T0137686

Intelligent Systems Reference Library

Volume 72

About this Series

The aim of this series is to publish a Reference Library, including novel advances and developments in all aspects of Intelligent Systems in an easily accessible and well structured form. The series includes reference works, handbooks, compendia, textbooks, well-structured monographs, dictionaries, and encyclopedias. It contains well integrated knowledge and current information in the field of Intelligent Systems. The series covers the theory, applications, and design methods of Intelligent Systems. Virtually all disciplines such as engineering, computer science, avionics, business, e-commerce, environment, healthcare, physics and life science are included.

More information about this series at http://www.springer.com/series/8578

Salvador García · Julián Luengo
Francisco Herrera

Data Preprocessing in Data Mining

Salvador García
Department of Computer Science
University of Jaén
Jaén
Spain

Julián Luengo
Department of Civil Engineering
University of Burgos
Burgos
Spain

Francisco Herrera
Department of Computer Science
and Artificial Intelligence
University of Granada
Granada
Spain

ISSN 1868-4394 ISSN 1868-4408 (electronic)
ISBN 978-3-319-37731-5 ISBN 978-3-319-10247-4 (eBook)
DOI 10.1007/978-3-319-10247-4

Springer Cham Heidelberg New York Dordrecht London

Softcover reprint of the hardcover 1st edition 2015

Printed on acid-free paper

Springer is part of Springer Science+Business Media (www.springer.com)

This book is dedicated to all people with whom we have worked over the years and have made it possible to reach this moment. Thanks to the members of the research group "Soft Computing and Intelligent Information Systems"

To our families.

Preface

Data preprocessing is an often neglected but major step in the data mining process. The data collection is usually a process loosely controlled, resulting in out of range values, e.g., impossible data combinations (e.g., Gender: Male; Pregnant: Yes), missing values, etc. Analyzing data that has not been carefully screened for such problems can produce misleading results. Thus, the representation and quality of data is first and foremost before running an analysis. If there is much irrelevant and redundant information present or noisy and unreliable data, then knowledge discovery is more difficult to conduct. Data preparation can take considerable amount of processing time.

Data preprocessing includes data preparation, compounded by integration, cleaning, normalization and transformation of data; and data reduction tasks; such as feature selection, instance selection, discretization, etc. The result expected after a reliable chaining of data preprocessing tasks is a final dataset, which can be considered correct and useful for further data mining algorithms.

This book covers the set of techniques under the umbrella of data preprocessing, being a comprehensive book devoted completely to the field of Data Mining, including all important details and aspects of all techniques that belonging to this families. In recent years, this area has become of great importance because the data mining algorithms require meaningful and manageable data to correctly operate and to provide useful knowledge, predictions or descriptions. It is well known that most of the efforts made in a knowledge discovery application is dedicated to data preparation and reduction tasks. Both theoreticians and practitioners are constantly searching for data preprocessing techniques to ensure reliable and accurate results together trading off with efficiency and time-complexity. Thus, an exhaustive and updated background in the topic could be very effective in areas such as data mining, machine learning, and pattern recognition. This book invites readers to explore the many advantages the data preparation and reduction provide:

- To adapt and particularize the data for each data mining algorithm.
- To reduce the amount of data required for a suitable learning task, also decreasing its time-complexity.
- To increase the effectiveness and accuracy in predictive tasks.
- To make possible the impossible with raw data, allowing data mining algorithms to be applied over high volumes of data.
- To support to the understanding of the data.
- Useful for various tasks, such as classification, regression and unsupervised learning.

The target audience for this book is anyone who wants a better understanding of the current state-of-the-art in a crucial part of the knowledge discovery from data: the data preprocessing. Practitioners in industry and enterprise should find new insights and possibilities in the breadth of topics covered. Researchers and data scientist and/or analysts in universities, research centers, and government could find a comprehensive review in the topic addressed and new ideas for productive research efforts.

Granada, Spain, June 2014

Salvador García
Julián Luengo
Francisco Herrera

Contents

Acronyms

ANN	Artificial Neural Network
CV	Cross Validation
DM	Data Mining
DR	Dimensionality Reduction
EM	Expectation-Maximization
FCV	Fold Cross Validation
FS	Feature Selection
IS	Instance Selection
KDD	Knowledge Discovery in Data
KEEL	Knowledge Extraction based on Evolutionary Learning
KNN	K-Nearest Neighbors
LLE	Locally Linear Embedding
LVQ	Learning Vector Quantization
MDS	Multi Dimensional Scaling
MI	Mutual Information
ML	Machine Learning
MLP	Multi-Layer Perceptron
MV	Missing Value
PCA	Principal Components Analysis
RBFN	Radial Basis Function Network
SONN	Self Organizing Neural Network
SVM	Support Vector Machine

Chapter 1
Introduction

Abstract The main background addressed in this book should be presented regarding Data Mining and Knowledge Discovery. Major concepts used throughout the contents of the rest of the book will be introduced, such as learning models, strategies and paradigms, etc. Thus, the whole process known as Knowledge Discovery in Data is provided in Sect. 1.1. A review on the main models of Data Mining is given in Sect. 1.2, accompanied a clear differentiation between Supervised and Unsupervised learning (Sects. 1.3 and 1.4, respectively). In Sect. 1.5, apart from the two classical data mining tasks, we mention other related problems that assume more complexity or hybridizations with respect to the classical learning paradigms. Finally, we establish the relationship between Data Preprocessing with Data Mining in Sect. 1.6.

1.1 Data Mining and Knowledge Discovery

Vast amounts of data are around us in our world, raw data that is mainly intractable for human or manual applications. So, the analysis of such data is now a necessity. The World Wide Web (WWW), business related services, society, applications and networks for science or engineering, among others, are continuously generating data in exponential growth since the development of powerful storage and connection tools. This immense data growth does not easily allow to useful information or organized knowledge to be understood or extracted automatically. This fact has led to the start of Data Mining (DM), which is currently a well-known discipline increasingly preset in the current world of the Information Age.

DM is, roughly speaking, about solving problems by analyzing data present in real databases. Nowadays, it is qualified as science and technology for exploring data to discover already present unknown patterns. Many people distinguish DM as synonym of the Knowledge Discovery in Databases (KDD) process, while others view DM as the main step of KDD [16, 24, 32].

There are various definitions of KDD. For instance, [10] define it as "the nontrivial process of identifying valid, novel, potentially useful, and ultimately understandable patterns in data" [11] considers the KDD process as an automatic exploratory data

S. García et al., *Data Preprocessing in Data Mining*,
Intelligent Systems Reference Library 72, DOI 10.1007/978-3-319-10247-4_1

analysis of large databases. A key aspect that characterizes the KDD process is the way it is divided into stages according the agreement of several important researchers in the topic. There are several methods available to make this division, each with advantages and disadvantages [16]. In this book, we adopt a hybridization widely used in recent years that categorizes these stages into six steps:

1. **Problem Specification:** Designating and arranging the application domain, the relevant prior knowledge obtained by experts and the final objectives pursued by the end-user.
2. **Problem Understanding:** Including the comprehension of both the selected data to approach and the expert knowledge associated in order to achieve high degree of reliability.
3. **Data Preprocessing:** This stage includes operations for data cleaning (such as handling the removal of noise and inconsistent data), data integrationdata integration (where multiple data sources may be combined into one), data transformation (where data is transformed and consolidated into forms which are appropriate for specific DM tasks or aggregation operations) and data reduction, including the selection and extraction of both features and examples in a database. This phase will be the focus of study throughout the book.
4. **Data Mining:** It is the essential process where the methods are used to extract valid data patterns. This step includes the choice of the most suitable DM task (such as classification, regression, clustering or association), the choice of the DM algorithm itself, belonging to one of the previous families. And finally, the employment and accommodation of the algorithm selected to the problem, by tuning essential parameters and validation procedures.
5. **Evaluation:** Estimating and interpreting the mined patterns based on interestingness measures.
6. **Result Exploitation:** The last stage may involve using the knowledge directly; incorporating the knowledge into another system for further processes or simply reporting the discovered knowledge through visualization tools.

Figure 1.1 summarizes the KDD process and reveals the six stages mentioned previously. It is worth mentioning that all the stages are interconnected, showing that the KDD process is actually a self-organized scheme where each stage conditions the remaining stages and reverse path is also allowed.

1.2 Data Mining Methods

A large number of techniques for DM are well-known and used in many applications. This section provides a short review of selected techniques considered the most important and frequent in DM. This review only highlights some of the main features of the different techniques and some of the influences related to data preprocessing procedures presented in the remaining chapters of this book. Our intention is not to

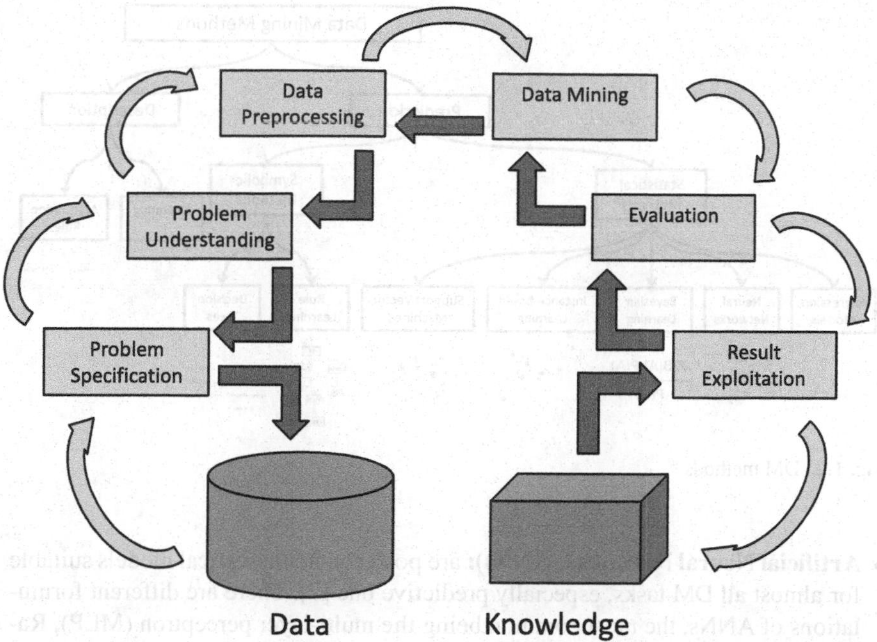

Fig. 1.1 KDD process

provide a complete explanation on how these techniques operate with detail, but to stay focused on the data preprocessing step.

Figure 1.2 shows a division of the main DM methods according to two methods of obtaining knowledge: prediction and description. In the following, we will give a short description for each method, including references for some representative and concrete algorithms and major considerations from the point of view of data preprocessing.

Within the prediction family of methods, two main groups can be distinguished: statistical methods and symbolic methods [4]. Statistical methods are usually characterized by the representation of knowledge through mathematical models with computations. In contrast, symbolic methods prefer to represent the knowledge by means of symbols and connectives, yielding more interpretable models for humans.

The most applied statistical methods are:

- **Regression Models:** being the oldest DM models, they are used in estimation tasks, requiring the class of equation modelling to be used [24]. Linear, quadratic and logistic regression are the most well known regression models in DM. There are basic requirement that they impose on the data. Among them, the use of numerical attributes are not designed for dealing with missing svalues, they try to fit outliers to the models and use all the features independently whether or not they are useful or dependent on one another.

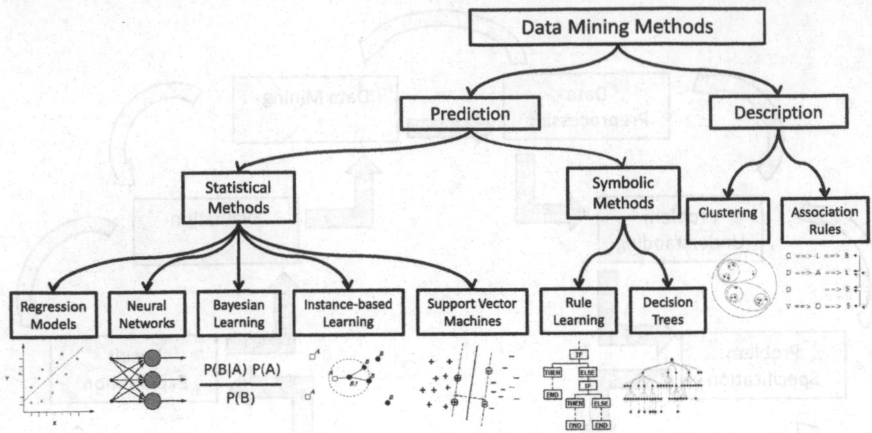

Fig. 1.2 DM methods

- **Artificial Neural Networks (ANNs):** are powerful mathematical models suitable for almost all DM tasks, especially predictive one [7]. There are different formulations of ANNs, the most common being the multi-layer perceptron (MLP), Radial Basis Function Networks (RBFNs) and Learning Vector Quantization (LVQ). ANNs are based on the definition of neurons, which are atomic parts that compute the aggregation of their input to an output according to an activation function. They usually outperform all other models because of their complex structure; however, the complexity and suitable configuration of the networks make them not very popular when regarding other methods, being considered as the typical example of black box models. Similar to regression models, they require numeric attributes and no MVs. However, if they are appropriately configured, they are robust against outliers and noise.
- **Bayesian Learning:** positioned using the probability theory as a framework for making rational decisions under uncertainty, based on Bayes' theorem. [6]. The most applied bayesian method is Naïve Bayes, which assumes that the effect of an attribute value of a given class is independent of the values of other attributes. Initial definitions of these algorithms only work with categorical attributes, due to the fact that the probability computation can only be made in discrete domains. Furthermore, the independence assumption among attributes causes these methods to be very sensitive to the redundancy and usefulness of some of the attributes and examples from the data, together with noisy and outliers examples. They cannot deal with MVs. Besides Naïve Bayes, there are also complex models based on dependency structures such as Bayesian networks.
- **Instance-based Learning:** Here, the examples are stored verbatim, and a distance function is used to determine which members of the database are closest to a new example with a desirable prediction. Also called lazy learners [3], the difference among them lies in the distance function used, the number of examples taken to

make the prediction, their influence when using voting or weighting mechanisms and the use of efficient algorithms to find the nearest examples, as KD-Trees or hashing schemes. The K-Nearest Neighbor (KNN) is the most applied, useful and known method in DM. Nevertheless, it suffers from several drawbacks such as high storage requirements, low efficiency in prediction response, and low noise tolerance. Thus, it is a good candidate to be improved through data reduction procedures.

- **Support Vector Machines:** SVMs are machine learning algorithms based on learning theory [30]. They are similar to ANNs in the sense that they are used for estimation and perform very well when data is linearly separable. SVMs usually do not require the generation of interaction among variables, as regression methods do. This fact should save some data preprocessing steps. Like ANNs, they require numeric non-missing data and are commonly robust against noise and outliers.

Regarding symbolic methods, we mention the following:

- **Rule Learning:** also called separate-and-conquer or covering rule algorithms [12]. All methods share the main operation. They search for a rule that explains some part of the data, separate these examples and recursively conquer the remaining examples. There are many ways for doing this, and also many ways to interpret the rules yielded and to use them in the inference mechanism. From the point of view of data preprocessing, generally speaking, they require nominal or discretized data (although this task is frequently implicit in the algorithm) and dispose of an innate selector of interesting attributes from data. However, MVs, noisy examples and outliers may prejudice the performance of the final model. Good examples of these models are the algorithms AQ, CN2, RIPPER, PART and FURIA.
- **Decision Trees:** comprising predictive models formed by iterations of a divide-and-conquer scheme of hierarchical decisions [28]. They work by attempting to split the data using one of the independent variables to separate data into homogeneous subgroups. The final form of the tree can be translated to a set of If-Then-Else rules from the root to each of the leaf nodes. Hence, they are closely related to rule learning methods and suffer from the same disadvantage as them. The most well known decision trees are CART, C4.5 and PUBLIC.

Considering the data descriptive task, we prefer to categorize the usual problems instead of the methods, due to the fact that both are intrinsically related to the case of predictive learning.

- **Clustering:** it appears when there is no class information to be predicted but the examples must be divided into natural groups or clusters [2]. These clusters reflect subgroups of examples that share some properties or have some similarities. They work by calculating a multivariate distance measure between observations, the observations that are more closely related. Roughly speaking, they belong to three broad categories: Agglomerative clustering, divisive clustering and partitioning clustering. The former two are hierarchical types of clustering opposite one another. The divisive one applies recursive divisions the entire data set whereas agglomerative ones start by considering each example as a cluster and performing an iterative merging of clusters until a criterion is satisfied. Partitioning based clustering, with k-Means algorithms as the most representative, starts with a fixed k number of clusters and iteratively adds or removes examples to and from them until no improvement is achieved based on a minimization of intra and/or inter cluster distance measure. As usual when distance measures are involved, numeric data is preferable together with no-missing data and the absence of noise and outliers. Other well known examples of clustering algorithms are COBWEB and Self Organizing Maps.

- **Association Rules:** they are a set of techniques that aim to find association relationships in the data. The typical application of these algorithms is the analysis of retail transaction data [1]. For example, the analysis would aim to find the likelihood that when a customer buys product X, she would also buy product Y. Association rule algorithms can also be formulated to look for sequential patterns. As a result of the data usually needed for association analysis is transaction data, the data volumes are very large. Also, transactions are expressed by categorical values, so the data must be discretized. Data transformation and reduction is often needed to perform high quality analysis in this DM problem. The Apriori technique is the most emblematic technique to address this problem.

1.3 Supervised Learning

In the DM community, prediction methods are commonly referred to as supervised learning. Supervised methods are thought to attempt the discovery of the relationships between input attributes (sometimes called variables or features) and a target attribute (sometimes referred to as class). The relationship which is sought after is represented in a structure called a model. Generally, a model describes and explains experiences, which are hidden in the data, and which can be used in the prediction of the value of the target attribute, when the values of the input attributes are known. Supervised learning is present in many application domains, such as finance, medicine and engineering.

In a typical supervised learning scenario, a training set is given and the objective is to form a description that can be used to predict unseen examples. This training set can be described in a variety of ways. The most common is to describe it by a set of instances, which is basically a collection of tuples that may contain duplicates. Each tuple is described by a vector of attribute values. Each attribute has an associate domain of values which are known prior to the learning task. Attributes are typically one of two types: nominal or categorical (whose values are members of an unordered set), or numeric (values are integer or real number, and an order is assumed). The nominal attributes have a finite cardinality, whereas numeric attributes domains are delimitated by lower and upper bounds. The instance space (the set of possible examples) is defined as a cartesian product of all the input attributes domains. The universal instance space is defined as a cartesian product of all input attribute domain and the target attribute domain.

The two basic and classical problems that belong to the supervised learning category are classification and regression. In classification, the domain of the target attribute is finite and categorical. That is, there are a finite number of classes or categories to predict a sample and they are known by the learning algorithm. A classifier must assign a class to a unseen example when it is trained by a set of training data. The nature of classification is to discriminate examples from others, attaining as a main application a reliable prediction: once we have a model that fits the past data, if the future is similar to the past, then we can make correct predictions for new instances. However, when the target attribute is formed by infinite values, such as in the case of predicting a real number between a certain interval, we are referring to regression problems. Hence, the supervised learning approach here has to fit a model to learn the output target attribute as a function of input attributes. Obviously, the regression problem present more difficulties than the classification problem and the required computation resources and the complexity of the model are higher.

There is another type of supervised learning that involves time data. Time series analysis is concerned with making predictions in time. Typical applications include analysis of stock prices, market trends and sales forecasting. Due to the time dependence of the data, the data preprocessing for time series data is different from the main theme of this book. Nevertheless, some basic procedures may be of interest and will be also applicable in this field.

1.4 Unsupervised Learning

We have seen that in supervised learning, the aim is to obtain a mapping from the input to an output whose correct and definite values are provided by a supervisor. In unsupervised learning, there is no such supervisor and only input data is available. Thus, the aim is now to find regularities, irregularities, relationships, similarities and associations in the input. With unsupervised learning, it is possible to learn larger and more complex models than with supervised learning. This is because in supervised learning one is trying to find the connection between two sets of observations. The difficulty of the learning task increases exponentially with the number of steps between the two sets and that is why supervised learning cannot, in practice, learn models with deep hierarchies.

Apart from the two well-known problems that belong to the unsupervised learning family, clustering and association rules, there are other related problems that can fit into this category:

1.4.1 Pattern Mining [25]

It is adopted as a more general term than frequent pattern mining or association mining since pattern mining also covers rare and negative patterns as well. For example, in pattern mining, the search of rules is also focused on multilevel, multidimensional, approximate, uncertain, compressed, rare/negative and high-dimensional patterns. The mining methods do not only involve candidate generation and growth, but also interestingness, correlation and exception rules, distributed and incremental mining, etc.

1.4.2 Outlier Detection [9]

Also known as anomaly detection, it is the process of finding data examples with behaviours that are very different from the expectation. Such examples are called outliers or anomalies. It has a high relation with clustering analysis, because the latter finds the majority patterns in a data set and organizes the data accordingly, whereas outlier detection attempts to catch those exceptional cases that present significant deviations from the majority patterns.

1.5 Other Learning Paradigms

Some DM problems are being clearly differentiated from the classical ones and some of them even cannot be placed into one of the two mentioned learning categories, neither supervised or unsupervised learning. As a result, this section will supply a brief description of other major learning paradigms which are widespread and recent challenges in the DM research community.

We establish a general division based on the nature of the learning paradigm. When the paradigm presents extensions on data acquirement or distribution, imposed restrictions on models or the implication of more complex procedures to obtain suitable knowledge, we refer to extended paradigm. On the other hand, when the paradigm can only be understood as an mixture of supervised and unsupervised learning, we refer to hybrid paradigm. Note that we only mention some learning paradigms out of the universe of possibilities and its interpretations, assuming that this section is just intended to introduce the issue.

1.5.1 Imbalanced Learning [22]

It is an extended supervised learning paradigm, a classification problem where the data has exceptional distribution on the target attribute. This issue occurs when the number of examples representing the class of interest is much lower than that of the other classes. Its presence in many real-world applications has brought along a growth of attention from researchers.

1.5.2 *Multi-instance Learning [5]*

This paradigm constitutes an extension based on imposed restrictions on models in which each example consists of a bag of instances instead of an unique instance. There are two main ways of addressing this problem, either converting multi-instance into single-instance by data transformations or by means of upgrade of single-case algorithms.

1.5.3 *Multi-label Classification [8]*

It is generalization of traditional classification, in which each processed instance is associated not with a class, but with a subset of them. In recent years different techniques have appeared which, through the transformation of the data or the adaptation of classic algorithms, aim to provide a solution to this problem.

1.5.4 *Semi-supervised Learning [33]*

This paradigm arises as an hybrid between the classification predictive task and the clustering descriptive analysis. It is a learning paradigm concerned with the design of models in the presence of both labeled and unlabeled data. Essentially, the developments in this field use unlabeled samples to either modify or re-prioritize the hypothesis obtained from the labeled samples alone. Both semi-supervised classification and semi-supervised clustering have emerged extending the traditional paradigms by including unlabeled or labeled examples, respectively. Another paradigm called Active Learning, with the same objective as Semi-supervised Learning, tries to select the most important examples from a pool of unlabeled data, however these examples are queried by an human expert.

1.5.5 *Subgroup Discovery [17]*

Also known as Contrast Set Mining and Emergent Pattern Mining, it is formed as the result of another hybridization between supervised and unsupervised learning tasks, specifically classification and association mining. A subgroup discovery method aims to extract interesting rules with respect to a target attribute.

1.5.6 Transfer Learning [26]

Aims to extract the knowledge from one or more source tasks and apply the knowledge to a target task. In this paradigm, the algorithms apply knowledge about source tasks when building a model for a new target task. Traditional learning algorithms assume that the training data and test data are drawn from the same distribution and feature space, but if the distribution changes, such methods need to rebuild or adapt the model in order to perform well. The so-called data shift problem is closely related to transfer learning.

1.5.7 Data Stream Learning [13]

In some situations, all data is not available at a specific moment, so it is necessary to develop learning algorithms that treat the input as a continuous data stream. Its core assumption is that each instance can be inspected only once and must then be discarded to make room for subsequent instances. This paradigm is an extension of data acquirement and it is related to both supervised and unsupervised learning.

1.6 Introduction to Data Preprocessing

Once some basic concepts and processes of DM have been reviewed, the next step is to question the data to be used. Input data must be provided in the amount, structure and format that suit each DM task perfectly. Unfortunately, real-world databases are highly influenced by negative factors such the presence of noise, MVs, inconsistent and superfluous data and huge sizes in both dimensions, examples and features. Thus, low-quality data will lead to low-quality DM performance [27].

In this section, we will describe the general categorization in which we can divide the set of data preprocessing techniques. More details will be given in the rest of chapters of this book, but for now, our intention is to provide a brief summary of the preprocessing techniques that we should be familiar with after reading this book. For this purpose, several subsections will be presented according to the type and set of techniques that belong to each category.

1.6.1 Data Preparation

Throughout this book, we refer to data preparation as the set of techniques that initialize the data properly to serve as input for a certain DM algorithm. It is worth mentioning that we prefer the data preparation notation to design parts of data preprocessing, which is a confusing nomenclature used in previous texts as the whole set of processes that perform data preprocessing tasks. This is not incorrect and we respect this nomenclature, however we prefer to clearly distinguish between data preparation and data reduction due to raised importance that the latter set of techniques have been achieving in recent years and some of the clear differentiations that can be extracted from this understanding.

Data preparation is normally a mandatory step. It converts prior useless data into new data that fits a DM process. First of all, if data is not prepared, the DM algorithm might not receive ir in order to operate or surely it will report errors during its runtime. In the best of cases, the algorithm will work, but the results offered will not make sense or will not be considered as accurate knowledge.

Thus, what are the basic issues that must be resolved in data preparation? Here, we provide a list of questions accompanied with the correct answers involving each type of process that belongs to the data preparation family of techniques:

- How do I clean up the data?—Data Cleaning.
- How do I provide accurate data?—Data Transformation.
- How do I incorporate and adjust data?—Data Integration.
- How do I unify and scale data?—Data Normalization.
- How do I handle missing data?—Missing Data Imputation.
- How do I detect and manage noise?—Noise Identification.

Next, we will briefly describe each one of these techniques listed above. Figure 1.3 shows an explanatory illustration of the forms of data preparation. We recall that they will be studied more in-depth in the following chapters of this book.

1.6.1.1 Data Cleaning

Or data cleansing, includes operations that correct bad data, filter some incorrect data out of the data set and reduce the unnecessary detail of data. It is a general concept that comprises or overlaps other well-known data preparation techniques. Treatment of missing and noise data is included here, but both categories have been separated in order to devote a deeper analysis of the intelligent proposals to them further into this book. Other cleaning data tasks involve the detection of discrepancies and dirty data (fragments of the original data which do not make sense). The latter tasks are more related to the understanding of the original data and they generally require human audit.

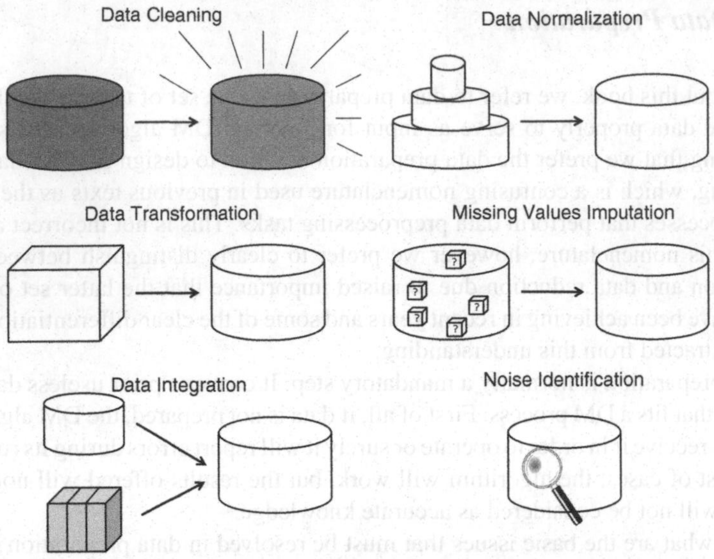

Fig. 1.3 Forms of data preparation

1.6.1.2 Data Transformation

In this preprocessing step, the data is converted or consolidated so that the mining process result could be applied or may be more efficient. Subtasks inside data transformation are the smoothing, the feature construction, aggregation or summarization of data, normalization, discretization and generalization. Most of them will be segregated as independent tasks, due to the fact that data transformation, such as the case of data cleaning, is referred to as a general data preprocessing family of techniques. Those tasks that require human supervision and are more dependent on the data are the classical data transformation techniques, such as the report generation, new attributes that aggregate existing ones and generalization of concepts especially in categorical attributes, such as the replacing complete dates in the database with year numbers only.

1.6.1.3 Data Integration

It comprises the merging of data from multiple data stores. This process must be carefully performed in order to avoid redundancies and inconsistencies in the resulting data set. Typical operations accomplished within the data integration are the identification and unification of variables and domains, the analysis of attribute correlation, the duplication of tuples and the detection of conflicts in data values of different sources.

1.6.1.4 Data Normalization

The measurement unit used can affect the data analysis. All the attributes should be expressed in the same measurement units and should use a common scale or range. Normalizing the data attempts to give all attributes equal weight and it is particularly useful in statistical learning methods.

1.6.1.5 Missing Data Imputation [23]

It is a form of data cleaning, where the purpose is to fill the variables that contain MVs with some intuitive data. In most of the cases, adding a reasonable estimate of a suitable data value is better than leaving it blank.

1.6.1.6 Noise Identification [29]

Included as a step of data cleaning and also known as the smoothing in data transformation, its main objective is to detect random errors or variances in a measured variable. Note that we refer to the detection of noise instead of the removal of noise, which is more related to the IS task within data reduction. Once a noisy example is detected, we can apply a correction-based process that could involve some kind of underlying operation.

1.6.2 Data Reduction

Data reduction comprises the set of techniques that, in one way or another, obtain a reduced representation of the original data. For us, the distinction of data preparation techniques is those that are needed to appropriately suit the data as input of a DM task. As we have mentioned before, this means that if data preparation is not properly conducted, the DM algorithms will not be run or will surely report wrong results after running. In the case of data reduction, the data produced usually maintains the essential structure and integrity of the original data, but the amount of data is downsized. So, can the original data be used, without applying a data reduction process, as input of a DM process? The answer is yes, but other major issues must be taken into account, being just as crucial as the issues addressed by data preparation.

Hence, at a glance, it can be considered as an optional step. However, this affirmation may be conflictive. Although the integrity of the data is maintained, it is well known that any algorithm has a certain time complexity that depends on several parameters. In DM, one of these parameters is somehow directly proportional to the size of the input database. If the size exceeds the limit, the limit being very dependant on the type of DM algorithms, the running of the algorithm can be prohibitive, and then the data reduction task is as crucial as data preparation is. Regarding other

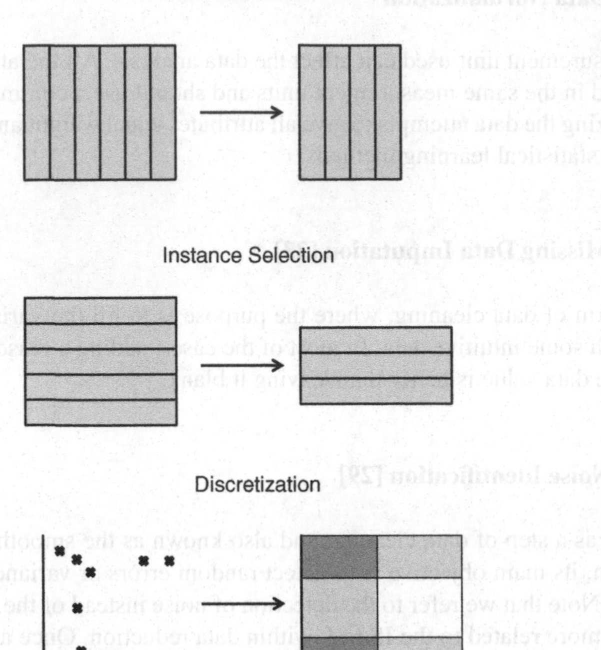

Fig. 1.4 Forms of data reduction

factors, such as the decreasing of the complexity and improvement of the quality of the models yielded, the role of data reduction is again decisive.

As mentioned before, what are the basic issues that must be resolved in data reduction? Again, we provide a series of questions associated with the correct answer related to each type of task that belongs to the data reduction techniques:

- How do I reduce the dimensionality of data?—Feature Selection (FS).
- How do I remove redundant and/or conflictive examples?—Instance Selection (IS).
- How do I simplify the domain of an attribute?—Discretization.
- How do I fill in gaps in data?—Feature Extraction and/or Instance Generation.

In the following, we provide a concise explanation of the four techniques enumerated before. Figure 1.4 shows an illustrative picture that reflects the forms of data reduction. All of them will be extended, studied and analyzed throughout the various chapters of the book.

1.6.2.1 Feature Selection [19, 21]

Achieves the reduction of the data set by removing irrelevant or redundant features (or dimensions). The goal of FS is to find a minimum set of attributes, such as the resulting probability distribution of the data output attributes, (or classes) is as close as possible to the original distribution obtained using all attributes. It facilitates the understanding of the pattern extracted and increases the speed of the learning stage.

1.6.2.2 Instance Selection [14, 20]

Consists of choosing a subset of the total available data to achieve the original purpose of the DM application as if the whole data had been used. It constitutes the family of oriented methods that perform in a somewhat intelligent way the choice of the best possible subset of examples from the original data by using some rules and/or heuristics. The random selection of examples is usually known as Sampling and it is present in a very large number of DM models for conducting internal validation and for avoiding over-fitting.

1.6.2.3 Discretization [15]

This procedure transforms quantitative data into qualitative data, that is, numerical attributes into discrete or nominal attributes with a finite number of intervals, obtaining a non-overlapping partition of a continuous domain. An association between each interval with a numerical discrete value is then established. Once the discretization is performed, the data can be treated as nominal data during any DM process.

It is noteworthy that discretization is actually a hybrid data preprocessing technique involving both data preparation and data reduction tasks. Some sources include discretization in the data transformation category and another sources consider a data reduction process. In practice, discretization can be viewed as a data reduction method since it maps data from a huge spectrum of numeric values to a greatly reduced subset of discrete values. Our decision is to mostly include it in data reduction although we also agree with the other trend. The motivation behind this is that recent discretization schemes try to reduce the number of discrete intervals as much as possible while maintaining the performance of the further DM process. In other words, it is often very easy to perform basic discretization with any type of data, given that the data is suitable for a certain algorithm with a simple map between continuous and categorical values. However, the real difficulty is to achieve good reduction without compromising the quality of data, and much of the effort expended by researchers follows this tendency.

1.6.2.4 Feature Extraction/Instance Generation [18, 20, 31]

Extends both the feature and IS by allowing the modification of the internal values that represent each example or attribute. In feature extraction, apart from the removal operation of attributes, subsets of attributes can be merged or can contribute to the creation of artificial substitute attributes. Regarding instance generation, the process is similar in the sense of examples. It allows the creation or adjustment of artificial substitute examples that could better represent the decision boundaries in supervised learning.

References

1. Adamo, J.M.: Data Mining for Association Rules and Sequential Patterns: Sequential and Parallel Algorithms. Springer, New York (2001)
2. Aggarwal, C., Reddy, C.: Data Clustering: Recent Advances and Applications. Data Mining and Knowledge Discovery Series. Chapman and Hall/CRC, Taylor & Francis Group, Boca Raton (2013)
3. Aha, D.W., Kibler, D., Albert, M.K.: Instance-based learning algorithms. Mach. Learn. **6**(1), 37–66 (1991)
4. Alpaydin, E.: Introduction to Machine Learning, 2nd edn. MIT Press, Cambridge (2010)
5. Amores, J.: Multiple instance classification: review, taxonomy and comparative study. Artif. Intell. **201**, 81–105 (2013)
6. Barber, D.: Bayesian Reasoning and Machine Learning. Cambridge University Press, New York (2012)
7. Bishop, C.M.: Neural Networks for Pattern Recognition. Oxford University Press, New York (1995)
8. Boutell, M.R., Luo, J., Shen, X., Brown, C.M.: Learning multi-label scene classification. Pattern Recogn. **37**(9), 1757–1771 (2004)
9. Chandola, V., Banerjee, A., Kumar, V.: Anomaly detection: a survey. ACM Comput. Surv. **41**(3), 15:1–15:58 (2009)
10. Fayyad, U.M., Piatetsky-Shapiro, G., Smyth, P. (eds.): From data mining to knowledge discovery: an overview as chapter. Advances in Knowledge Discovery and Data Mining. American Association for Artificial Intelligence, San Francisco (1996)
11. Friedman, J.H.: Data mining and statistics: What is the connection The Data Administrative Newsletter? (1997)
12. Frunkranz, J., Gamberger, D., Lavrac, N.: Foundations of Rule Learning. Springer, New York (2012)
13. Gama, J.: Knowledge Discovery from Data Streams, 1st edn. Chapman & Hall/CRC, Taylor and Francis, Boca Raton (2010)
14. García, S., Derrac, J., Cano, J.R., Herrera, F.: Prototype selection for nearest neighbor classification: Taxonomy and empirical study. IEEE Trans. Pattern Anal. Mach. Intell. **34**(3), 417–435 (2012)
15. García, S., Luengo, J., Sáez, J.A., López, V., Herrera, F.: A survey of discretization techniques: taxonomy and empirical analysis in supervised learning. IEEE Trans. Knowl. Data Eng. **25**(4), 734–750 (2013)
16. Han, J.: Data Mining: Concepts and Techniques. Morgan Kaufmann, San Francisco (2011)
17. Herrera, F., Carmona, C.J., González, P., del Jesus, M.J.: An overview on subgroup discovery: foundations and applications. Knowl. Inf. Syst. **29**(3), 495–525 (2011)

18. Liu, H., Motoda, H.: Feature Extraction, Construction and Selection: A Data Mining Perspective. Kluwer International Series in Engineering and Computer Science. Kluwer Academic, Boston (1998)
19. Liu, H., Motoda, H.: Feature Selection for Knowledge Discovery and Data Mining. Kluwer International Series in Engineering and Computer Science. Kluwer Academic, Boston (1998)
20. Liu, H., Motoda, H.: Instance Selection and Construction for Data Mining. Kluwer Academic, Norwell (2001)
21. Liu, H., Motoda, H.: Computational Methods of Feature Selection. Data Mining and Knowledge Discovery Series. Chapman & Hall/CRC, Taylor and Francis, Boca Raton (2007)
22. López, V., Fernández, A., García, S., Palade, V., Herrera, F.: An insight into classification with imbalanced data: Empirical results and current trends on using data intrinsic characteristics. Inf. Sci. **250**, 113–141 (2013)
23. Luengo, J., García, S., Herrera, F.: On the choice of the best imputation methods for missing values considering three groups of classification methods. Knowl. Inf. Syst. **32**(1), 77–108 (2012)
24. Nisbet, R., Elder, J., Miner, G.: Handbook of Statistical Analysis and Data Mining Applications. Academic Press, Boston (2009)
25. Ong, K.: Frequent Pattern Mining. VDM Publishing, Germany (2010)
26. Pan, S.J., Yang, Q.: A survey on transfer learning. IEEE Trans. Knowl. Data Eng. **22**(10), 1345–1359 (2010)
27. Pyle, D.: Data Preparation for Data Mining. Morgan Kaufmann, San Francisco (1999)
28. Rokach, L.: Data Mining with Decision Trees: Theory and Applications. Series in Machine Perception and Artificial Intelligence. World Scientific, Singapore (2007)
29. Sáez, J.A., Luengo, J., Herrera, F.: Predicting noise filtering efficacy with data complexity measures for nearest neighbor classification. Pattern Recogn. **46**(1), 355–364 (2013)
30. Schölkopf, B., Smola, A.J.: Learning with Kernels: Support Vector Machines, Regularization, Optimization, and Beyond. Adaptive Computation and Machine Learning. MIT Press, Cambridge (2002)
31. Triguero, I., Derrac, J., García, S., Herrera, F.: A taxonomy and experimental study on prototype generation for nearest neighbor classification. IEEE Trans. Syst. Man Cybern. Part C **42**(1), 86–100 (2012)
32. Witten, I.H., Frank, E.: Data Mining: Practical Machine Learning Tools and Techniques. Morgan Kaufmann Series in Data Management Systems, 2nd edn. Morgan Kaufmann, San Francisco (2005)
33. Zhu, X., Goldberg, A.B., Brachman, R.: Introduction to Semi-Supervised Learning. Morgan and Claypool, California (2009)

Chapter 2
Data Sets and Proper Statistical Analysis of Data Mining Techniques

Abstract Presenting a Data Mining technique and analyzing it often involves using a data set related to the domain. In research fortunately many well-known data sets are available and widely used to check the performance of the technique being considered. Many of the subsequent sections of this book include a practical experimental comparison of the techniques described in each one as a exemplification of this process. Such comparisons require a clear bed test in order to enable the reader to be able to replicate and understand the analysis and the conclusions obtained. First we provide an insight of the data sets used to study the algorithms presented as representative in each section in Sect. 2.1. In this section we elaborate on the data sets used in the rest of the book indicating their characteristics, sources and availability. We also delve in the partitioning procedure and how it is expected to alleviate the problematic associated to the validation of any supervised method as well as the details of the performance measures that will be used in the rest of the book. Section 2.2 takes a tour of the most common statistical techniques required in the literature to provide meaningful and correct conclusions. The steps followed to correctly use and interpret the statistical test outcome are also given.

2.1 Data Sets and Partitions

The ultimate goal of any DM process is to be applied to real life problems. As testing a technique in every problem is unfeasible, the common procedure is to evaluate such a technique in a set of standard DM problems (or data sets) publicly available. In this book we will mainly use the KEEL DM tool which is also supported by the KEEL-Dataset repository[1] where data sets from different well-known sources as UCI [2] and others have been converted to KEEL ARFF format and partitioned. This enables the user to replicate all the experiments presented in this book with ease.

As this book focus on supervised learning, we will provide a list with the data sets enclosed in this paradigm. The representative data sets that will be used in classification are shown in Table 2.1. The table includes the most relevant information about the data set:

[1] http://keel.es/datasets.php.

S. García et al., *Data Preprocessing in Data Mining*,
Intelligent Systems Reference Library 72, DOI 10.1007/978-3-319-10247-4_2

Table 2.1 Classification data sets to be used in the remainder of this book

Acronym	Data set	#Attributes (R/I/N)	#Examples	#Classes	Miss val.
ABA	Abalone	8 (7/0/1)	4174	28	No
APP	Appendicitis	7 (7/0/0)	106	2	No
AUD	Audiology	71 (0/0/71)	226	24	Yes
AUS	Australian	14 (3/5/6)	690	2	No
AUT	Autos	25 (15/0/10)	205	6	Yes
BAL	Balance	4 (4/0/0)	625	3	No
BAN	Banana	2 (2/0/0)	5300	2	No
BND	Bands	19 (13/6/0)	539	2	Yes
BRE	Breast	9 (0/0/9)	286	2	Yes
BUP	Bupa	6 (1/5/0)	345	2	No
CAR	Car	6 (0/0/6)	1728	4	No
CLE	Cleveland	13 (13/0/0)	303	5	Yes
CTR	Contraceptive	9 (0/9/0)	1473	3	No
CRX	Crx	15 (3/3/9)	690	2	Yes
DER	Dermatology	34 (0/34/0)	366	6	Yes
ECO	Ecoli	7 (7/0/0)	336	8	No
FLA	Flare	11 (0/0/11)	1066	6	No
GER	German	20 (0/7/13)	1000	2	No
GLA	Glass	9 (9/0/0)	214	7	No
HAB	Haberman	3 (0/3/0)	306	2	No
HAY	Hayes	4 (0/4/0)	160	3	No
HEA	Heart	13 (1/12/0)	270	2	No
HEP	Hepatitis	19 (2/17/0)	155	2	Yes
HOC	Horse colic	23 (7/1/15)	368	2	Yes
HOU	Housevotes	16 (0/0/16)	435	2	Yes
IRI	Iris	4 (4/0/0)	150	3	No
LED	Led7digit	7 (7/0/0)	500	10	No
LUN	Lung cancer	57 (0/0/57)	32	3	Yes
LYM	Lymphography	18 (0/3/15)	148	4	No
MAM	Mammographic	5 (0/5/0)	961	2	Yes
MON	Monk-2	6 (0/6/0)	432	2	No
MOV	Movement	90 (90/0/0)	360	15	No
NTH	Newthyroid	5 (4/1/0)	215	3	No
PAG	Pageblocks	10 (4/6/0)	5472	5	No
PEN	Penbased	16 (0/16/0)	10992	10	No
PHO	Phoneme	5 (5/0/0)	5404	2	No
PIM	Pima	8 (8/0/0)	768	2	No
PRT	Primary tumor	18 (0/0/18)	339	21	Yes
SAH	Saheart	9 (5/3/1)	462	2	No

(continued)

Table 2.1 (continued)

Acronym	Data set	#Attributes (R/I/N)	#Examples	#Classes	Miss val.
SAT	Satimage	36 (0/36/0)	6435	7	No
SEG	Segment	19 (19/0/0)	2310	7	No
SON	Sonar	60 (60/0/0)	208	2	No
SPO	Sponge	45 (0/3/42)	76	12	Yes
SPA	Spambase	57 (57/0/0)	4597	2	No
SPH	Specfheart	44 (0/44/0)	267	2	No
TAE	Tae	5 (0/5/0)	151	3	No
TNC	Titanic	3 (3/0/0)	2201	2	No
VEH	Vehicle	18 (0/18/0)	846	4	No
VOW	Vowel	13 (10/3/0)	990	11	No
WAT	Water treatment	38 (38/0/0)	526	13	Yes
WIN	Wine	13 (13/0/0)	178	3	No
WIS	Wisconsin	9 (0/9/0)	699	2	Yes
YEA	Yeast	8 (8/0/0)	1484	10	No

- The name of the data set and the abbreviation that will be used as future reference.
- *#Attributes* is the number of attributes/features and their type. *R* stands for real valued attributes, *I* means integer attributes and *N* indicates the number of nominal attributes.
- *#Examples* is the number of examples/instances contained in the data set.
- *#Classes* is the quantity of different class labels the problem presents.
- Whether the data set contains MVs or not.

2.1.1 Data Set Partitioning

The benchmark data sets presented are used with one goal: to evaluate the performance of a given model over a set of well-known standard problems. Thus the results can be replicated by other users and compared to new proposals. However the data must be correctly used in order to avoid bias in the results.

If the whole data set is used for both build and validate the model generated by a ML algorithm, we have no clue about how the model will behave with new, unseen cases. Two main problems may arise by using the same data to train and evaluate the model:

- **Underfitting** is the easiest problem to understand. It happens when the model is poorly adjusted to the data, suffering from high error both in training and test (unseen) data.

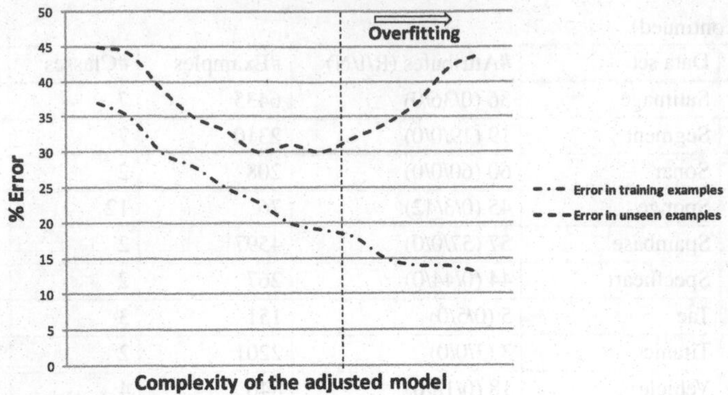

Fig. 2.1 Typical evolution of % error when adjusting a supervised model. Underfitting is noticeable in the *left side* of the figure

- **Overfitting** happens when the model is too tightly adjusted to data offering high precision to known cases but behaving poorly with unseen data.

By using the whole data we may be aware of underfitting problems due to a low performance of the model. Adjusting such a model to better fit the data may lead to overfitting but the lack of unseen case makes impossible to notice this situation. Please also note that taking this procedure to an extreme may lead to overfitting as represented in Fig. 2.1. According to Occam's Razor reasoning given two models of similar generalization errors, one should prefer the simpler model over the more complex model.

Overfitting may also appear due other reasons like noise as it may force the model to be wrongly adjusted to false regions of the problem space. The lack of data will also cause underfitting, as the inner measures followed by the ML algorithm can only take into account known examples and their distribution in the space.

In order to control the model's performance, avoid overfitting and to have a generalizable estimation of the quality of the model obtained several partitioning schemes are introduced in the literature. The most common one is k-Fold Cross Validation (k-FCV) [17]:

1. In k-FCV, the original data set is randomly partitioned into k equal size folds or *partitions*.
2. From the k partitions, one is retained as the validation data for testing the model, and the remaining $k - 1$ subsamples are used to build the model.
3. As we have k partitions, the process is repeated k times with each of the k subsamples used exactly once as the validation data.

Finally the k results obtained from each one of the test partitions must be combined, usually by averaging them, to produce a single value as depicted in Fig. 2.2. This procedure is widely used as it has been proved that these schemes asymptotically converge to a stable value, which allows realistic comparisons between classifiers

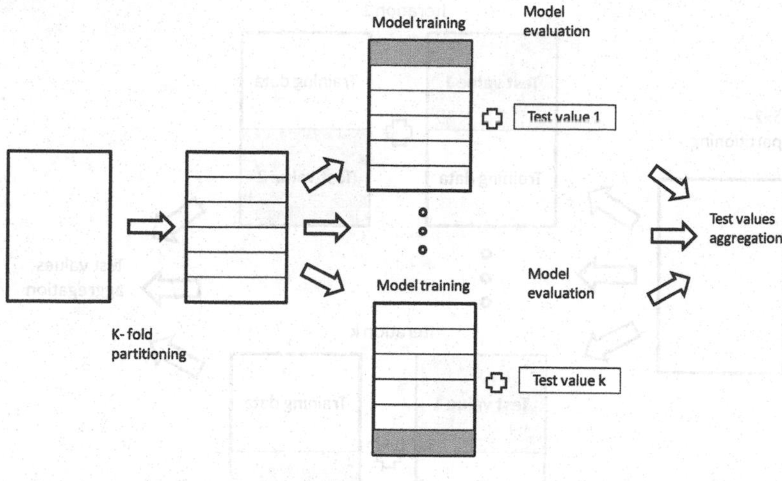

Fig. 2.2 K-fold process

[6, 26]. The value of k may vary, 5 and 10 being the most common ones. Such a value needs to be adjusted to avoid to generate a small test partition poorly populated with examples that may bias the performance measures used. If big data sets are being used, 10-FCV is usually utilized while for smaller data sets 5-FCV is more frequent.

Simple k-FCV may also lead to disarranging the proportion of examples from each class in the test partition. The most commonly employed method in the literature to avoid this problem is stratified k-FCV . It places an equal number of samples of each class on each partition to maintain class distributions equal in all partitions

Other popular validation schemes are:

- In 5×2 *CV* [22] the whole data set is randomly partitioned in two subsets A and B. Then the model is first built using A and validated with B and then the process is reversed with the model built with B and tested with A. This partitioning process is repeated as desired aggregating the performance measure in each step. Figure 2.3 illustrates the process. Stratified 5×2 cross-validation is the variation most commonly used in this scheme.
- *Leave one out* is an extreme case of k-FCV, where k equals the number of examples in the data set. In each step only one instance is used to test the model whereas the rest of instances are used to learn it.

How to partition the data is a key issue as it will largely influence in the performance of the methods and in the conclusions extracted from that point on. Performing a bad partitioning will surely lead to incomplete and/or biased behavior data about the model being evaluated. This issue is being actively investigated nowadays, with special attention being paid to data set shift [21] as a decisive factor that impose large k values in k-FCV to reach performance stability in the model being evaluated.

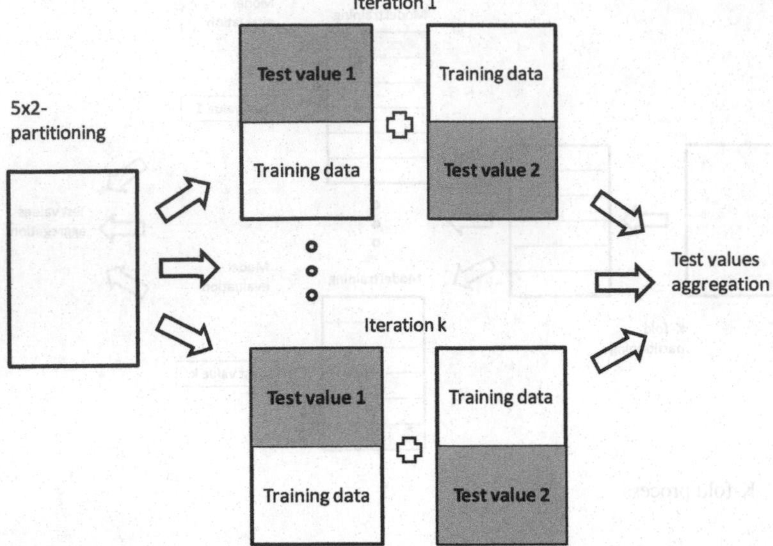

Fig. 2.3 5 × 2-fold process

2.1.2 Performance Measures

Most of the preprocessing stages aim to improve the quality of the data. Such improvement is later measured by analyzing the model constructed over the data and it depends on the type of the DM process carried out afterwards. Predictive processes like classification and regression rely in a measure of how well the model fits the data, resulting in a series of measures that work over the predictions made.

In classification literature we can observe that most of the performance measures are designed for binary-class problems [25]. Well-known accuracy measures for binary-class problems are: classification rate , precision, sensitivity, specificity, G-mean [3], F-score, AUC [14], Youden's index γ [31] and Cohen's Kappa [4].

Some of the two-class accuracy measures have been adapted for multi-class problems. For example, in a recent paper [18], the authors propose an approximating multi-class ROC analysis, which is theoretically possible but its computation is still restrictive. Only two measures are widely used because of their simplicity and successful application when the number of classes is large enough. We refer to classification rate and Cohen's kappa measures, which will be explained in the following.

- *Classification rate (also known as accuracy)*: is the number of successful hits relative to the total number of classifications. It has been by far the most commonly used metric for assessing the performance of classifiers for years [1, 19, 28].
- *Cohen's kappa*: is an alternative to classification rate, a method, known for decades, that compensates for random hits (Cohen 1960). Its original purpose was to measure the degree of agreement or disagreement between two people observing the

same phenomenon. Cohen's kappa can be adapted to classification tasks and it is recommended to be employed because it takes random successes into consideration as a standard, in the same way as the AUC measure (Ben-David 2007). Also, it is used in some well-known software packages, such as Weka [28], SAS, SPSS, etc.

An easy way of computing Cohen's kappa is to make use of the resulting confusion matrix in a classification task. Specifically, the Cohen's kappa measure can be obtained using the following expression:

$$kappa = \frac{n \sum_{i=1}^{C} x_{ii} - \sum_{i=1}^{C} x_{i\cdot} x_{\cdot i}}{n^2 - \sum_{i=1}^{C} x_{i\cdot} x_{\cdot i}} \tag{2.1}$$

where x_{ii} is the cell count in the main diagonal, n is the number of examples in the data set, C is the number of class labels and $x_{i\cdot}$, $x_{\cdot i}$ are the rows and columns total counts respectively.

Cohen's kappa ranges from -1 (total disagreement) through 0 (random classification) to 1 (perfect agreement). Being a scalar, it is less expressive than ROC curves when applied to binary-classification. However, for multi-class problems, kappa is a very useful, yet simple, meter for measuring the accuracy of the classifier while compensating for random successes.

The main difference between classification rate and Cohen's kappa is the scoring of the correct classifications. Classification rate scores all the successes over all classes, whereas Cohen's kappa scores the successes independently for each class and aggregates them. The second way of scoring is less sensitive to randomness caused by different number of examples in each class, which causes a bias in the learner towards the obtention of data-dependent models.

2.2 Using Statistical Tests to Compare Methods

Using the raw performance measures to compare different ML methods and to establish a ranking is discouraged. It has been recently analyzed [5, 10] that other tools of statistical nature must be utilized in order to obtain meaningful and durable conclusions.

In recent years, there has been a growing interest for the experimental analysis in the field of DM. It is noticeable due to the existence of numerous papers which analyze and propose different types of problems, such as the basis for experimental comparisons of algorithms, proposals of different methodologies in comparison or proposals of use of different statistical techniques in algorithms' comparison.

The "No free lunch" theorem [29] demonstrates that it is not possible to find one algorithm behaving better for any problem. On the other hand, we know that we can work with different degrees of knowledge of the problem which we expect to solve, and that it is not the same to work without knowledge of the problem (hypothesis of the "no free lunch" theorem) than to work with partial knowledge about the problem, knowledge that allows us to design algorithms with specific characteristics which can make them more suitable to solve of the problem.

2.2.1 Conditions for the Safe Use of Parametric Tests

In [24] the distinction between parametric and non-parametric tests is based on the level of measure represented by the data to be analyzed. That is, a parametric test usually uses data composed by real values.

However the latter does not imply that when we always dispose of this type of data, we should use a parametric test. Other initial assumptions for a safe usage of parametric tests must be fulfilled. The non fulfillment of these conditions might cause a statistical analysis to lose credibility.

The following conditions are needed in order to safely carry out parametric tests [24, 32]:

- **Independence**: In statistics, two events are independent when the fact that one occurs does not modify the probability of the other one occurring.
- **Normality**: An observation is normal when its behaviour follows a normal or Gauss distribution with a certain value of average μ and variance σ. A normality test applied over a sample can indicate the presence or absence of this condition in observed data. Three normality tests are usually used in order to check whether normality is present or not:
  - *Kolmogorov–Smirnov*: compares the accumulated distribution of observed data with the accumulated distribution expected from a Gaussian distribution, obtaining the p-value based on both discrepancies.
  - *Shapiro–Wilk*: analyzes the observed data to compute the level of symmetry and kurtosis (shape of the curve) in order to compute the difference with respect to a Gaussian distribution afterwards, obtaining the p-value from the sum of the squares of the discrepancies.
  - *D'Agostino–Pearson*: first computes the skewness and kurtosis to quantify how far from Gaussian the distribution is in terms of asymmetry and shape. It then calculates how far each of these values differs from the value expected with a Gaussian distribution, and computes a single p-value from the sum of the discrepancies.
- **Heteroscedasticity**: This property indicates the existence of a violation of the hypothesis of equality of variances. Levene's test is used for checking whether or not k samples present this homogeneity of variances (homoscedasticity). When observed data does not fulfill the normality condition, this test's result is more reliable than Bartlett's test [32], which checks the same property.

With respect to the independence condition, Demsăr suggests in [5] that independency is not truly verified in k-FCV and 5×2CV (a portion of samples is used either for training and testing in different partitions). Hold-out partitions can be safely take as independent, since training and tests partitions do not overlap.

The independence of the events in terms of getting results is usually obvious, given that they are independent runs of the algorithm with randomly generated initial seeds. In the following, we show a normality analysis by using KolmogorovSmirnov's, ShapiroWilk's and D'AgostinoPearson's tests, together with a heteroscedasticity analysis by using Levene's test in order to show the reader how to check such property.

2.2.2 Normality Test over the Group of Data Sets and Algorithms

Let us consider an small case of study, where we take into account an stochastic algorithm that needs a seed to generate its model. A classic example of these types of algorithms is the MLP. Using a small set of 6 well-known classification problems, we aim to analyze whether the conditions required to safely perform a parametric statistical analysis are held. We have used a 10-FCV validation scheme in which MLP is run 5 times per fold, thus obtaining 50 results per data set. Please note that using a k-FCV will mean that independence is not held but it is the most common validation scheme used in classification so this study case turns out to be relevant.

First of all, we want to check if our samples follow a normal distribution. In Table 2.2 the p-values obtained for the normality test were described in the previous section. As we can observe, in many cases the normality assumption is not held (indicated by an "a" in the table).

In addition to this general study, we show the sample distribution in three cases, with the objective of illustrating representative cases in which the normality tests obtain different results.

From Fig. 2.4 to 2.6, different examples of graphical representations of histograms and Q-Q graphics are shown. A histogram represents a statistical variable by using bars, so that the area of each bar is proportional to the frequency of the represented values. A Q-Q graphic represents a confrontation between the quartiles from data observed and those from the normal distributions.

In Fig. 2.4 we can observe a general case in which the property of abnormality is clearly presented. On the contrary, Fig. 2.5 is the illustration of a sample whose distribution follows a normal shape, and the three normality tests employed verified

Table 2.2 Normality test applied to a sample case

	Cleveland	Glass	Iris	Pima	Wine	Wisconsin
Kolmogorov–Smirnov	0.09	0.00[a]	0.00[a]	0.20	0.00[a]	0.09[a]
Shapiro–Wilk	0.04	0.00[a]	0.00[a]	0.80	0.00[a]	0.02[a]
D'Agostino–Pearson	0.08	0.01[a]	0.02[a]	0.51	0.00[a]	0.27

[a] indicates that the normality is not satisfied

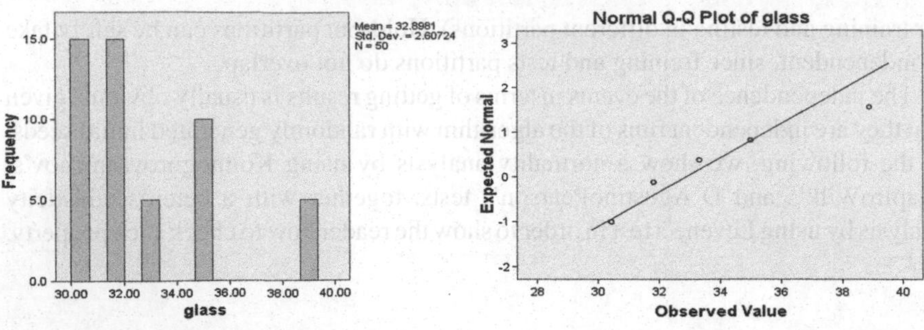

Fig. 2.4 Example of non-normal distribution: *glass* data set for MLP: Histogram and Q-Q graphic

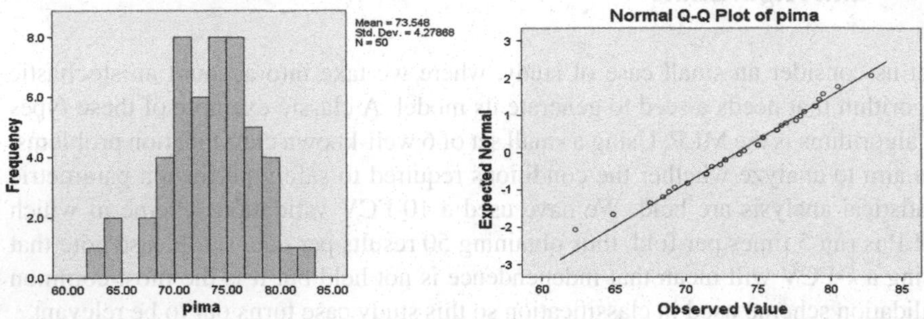

Fig. 2.5 Example of normal distribution: *pima* data set for MLP: Histogram and Q-Q graphic

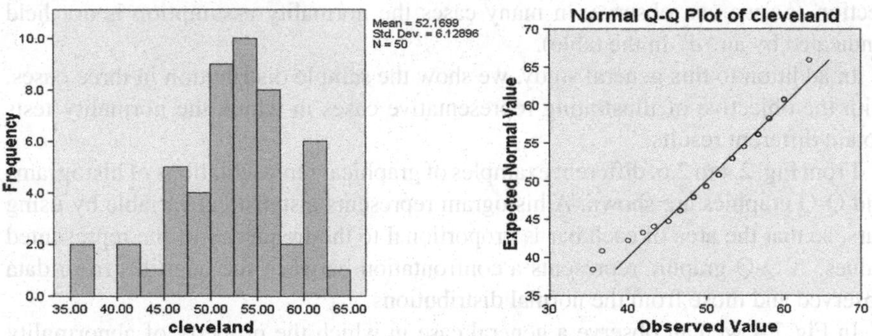

Fig. 2.6 Example of a special case: *cleveland* data set for MLP: Histogram and Q-Q graphic

this fact. Finally, Fig. 2.6 shows a special case where the similarity between both distributions, the sample of results and the normal distribution, is not confirmed by all normality tests. In this case, one normality test could work better than another, depending on types of data, number of ties or number of results collected. Due to this fact, we have employed three well-known normality tests for studying the normality condition. The choice of the most appropriate normality test depending on the problem is out of the scope of this book.

Table 2.3 Test of heteroscedasticity of levene (based on means)

Cleveland	Glass	Iris	Pima	Wine	Wisconsin
(0.000)[a]	(0.000)[a]	(0.000)[a]	(0.003)[a]	(0.000)[a]	(0.000)[a]

[a] indicates that homocedasticity is not satisfied

The third condition needing to be fulfilled is heteroscedasticity. Applying Levene's test to the samples of the six data sets results in Table 2.3.

Clearly, in both cases, the non fulfillment of the normality and homoscedasticity conditions is perfectible. In most functions, the normality condition is not verified in a single-problem analysis. The homoscedasticity is also dependent of the number of algorithms studied, because it checks the relationship of the variances of all population samples. Even though in this case we only analyze this condition in results for two algorithms, the condition is also not fulfilled in many other cases.

Obtaining results in a single data set analysis when using stochastics ML algorithms is a relatively easy task, due to the fact that new results can be yielded in new runs of the algorithms. In spite of this fact, a sample of 50 results that should be large enough to fulfill the parametric conditions does not always verify the necessary precepts for applying parametric tests, as we could see in the previous section.

On the other hand, other ML approaches are not stochastic and it is not possible to obtain a larger sample of results. This makes the comparison between stochastic ML methods and deterministic ML algorithms difficult, given that the sample of results might not be large enough or it might be necessary to use procedures which can operate with samples of different size.

For all these reasons, the use of non-parametric test for comparing ML algorithms is recommended [5].

2.2.3 *Non-parametric Tests for Comparing Two Algorithms in Multiple Data Set Analysis*

The authors are usually familiarized with parametric tests for pairwise comparisons. ML approaches have been compared through parametric tests by means of paired *t* tests.

In some cases, the *t* test is accompanied with the non-parametric Wilcoxon test applied over multiple data sets. The use of these types of tests is correct when we are interested in finding the differences between two methods, but they must not be used when we are interested in comparisons that include several methods. In the case of repeating pairwise comparisons, there is an associated error that grows as the number of comparisons done increases, called the family-wise error rate (FWER), defined as the probability of at least one error in the family of hypotheses. To solve this problem, some authors use the Bonferroni correction for applying paired *t* test in their works [27] although is not recommended.

In this section our interest lies in presenting a methodology for analyzing the results offered by a pair algorithms in a certain study, by using non-parametric tests in a multiple data set analysis. Furthermore, we want to comment on the possibility of comparison with other deterministic ML algorithms. Non-parametric tests could be applied to a small sample of data and their effectiveness have been proved in complex experiments. They are preferable to an adjustment of data with transformations or to a discarding of certain extreme observations (outliers) [16].

This section is devoted to describing a non-parametric statistical procedure for performing pairwise comparisons between two algorithms, also known as the Wilcoxon signed-rank test, Sect. 2.2.3.1; and to show the operation of this test in the presented case study, Sect. 2.2.3.2.

2.2.3.1 Wilcoxon Signed-Ranks Test

This is the analogue of the paired t-test in non-parametric statistical procedures; therefore, it is a pairwise test that aims to detect significant differences between two sample means, that is, the behavior of two algorithms. Let d_i be the difference between the performance scores of the two classifiers on i th out of N_{ds} data sets. The differences are ranked according to their absolute values; average ranks are assigned in case of ties. Let R^+ be the sum of ranks for the data sets on which the first algorithm outperformed the second, and R^- the sum of ranks for the opposite. Ranks of $d_i = 0$ are evenly split among the sums; if there is an odd number of them, one is ignored:

$$R^+ = \sum_{d_i>0} rank(d_i) + \frac{1}{2}\sum_{d_i=0} rank(d_i)$$

$$R^+ = \sum_{d_i<0} rank(d_i) + \frac{1}{2}\sum_{d_i=0} rank(d_i)$$

Let T be the smaller of the sums, $T = min(R^+, R^-)$. If T is less than or equal to the value of the distribution of Wilcoxon for N_{ds} degrees of freedom ([32], Table B.12), the null hypothesis of equality of means is rejected.

Wilcoxon signed ranks test is more sensible than the t-test. It assumes commensurability of differences, but only qualitatively: greater differences still count more, which is probably desired, but the absolute magnitudes are ignored. From the statistical point of view, the test is safer since it does not assume normal distributions. Also, the outliers (exceptionally good/bad performances on a few data sets) have less effect on the Wilcoxon than on the t test. The Wilcoxon test assumes continuous differences d_i, therefore they should not be rounded to one or two decimals, since this would decrease the power of the test due to a high number of ties.

Please note when the assumptions of the paired t test are met, Wilcoxon signed-ranks test is less powerful than the paired t test. On the other hand, when the assumptions are violated, the Wilcoxon test can be even more powerful than the t test. This allows us to apply it over the means obtained by the algorithms in each data set, without any assumptions about the sample of results obtained.

2.2.3.2 A Case Study: Performing Pairwise Comparisons

In the following, we will perform the statistical analysis by means of pairwise comparisons by using the results of performance measures obtained by the algorithms taken as reference for this section: MLP, RBFN, SONN and LVQ. A similar yet more detailed study can be found in [20].

In order to compare the results between two algorithms and to stipulate which one is the best, we can perform a Wilcoxon signed-rank test for detecting differences in both means. This statement must be enclosed by a probability of error, that is the complement of the probability of reporting that two systems are the same, called the p value [32]. The computation of the p value in the Wilcoxon distribution could be carried out by computing a normal approximation [24]. This test is well known and it is usually included in standard statistics packages (such as SPSS, R, SAS, etc.) as well as in open source implementations like as in KEEL [9] (see Chap. 10).

In Table 2.4 the ranks obtained by each method are shown. Table 2.5 shows the results obtained in all possible comparisons of the algorithms considered in the example study. We stress with bullets the winning algorithm in each row/column when the p value associated is below 0.1 and/or 0.05.

The comparisons performed in this study are independent, so they never have to be considered in a whole. If we try to extract a conclusion which involves more than one comparison from the previous tables, we will lose control of the FWER. For instance,

Table 2.4 Ranks computed by the Wilcoxon test

	(1)	(2)	(3)	(4)	(5)
MLP-CG-C (2)	19.0	-	16.0	0.0	15.0
RBFN-C (3)	11.0	5.0	-	0.0	5.0
SONN-C (4)	21.0	21.0	21.0	-	21.0
LVQ-C (5)	16.0	6.0	16.0	0.0	-

Row algorithm is the reference

Table 2.5 Summary of the Wilcoxon test

	(1)	(2)	(3)	(4)	(5)
MLP-CG-C (2)		-		○	
RBFN-C (3)			-	○	
SONN-C (4)	•	•	•	-	•
LVQ-C (5)				○	-

• = the method in the row improves the method of the column
○ = the method in the column improves the method of the row. Upper diagonal of level significance $\alpha = 0.9$, Lower diagonal level of significance $\alpha = 0.95$

the statement: "SONN obtains a classification rate better than RBFN, LVQ and MLP algorithms with a p value lower than 0.05" is incorrect, since we do not prove the control of the FWER. The SONN algorithm really outperforms MLP, RBFN and LVQ algorithms considering classification rate in independent comparisons.

The true statistical signification for combining pairwise comparisons is given by Eq. 2.2:

$$
\begin{aligned}
p &= P(reject\ H_0|H_0\ true) \\
&= 1 - P(Accept\ H_0|H_0\ true) \\
&= 1 - P(Accept\ A_k = A_i, i = 1,\ldots,k+1|H_0\ true) \\
&= 1 - \prod_{i=1}^{k-1} P(Accept\ A_k = A_i|H_0\ true) \\
&= 1 - \prod_{i=1}^{k-1} [1 - P(Reject\ A_k = A_i|H_0\ true)] \\
&= 1 - \prod_{i=1}^{k-1} (1 - p_{H_i})
\end{aligned}
\tag{2.2}
$$

2.2.4 *Non-parametric Tests for Multiple Comparisons Among More than Two Algorithms*

When a new ML algorithm proposal is developed or just being taking as reference, it could be interesting to compare it with previous approaches. Making pairwise comparisons allows this analysis to be conducted, but the experiment wise error can not be previously controlled. Furthermore, a pairwise comparison is not influenced by any external factor, whereas in a multiple comparison, the set of algorithms chosen can determine the results of the analysis.

Multiple comparison procedures are designed for allowing the FWER to be fixed before performing the analysis and for taking into account all the influences that can exist within the set of results for each algorithm. Following the same structure as in the previous section, the basic and advanced non-parametrical tests for multiple comparisons are described in Sect. 2.2.4.1 and their application on the case study is conducted in Sect. 2.2.4.2.

2.2.4.1 Friedman Test and Post-hoc Tests

One of the most frequent situations where the use of statistical procedures is requested is in the joint analysis of the results achieved by various algorithms. The groups of differences between these methods (also called blocks) are usually associated with the problems met in the experimental study. For example, in a multiple problem comparison, each block corresponds to the results offered for a specific problem. When referring to multiple comparisons tests, a block is composed of three or more subjects or results, each one corresponding to the performance evaluation of the algorithm for the problem.

Please remember that in pairwise analysis, if we try to extract a conclusion involving more than one pairwise comparisons, we will obtain an accumulated error coming from its combination losing the control on the Family-Wise Error Rate (FWER) (see Eq. 2.2).

This section is devoted to describing the use of several procedures for multiple comparisons considering a control method. In this sense, a control method can be defined as the most interesting algorithm for the researcher of the experimental study (usually its new proposal). Therefore, its performance will be contrasted against the rest of the algorithms of the study.

The best-known procedure for testing the differences between more than two related samples, the Friedman test, will be introduced in the following.

Friedman test

The Friedman test [7, 8] (Friedman two-way analysis of variances by ranks) is a nonparametric analog of the parametric two-way analysis of variance. It can be used to answer the following question: in a set of k samples (where $k \geq 2$), do at least two of the samples represent populations with different median values?. The Friedman test is the analog of the repeated measures ANOVA in non-parametric statistical procedures; therefore, it is a multiple comparisons test that aims to detect significant differences between the behavior of two or more algorithms.

The null hypothesis for Friedman's test states equality of medians between the populations. The alternative hypothesis is defined as the negation of the null hypothesis, so it is nondirectional.

The Friedman test method is described as follows: It ranks the algorithms for each data set separately, the best performing algorithm getting the rank of 1, the second best rank 2, and so on. In case of ties average ranks are assigned. Let r_i^j be the rank of the jth of k algorithms on the ith of N data sets. The Friedman test compares the average ranks of algorithms, $R_j = \frac{1}{N}\sum_i r_i^j$. Under the null hypothesis, which states that all the algorithms are equivalent and so their ranks R_j should be equal, the Friedman statistic:

$$\chi_F^2 = \frac{12\,N}{k(k+1)}\left[\sum jR_j^2 - \frac{k(k+1)^2}{4}\right] \quad (2.3)$$

which is distributed according to a χ^2 distribution with $k-1$ degrees of freedom,when n and k are big enough(as a rule of a thumb, $n > 10$ and $k > 5$).

Iman–Davenport test

Iman and Davenport [15] proposed a derivation from the Friedman statistic given that this last metric often produces a conservative effect not desired. The proposed statistic is

$$F_{ID} = \frac{(n-1)\chi_F^2}{n(k-1)\chi_F^2} \tag{2.4}$$

which is distributed according to an F distribution with $k1$ and $(k1)(N1)$ degrees of freedom. See Table A10 in [24] to find the critical values.

A drawback of the ranking scheme employed by the Friedman test is that it allows for intra-set comparisons only. When the number of algorithms for comparison is small, this may pose a disadvantage, since inter-set comparisons may not be meaningful. In such cases, comparability among problems is desirable. The method of aligned ranks [12] for the Friedman test overcomes this problem but for the sake of simplicity we will not elaborate on such an extension.

Post-hoc procedures

The rejection of the null hypothesis in both tests described above does not involve the detection of the existing differences among the algorithms compared. They only inform us of the presence of differences among all samples of results compared. In order to conducting pairwise comparisons within the framework of multiple comparisons, we can proceed with a post-hoc procedure. In this case, a control algorithm (maybe a proposal to be compared) is usually chosen. Then, the post-hoc procedures proceed to compare the control algorithm with the remain $k-1$ algorithms. In the following, we describe three post-hoc procedures:

- Bonferroni-Dunn's procedure [32]: it is similar to Dunnet's test for ANOVA designs. The performance of two algorithms is significantly different if the corresponding average of rankings is at least as great as its critical difference (CD).

$$CD = q_\alpha \sqrt{\frac{k(k+1)}{6N}} \tag{2.5}$$

  The value of q_α is the critical value of Q' for a multiple non-parametric comparison with a control (Table B.16 in [32]).
- Holm [13] procedure: for contrasting the procedure of Bonferroni-Dunn, we dispose of a procedure that sequentially checks the hypotheses ordered according to their significance. We will denote the p-values ordered by $p1, p2, \ldots$, in the way that $p1 \leq p2 \leq \cdots \leq p1$. Holm's method compares each p_i with $\alpha/(ki)$ starting from the most significant p-value. If p_1 is lower than $\alpha/(k1)$, the corresponding hypothesis is rejected and it leaves us to compare p_2 with $\alpha/(k2)$. If the second

hypothesis is rejected, we continue with the process. As soon as a certain hypothesis can not be rejected, all the remaining hypotheses are maintained as supported. The statistic for comparing the i algorithm with the j algorithm is:

$$z = (R_i - R_j)/\sqrt{\frac{k(k+1)}{6N}}$$

The value of z is used for finding the corresponding probability from the table of the normal distribution (p-value), which is compared with the corresponding value of α. Holm's method is more powerful than Bonferroni-Dunn's and it does no additional assumptions about the hypotheses checked.

- Hochberg [11] procedure: It is a step-up procedure that works in the opposite direction to Holm's method, comparing the largest p-value with ?, the next largest with $\alpha/2$ and so forth until it encounters a hypothesis it can reject. All hypotheses with smaller p values are then rejected as well. Hochberg's method is more powerful than Holm's [23].

The post-hoc procedures described above allow us to know whether or not a hypothesis of comparison of means could be rejected at a specified level of significance α. However, it is very interesting to compute the p-value associated to each comparison, which represents the lowest level of significance of a hypothesis that results in a rejection. In this manner, we can know whether two algorithms are significantly different and also get a metric of how different they are.

In the following, we will describe the method for computing these exact p-values for each test procedure, which are called "adjusted p-values" [30].

- The adjusted p-value for BonferroniDunn's test (also known as the Bonferroni correction) is calculated by $p_{Bonf} = (k-1)p_i$.
- The adjusted p-value for Holm's procedure is computed by $p_{Holm} = (k-i)p_i$. Once all of them have been computed for all hypotheses, it will not be possible to find an adjusted p-value for the hypothesis i lower than that for the hypothesis j, $j < i$. In this case, the adjusted p-value for hypothesis i is set equal to the p-values associated to the hypothesis j.
- The adjusted p-value for Hochberg's method is computed with the same formula as Holm's, and the same restriction is applied in the process, but to achieve the opposite, that is, so that it will not possible to find an adjusted p-value for the hypothesis i lower than for the hypothesis j, $j > i$.

2.2.4.2 A Case Study: Performing Multiple Comparisons

In this section we carry out a toy example on the analysis of a multiple comparison using the same ML algorithms as Sect. 2.2.3.2: MLP, RBFN, SONN and LVQ.

In Table 2.6 we show the ranks obtained by each algorithm for Friedman test. From this table we can observe that SONN is the algorithm with the lowest rank and hence will act as the control algorithm.

Table 2.6 Average rankings of the algorithms (Friedman)

Algorithm	Ranking
MLP-CG-C	2.5
RBFN-C	3.3333
SONN-C	1
LVQ-C	3.1667

Computing the Friedman and Iman–Davenport statistics as in Eqs. (2.3) and (2.4) the respective values are:

- *Friedman statistic* (distributed according to chi-squared with 3 degrees of freedom): 12.2. p-value computed by Friedman Test: **0.006729**.
- *Iman and Davenport statistic* (distributed according to F-distribution with 3 and 15 degrees of freedom): 10.517241. p-value computed by Iman and Daveport Test: **0.000561296469**.

In our case, both Friedman's and ImanDavenport's tests indicate that significant differences in the results are found in the three validations used in this study. Due to these results, a post-hoc statistical analysis is required. In this analysis, we choose the best performing method, SONN, as the control method for comparison with the rest of algorithms.

Post-hoc comparision

We will first present the results obtained for Bonferroni-Dunn's, Holm's and Hochberg's post-hoc tests with no adjustment of the p-values. Table 2.7 summarizes the unadjusted p-values for each algorithm when compared to SONN.

By computing Bonferroni-Dunn's CD as in 2.5 those hypotheses that have an unadjusted p-value $\leq$ 0.016667 are rejected. By using the z value indicated for Holm's and Hochberg's procedures, we can observe that they reject those hypotheses that have an unadjusted p-value $\leq$ 0.05. The reader may notice that Bonferroni-Dunn's is not able to reject the null-hypothesis for SONN versus MLP, while Holm's and Hochberg's are able to reject all null-hypothesis due to their higher statistical power.

The reader may usually refer directly to the adjusted p-values for the post-hoc methods, as they make searching for critical differences unnecessary and improve

Table 2.7 Post Hoc comparison table for $\alpha = 0.05$ (Friedman)

i	Algorithm	$z = (R_0 - R_i)/SE$	p	Holm Hochberg
3	RBFN-C	3.130495	0.001745	0.016667
2	LVQ-C	2.906888	0.00365	0.025
1	MLP-CG-C	2.012461	0.044171	0.05

Table 2.8 Adjusted p-values for the post-hoc tests

i	Algorithm	Unadjusted p	p_{Bonf}	p_{Holm}	$p_{Hochberg}$
1	RBFN-C	0.001745	0.005235	0.005235	0.005235
2	LVQ-C	0.00365	0.010951	0.007301	0.007301
3	MLP-CG-C	0.044171	0.132514	0.044171	0.044171

the readability. Table 2.8 contains all the adjusted p-values for Bonferroni-Dunn's, Holm's and Hochberg's test from the unadjusted values.

Taking a significance level of $\alpha = 0.05$ we can observe that the conclusions obtained from Table 2.8 are the same than those obtained in Table 2.7 without the need to establish the unadjusted p-value that acts as a threshold for the null-hypothesis rejection. The user only needs to observe those adjusted p-values that fall under the desired α significance level.

References

1. Alpaydin, E.: Introduction to Machine Learning, 2nd edn. MIT Press, Cambridge (2010)
2. Bache, K., Lichman, M.: UCI machine learning repository (2013). http://archive.ics.uci.edu/ml
3. Barandela, R., Sánchez, J.S., García, V., Rangel, E.: Strategies for learning in class imbalance problems. Pattern Recognit. **36**(3), 849–851 (2003)
4. Ben-David, A.: A lot of randomness is hiding in accuracy. Eng. Appl. Artif. Intell. **20**(7), 875–885 (2007)
5. Dĕmsar, J.: Statistical comparisons of classifiers over multiple data sets. J. Mach. Learn. Res. **7**, 1–30 (2006)
6. Efron, B., Gong, G.: A leisurely look at the bootstrap, the jackknife, and cross-validation. Am. Stat. **37**(1), 36–48 (1983)
7. Friedman, M.: The use of ranks to avoid the assumption of normality implicit in the analysis of variance. J. Am. Stat. Assoc. **32**(200), 675–701 (1937)
8. Friedman, M.: A comparison of alternative tests of significance for the problem of m rankings. Ann. Math. Stat. **11**(1), 86–92 (1940)
9. García, S., Fernández, A., Luengo, J., Herrera, F.: Advanced nonparametric tests for multiple comparisons in the design of experiments in computational intelligence and data mining: Experimental analysis of power. Inf. Sci. **180**(10), 2044–2064 (2010)
10. García, S., Herrera, F.: An extension on "statistical comparisons of classifiers over multiple data sets" for all pairwise comparisons. J. Mach. Learn. Res. **9**, 2677–2694 (2008)
11. Hochberg, Y.: A sharper bonferroni procedure for multiple tests of significance. Biometrika **75**(4), 800–802 (1988)
12. Hodges, J., Lehmann, E.: Rank methods for combination of independent experiments in analysis of variance. Ann. Math. Statist **33**, 482–497 (1962)
13. Holm, S.: A simple sequentially rejective multiple test procedure. Scand. J. Stat. **6**, 65–70 (1979)
14. Huang, J., Ling, C.X.: Using AUC and accuracy in evaluating learning algorithms. IEEE Trans. Knowl. Data Eng. **17**(3), 299–310 (2005)
15. Iman, R., Davenport, J.: Approximations of the critical region of the Friedman statistic. Commun. Stat. **9**, 571–595 (1980)

16. Koch, G.: The use of non-parametric methods in the statistical analysis of a complex split plot experiment. Biometrics **26**(1), 105–128 (1970)
17. Kohavi, R.: A study of cross-validation and bootstrap for accuracy estimation and model selection. Proceedings of the 14th international joint conference on Artificial intelligence. IJCAI'95, vol. 2, pp. 1137–1143. Morgan Kaufmann Publishers Inc., San Francisco, CA (1995)
18. Landgrebe, T.C., Duin, R.P.: Efficient multiclass ROC approximation by decomposition via confusion matrix perturbation analysis. IEEE Trans. Pattern Anal. Mach. Intell. **30**(5), 810–822 (2008)
19. Lim, T.S., Loh, W.Y., Shih, Y.S.: A comparison of prediction accuracy, complexity, and training time of thirty-three old and new classification algorithms. Mach. Learn. **40**(3), 203–228 (2000)
20. Luengo, J., García, S., Herrera, F.: A study on the use of statistical tests for experimentation with neural networks: Analysis of parametric test conditions and non-parametric tests. Expert Syst. Appl. **36**(4), 7798–7808 (2009)
21. Moreno-Torres, J.G., Sáez, J.A., Herrera, F.: Study on the impact of partition-induced dataset shift on k -fold cross-validation. IEEE Trans. Neural Netw. Learn. Syst. **23**(8), 1304–1312 (2012)
22. Salzberg, S.L.: On comparing classifiers: Pitfalls to avoid and a recommended approach. Data Min. Knowl. Discov. **1**(3), 317–328 (1997)
23. Shaffer, J.P.: Multiple hypothesis testing. Annu. Rev. Psychol. **46**(1), 561–584 (1995)
24. Sheskin, D.J.: Handbook of Parametric and Nonparametric Statistical Procedures. Chapman & Hall/CRC, Boca Raton (2007)
25. Sokolova, M., Japkowicz, N., Szpakowicz, S.: Beyond accuracy, f-score and roc: A family of discriminant measures for performance evaluation. In: A. Sattar, B.H. Kang (eds.) Australian Conference on Artificial Intelligence, Lecture Notes in Computer Science, vol. 4304, pp. 1015–1021. Springer (2006).
26. Stone, M.: Asymptotics for and against cross-validation. Biometrika **64**(1), 29–35 (1977)
27. Tan, K.C., Yu, Q., Ang, J.H.: A coevolutionary algorithm for rules discovery in data mining. Int. J. Syst. Sci. **37**(12), 835–864 (2006)
28. Witten, I.H., Frank, E.: Data Mining: Practical Machine Learning Tools and Techniques, Second Edition (Morgan Kaufmann Series in Data Management Systems). Morgan Kaufmann Publishers Inc., San Francisco (2005)
29. Wolpert, D.H., Macready, W.G.: No free lunch theorems for optimization. Trans. Evol. Comp. **1**(1), 67–82 (1997)
30. Wright, S.P.: Adjusted P-values for simultaneous inference. Biometrics **48**(4), 1005–1013 (1992)
31. Youden, W.J.: Index for rating diagnostic tests. Cancer **3**(1), 32–35 (1950)
32. Zar, J.: Biostatistical Analysis, 4th edn. Prentice Hall, Upper Saddle River (1999)

Chapter 3
Data Preparation Basic Models

Abstract The basic preprocessing steps carried out in Data Mining convert real-world data to a computer readable format. An overall overview related to this topic is given in Sect. 3.1. When there are several or heterogeneous sources of data, an integration of the data is needed to be performed. This task is discussed in Sect. 3.2. After the data is computer readable and constitutes an unique source, it usually goes through a cleaning phase where the data inaccuracies are corrected. Section 3.3 focuses in the latter task. Finally, some Data Mining applications involve some particular constraints like ranges for the data features, which may imply the normalization of the features (Sect. 3.4) or the transformation of the features of the data distribution (Sect. 3.5).

3.1 Overview

Data gathered in data sets can present multiple forms and come from many different sources. Data directly extracted from relational databases or obtained from the real world is completely raw: it has not been transformed, cleansed or changed at all. Therefore, it may contain errors due to wrong data entry procedures or missing data, or inconsistencies due to ill-handled merging data processes.

Three elements define data quality [15]: accuracy, completeness and consistency. Unfortunately real-world data sets often present the opposite conditions, and the reasons may vary as mentioned above. Many preprocessing techniques have been devised to overcome the problems present in such real-world data sets and to obtain a final, reliable and accurate data set to later apply a DM technique [35].

Gathering all the data elements together is not an easy task when the examples come from different sources and they have to be merged in a single data set. Integrating data from different databases is usually called *data integration*. Different attribute names or table schemes will produce uneven examples that need to be consolidated. Moreover, attribute values may represent the same concept but with different names creating inconsistencies in the instances obtained. If some attributes are calculated from the others, the data sets will present a large size but the information contained will not scale accordingly: detecting and eliminating redundant attributes is needed.

S. García et al., *Data Preprocessing in Data Mining*,
Intelligent Systems Reference Library 72, DOI 10.1007/978-3-319-10247-4_3

Having an uniform data set without measurable inconsistences does not mean that the data is clean. Errors like MVs or uncontrolled noise may be still present. A *data cleaning* step is usually needed to filter or correct wrong data. Otherwise, the knowledge extracted by a DM algorithm will be barely accurate or DM algorithms will not be able to handle the data.

Ending up with a consistent and almost error-free data set does not mean that the data is in the best form for a DM algorithm. Some algorithms in DM work much better with normalized attribute values, such as ANNs or distance-based methods. Others are not able to work with nominal valued attributes, or benefit from subtle transformations in the data. *Data normalization* and *data transformation* techniques have been devised to adapt a data set to the needs of the DM algorithm that will be applied afterwards. Note that eliminating redundant attributes and inconsistencies may still yield a large data set that will slow down the DM process. The use of *data reduction* techniques to transform the data set are quite useful, as they can reduce the number of attributes or instances while maintaining the information as whole as possible.

To sum up, real-world data is usually incomplete, dirty and inconsistent. Therefore data preprocessing techniques are needed to improve the accuracy and efficiency the subsequent DM technique used. The rest of the chapter further describes the basic techniques used to perform the preparation of the data set, while leading the reader to the chapters where advanced techniques are deeper described and presented.

3.2 Data Integration

One hard problem in DM is to collect a single data set with information coming from varied and different sources. If the integration process is not properly performed redundancies and inconsistencies will soon appear, resulting in a decrease of the accuracy and speed of the subsequent DM processes. Matching the schema from different sources presents a notorious problem that usually does not usually come alone: inconsistent and duplicated tuples as well as redundant and correlated attributes are problems that the data set creation process will probably show sooner or later.

An essential part of the integration process is to build a data map that establishes how each instance should be arranged in a common structure to present a real-world example taken from the real world. When data is obtained from relational databases, it is usually *flattened*, gathered together into one single record. Some database frameworks enable the user to provide a map to directly traverse the database through in-database access utilities. While using this in-database mining tools has the advantage of not having to extract and create an external file for the data, it is not the best option for its treatment with preprocessing techniques. In this case extracting the data is usually the best option. Preprocessing takes time, and when the data is kept in the database, preprocessing has to be applied repeatedly. In this way, if the data is extracted to an external file, then processing and modeling it can be faster if the data is already preprocessed and completely fits in memory.

Automatic approaches used to integrate the data can be found in the literature, from techniques that match and find the schemas of the data [7, 8], to automatic procedures that reconcile different schemas [6].

3.2.1 Finding Redundant Attributes

Redundancy is a problem that should be avoided as much as possible. It will usually cause an increment in the data set size, meaning that the modeling time of DM algorithms is incremented as well, and may also induce overfitting in the obtained model. An attribute is redundant when it can be derived from another attribute or set of them. Inconsistencies in dimension or attribute names can cause redundancies as well.

Redundancies in attributes can be detected using correlation analysis. By means of such analysis we can measure how strong is the implication of one attribute to the other. When the data is nominal and the set of values is thus finite, the χ^2 (chi-squared) test is commonly applied. In numeric attributes the use of the correlation coefficient and the covariance is typical.

3.2.1.1 χ^2 Correlation Test

Suppose that two nominal attributes, A and B, contain c and r distinct values each, namely $a_1, \ldots, a_c$ and $b_1, \ldots, a_r$. We can check the correlation between them using the χ^2 test. In order to do so, a contingency table, with the joint events (A_i, B_j) in which attribute A takes the value a_i and the attribute B takes the value b_j, is created. Every possible joint event (A_i, B_j) has its own entry in the table. The χ^2 value (or Pearson χ^2 statistic) is computed as:

$$\chi^2 = \sum_{i=1}^{c} \sum_{j=1}^{r} \frac{(o_{ij} - e_{ij})^2}{e_{ij}}, \tag{3.1}$$

where o_{ij} is the observed frequency of the joint event (A_i, B_j), and e_{ij} is the expected frequency of (A_i, B_j) computed as:

$$e_{ij} = \frac{count(A = a_i) \times count(B = b_j)}{m}, \tag{3.2}$$

where m is the number of instances in the data set, $count(A = a_i)$ is the number of instances with the value a_i for attribute A and $count(B = b_j)$ is the number of instances having the value b_j for attribute B.

The χ^2 test checks the hypothesis that A and B are independent, with $(r - 1) \times (c - 1)$ degrees of freedom. The χ^2 statistic obtained in Eq. (3.1) is compared against any

χ^2 table using the suitable degrees of freedom or any available software that is able to provide this value. If the significance level of such a table is below the established one (or the statistic value computed is above the needed one in the table), we can say that the null hypothesis is rejected and therefore, A and B are statistically correlated.

3.2.1.2 Correlation Coefficient and Covariance for Numeric Data

When we have two numerical attributes, checking whether they are highly correlated or not is useful to determine if they are redundant. The most well-known correlation coefficient is the Pearson's product moment coefficient, given by:

$$r_{A,B} = \frac{\sum_{i=1}^{m}(a_i - \overline{A})(b_i - \overline{B})}{m\sigma_A\sigma_B} = \frac{\sum_{i=1}^{m}(a_i b_i) - m\overline{A}\,\overline{B}}{m\sigma_A\sigma_B}, \tag{3.3}$$

where m is the number of instances, a_i and b_i are the values of attributes A and B in the instances, $\overline{A}$ and $\overline{B}$ are the mean values of A and B respectively, and σ_A and σ_B are the standard deviations of A and B.

Please note that $-1 \le r_{A,B} \le +1$. When $r_{A,B} > 0$ it means that the two attributes are positively correlated: when values of A are increased, then the values of B are incremented too. The higher the coefficient is, the higher the correlation between them is. Having a high value of $r_{A,B}$ could also indicate that one of the two attributes can be removed.

When $r_{A,B} = 0$, it implies that attributes A and B are independent and no correlation can be found between them. If $r_{A,B} < 0$, then attributes A and B are negatively correlated and when the values of one attribute are increased, the values of the other attribute are decreased. Scatter plots can be useful to examine how correlated the data is and to visually check the results obtained.

Similarly to correlation, covariance is an useful and widely used measure in statistics in order to check how much two variables change together. Considering that the mean values are the expected values of attributes A and B, namely $E(A) = \overline{A}$ and $E(B) = \overline{B}$, the covariance between both is defined as

$$Cov(A, B) = E((A - \overline{A})(B - \overline{B})) = \frac{\sum_{i=1}^{m}(a_i - \overline{A})(b_i - \overline{B})}{m}. \tag{3.4}$$

It is easy to see the relation between the covariance and the correlation coefficient $r_{A,B}$ given in Eq. (3.3) expressed as

$$r_{A,B} = \frac{Cov(A, B)}{\sigma_A\sigma_B} \tag{3.5}$$

If two attributes vary similarly, when $A > \overline{A}$ then probably $B > \overline{B}$ and thus the covariance is positive. On the other hand, when one attribute tends to be above its expected value whereas the other is below its expected value, the covariance is negative. When the two variables are independent, it is satisfied that $E(A \cdot B) = E(A) \cdot E(B)$, and thus the covariance verifies

$$Cov(A, B) = E(A \cdot B) - \overline{A}\,\overline{B} = E(A) \cdot E(B) - \overline{A}\,\overline{B} = 0. \quad (3.6)$$

The reader must be cautious, as having $Cov(A, B) = 0$ does not imply the two attributes being independent, as some random variables may present a covariance of 0 but still being dependent. Additional assumptions (like the data follows multivariate normal distributions) are necessary if covariance is 0 to determine whether the two attributes are independent.

3.2.2 Detecting Tuple Duplication and Inconsistency

It is interesting to check, when the tuples have been obtained, that there are not any duplicated tuple. One source of duplication is the use of denormalized tables, sometimes used to speed up processes involving join operations.

Having duplicate tuples can be troublesome, not only wasting space and computing time for the DM algorithm, but they can also be a source of inconsistency. Due to errors in the entry process, differences in some attribute values (for example the identifier value) may produce identical repeated instances but which are considered as different. These samples are harder to detect than simply scanning the data set for duplicate instances.

Please note that sometimes the duplicity is subtle. For example, if the information comes from different sources, the systems of measurement may be different as well, resulting in some instances being actually the same, but not identified like that. Their values can be represented using the metric system and the imperial system in different sources, resulting in a not-so-obvious duplication. The instances may also be inconsistent if attribute values are out of the established range (usually indicated in the associated metadata for the data set), but this is an easy to check condition.

One of the most common sources of mismatches in the instances are the nominal attributes [9]. Analyzing the similarity between nominal attributes is not trivial, as distance functions are not applied in a straightforward way and several alternatives do exist. Several character-based distance measures for nominal values can be found in the literature. These and can be helpful to determine whether two nominal values are similar (even with entry errors) or different [9]:

- The **edit distance** [23] between two strings σ_1 and σ_2 is the minimum number of string operations (or *edit operations*) needed to convert one string in the other. Three types of edit operations are usually considered: inserting a character, replacing a character or deleting a character. Using dynamic programming the number of operations can be established. Modern versions of this distance measure establish different costs for each edit operation, depending on the characters involved [31].

- The edit distance does not work well if the string values have been shortened. The **affine gap distance** adds two edit operations: opening gap and extending gap. With these operations the truncated values can be matched with their original full string counterparts.
- Jaro introduced in [17] a string comparison algorithm mainly aimed to compare first and last names in census data. It uses the string lengths $|\sigma_1|$ and $|\sigma_2|$, the common c characters between strings σ_1 and σ_2, and t the number of transpositions by comparing the ordering of the most common characters in the string and calling a transposition each mismatch in the order. The Jaro distance value is computed as
$$Jaro(\sigma_1, \sigma_2) = \frac{1}{3}\left(\frac{c}{|\sigma_1|} + \frac{c}{|\sigma_2|} + \frac{c - t/2}{c}\right). \tag{3.7}$$
- Originally formulated for speech-recognition [22], **q-grams** are substrings of length q that are commonly shared between strings σ_1 and σ_2. From a string, the q-gram is obtained using a window of size q over the string and adding a special character shared between both strings and not in the alphabet used in the original encoding to fill in the beginning and the end of the string if needed. By means of hashing algorithms, matching between q-grams can be speeded up [12].

Although these measures work well for typographical errors, they fail with rearrangements of the string (for example placing the last name before the first name in some of the strings). Several options are available to the user to overcome this problem. Token-based similarity metrics are devised to compensate this problem. Measures based on atomic strings [26] or the WHIRL distance [4] are examples of these. Alternatives based on phonetic similarities can also be found. They try to match two strings considering how they are pronounced instead of how they are encoded. Metaphone [27] and ONCA [11] are examples of phonetic-based string matching algorithms. Please note that these algorithms are limited to the particular language used to read the strings and they are very dependant on the dialects as well.

Trying to detect similarities in numeric data is harder. Some authors encode the numbers as strings or use range comparisons (numbers within a range threshold are considered equivalent). However, these approaches are quite naive. More sophisticated techniques are being proposed, such as considering the distribution of the numeric data [21] or the extension of the cosine similarity metric used in WHIRL for numeric data [1]. Nevertheless, discrepancies in numeric data are often detected in the data cleaning step as will be shown in the next section, thanks to outlier or noise detection algorithms.

After we have presented some measures to detect duplicity in each attribute of an instance, describing procedures to establish whether two instances are duplicated or not is the following step. There are different approaches to this task, and they can be summarized as follows:

- Probabilistic approaches. Fellegi and Sunter [10] formulated the duplicate instance detection problem as Bayesian inference problem, and thus the Fellegi-Sunter model is the most widely used in probabilistic approaches. If the density function of an instance differs when it is an unique record from when it is duplicated, then a Bayesian inference problem can be formulated if these density functions are known. The Bayes decision rule is a common approach [9] and several variants minimizing the error and the cost are well-known. They use an Expectation-Maximization algorithm to estimate the conditional probabilities needed [34].

- Supervised (and semisupervised) approaches. Well-known ML algorithms have been used to detect duplicity in record entries. For example, in [3] CART is used for this task, whereas in [18] a SVM is used to merge the matching results for different attributes of the instances. Clustering techniques are also applied, using graph partitioning techniques [25, 32], to establish those instances which are similar and thus suitable for removing.
- Distance-based techniques. Simple approaches like the use of the distance metrics described above to establish the similar instances have been long considered in the field [26]. Weighted modifications are also recurrent in the literature [5] and even other approaches like ranking the most similar weighted instances to a given one to detect the less duplicated tuple among all are also used [13].
- When data is unsupervised, clustering algorithms are the most commonly used option. Clustering bootstrapping[33] or hierarchical graph models encode the attributes as binary "match-does not match" attributes to generate two probability distributions for the observed values (instead of modeling the distributions as it is done in the probabilistic approaches) [29].

3.3 Data Cleaning

After the data is correctly integrated into a data set, it does not mean that the data is free from errors. The integration may result in an inordinate proportion of the data being dirty [20]. Broadly, dirty data include missing data, wrong data and non-standard representation of the same data. If a high proportion of the data is dirty, applying a DM process will surely result in a unreliable model. Dirty data has varying degrees of impact depending on the DM algorithm, but it is difficult to quantify such an impact.

Before applying any DM technique over the data, the data must be cleaned to remove or repair dirty data. The sources of dirty data include data entry errors, data update errors, data transmission errors and even bugs in the data processing system. As a result, dirty data usually is presented in two forms: missing data and wrong (noisy) data. The authors in [20] also include under this categorization inconsistent instances, but we assume that such kind of erroneous instances have been already addressed as indicated in Sect. 3.2.2.

The presence of a high proportion of dirty data in the training data set and/or the testing data set will likely produce a less reliable model. The impact of dirty data also depends on the particular DM algorithm applied. Decision trees are known to be susceptible to noise (specially if the trees are of higher order than two) [2]. ANNs and distance based algorithms (like the KNN algorithm) are known to be susceptible to noise. The use of distance measures is heavily dependent on the values of the data, and if the data used is dirty the results will be faulty. However, the detection of the noisy data in the data set is not a trivial task [16] and a wrong detection will result in damage to the correct data.

The way of handling MVs and noisy data is quite different:

- *MVs* are treated by routines prior to the DM algorithm application. The instances containing MVs can be ignored, or filled in manually or with a constant. Elaborated strategies that use estimations over the data are recommended in order to obtain reliable and more general results. This task is deeper studied in Chap. 4.
- The presence of noise in data is often defined as a random error in a measured variable, changing its value. Basic statistical and descriptive techniques as scatter plots can be used to identify outliers. Multiple linear regression is considered to estimate the tendency of the attribute values if they are numerical. However, the most recommended approach in the literature is the noisy detection and treatment, usually by filtering. Chapter 5 is completely devoted to noise identification and filtering.

3.4 Data Normalization

The data collected in a data set may not be useful enough for a DM algorithm. Sometimes the attributes selected are *raw attributes* that have a meaning in the original domain from where they were obtained, or are designed to work with the operational system in which they are being currently used. Usually these original attributes are not good enough to obtain accurate predictive models. Therefore, it is common to perform a series of manipulation steps to transform the original attributes or to generate new attributes with better properties that will help the predictive power of the model. The new attributes are usually named *modeling variables* or *analytic variables*.

In this section we will focus on the transformations that do not generate new attributes, but they transform the distribution of the original values into a new set of values with the desired properties.

3.4.1 Min-Max Normalization

The min-max normalization aims to scale all the numerical values v of a numerical attribute A to a specified range denoted by $[new-min_A, new-max_A]$. Thus a transformed value is obtained by applying the following expression to v in order to obtain the new value v':

$$v' = \frac{v - min_A}{max_A - min_A}(new - max_A - new - min_A) + new - min_A, \quad (3.8)$$

where max_A and min_A are the original maximum and minimum attribute values respectively.

In the literature "normalization" usually refers to a particular case of the min-max normalization in which the final interval is [0, 1], that is, $new - min_A = 0$ and $new - max_A = 1$. The interval $[-1, 1]$ is also typical when normalizing the data.

This type of normalization is very common in those data sets being prepared to be used with learning methods based on distances. Using a normalization to rescale all the data to the same range of values will avoid those attributes with a large $max_A - min_A$ difference dominating over the other ones in the distance calculation, misleading the learning process by giving more importance to the former attributes. This normalization is also known for speeding up the learning process in ANNs, helping the weights to converge faster.

An alternative, but equivalent, formulation for the min-max normalization is obtained by using a base value $new - min_A$ and the desired new range R in which the values will be mapped after the transformation. Some well-known software packages such as SAS or Weka [14] use this type of formulation for the min-max transformation:

$$v' = new - min_A + R\left(\frac{v - min_A}{max_A - min_A}\right). \tag{3.9}$$

3.4.2 Z-score Normalization

In some cases, the min-max normalization is not useful or cannot be applied. When the minimum or maximum values of attribute A are not known, the min-max normalization is infeasible. Even when the minimum and maximum values are available, the presence of outliers can bias the min-max normalization by grouping the values and limiting the digital precision available to represent the values.

If $\overline{A}$ is the mean of the values of attribute A and σ_A is the standard deviation, original value v of A is normalized to v' using

$$v' = \frac{v - \overline{A}}{\sigma_A}. \tag{3.10}$$

By applying this transformation the attribute values now present a mean equal to 0 and a standard deviation of 1.

If the mean and standard deviation associated to the probability distribution are not available, it is usual to use instead the sample mean and standard deviation:

$$\overline{A} = \frac{1}{n}\sum_{i=1}^{n} v_i, \tag{3.11}$$

and

$$\sigma_A = +\sqrt{\frac{1}{n}\sum_{i=1}^{n}(v_i - \overline{A})^2}. \tag{3.12}$$

A variation of the z-score normalization, described in [15], uses the mean absolute deviation s_A of A instead of the standard deviation. It is computed as

$$s_A = \frac{1}{n}\sum_{i=1}^{n}|v_i - \overline{A}|. \tag{3.13}$$

As a result the z-score normalization now becomes:

$$v' = \frac{v - \overline{A}}{s_A} \tag{3.14}$$

An advantage of the s_A mean absolute deviation is that it is more robust to outliers than the standard deviation σ_A as the deviations from the mean calculated by $|v_i - \overline{A}|$ are not squared.

3.4.3 Decimal Scaling Normalization

A simple way to reduce the absolute values of a numerical attribute is to normalize its values by shifting the decimal point using a power of ten division such that the maximum absolute value is always lower than 1 after the transformation. This transformation is commonly known as decimal scaling [15] and it is expressed as

$$v' = \frac{v}{10^j}, \tag{3.15}$$

where j is the smallest integer such that $new - max_A < 1$.

3.5 Data Transformation

In the previous Sect. 3.4 we have shown some basic transformation techniques to adapt the ranges of the attributes or their distribution to a DM algorithm's needs. In this section we aim to present the process to create new attributes, often called transforming the attributes or the attribute set. Data transformation usually combines the original raw attributes using different mathematical formulas originated in business models or pure mathematical formulas.

3.5.1 Linear Transformations

In the area of scientific discoveries and machine control, normalizations may not be enough to adapt the data to improve the generated model. In these cases aggregating the information contained in various attributes might be beneficial. A family of simple methods that can be used to this purpose are linear transformations. They are based on simple algebraic transformations such as sums, averages, rotations, translations and so on. Let $A = A_1, A_2, \ldots, A_n$ be a set of attributes, let $B = B_1, B_2, \ldots, B_m$ be a subset of the complete set of attributes A. If the following expression is applied

$$Z = r_1 B_1 + r_2 B2 + \cdots + r_m B_M \tag{3.16}$$

a new derived attribute is constructed by taking a linear combination of attributes in B.

A special case arises when the m values are set to $r_1 = r_2 = \cdots = r_m = 1/m$, that situation averages the considered attributes in B:

$$Z = (B_1 + B_2 + \cdots + B_m)/m. \tag{3.17}$$

3.5.2 Quadratic Transformations

In quadratic transformations a new attribute is built as follows:

$$Z = r_{1,1} B_1^2 + r_{1,2} B_1 B_2 + \cdots + r_{m-1,m} B_{m-1} B_m + r_{m,m} B_m^2, \tag{3.18}$$

where $r_{i,j}$ is a real number. These kinds of transformations have been thoroughly studied and can help to transform data to make it separable.

In Table 3.1 we show an example of how quadratic transformations can help us to reveal knowledge that it is not explicitly present using the initial attributes of the data set. Let us consider a set of conic sections described by the coefficients of the algebraic expression that follows:

$$A_1 x^2 + A_2 xy + A_3 y^2 + A_4 x + A_5 y + A_6 = 0. \tag{3.19}$$

From Eq. (3.19) there is no obvious interpretation that can be used to label the tuples to any type of conic section. However, computing the quadratic transformation known as the discriminant given by

$$Z = (A_2^2 - 4A_1 A_3), \tag{3.20}$$

the sign of Z provides enough information to correctly label the tuples. Without the new derived attribute Z we could not be able to classify them.

Table 3.1 Applying quadratic transformations to identify the implicit conic figure

A_1	A_2	A_3	A_4	A_5	A_6	Z	Sign	Conic
1	0	42	2	34	−33	−168	−	Hyperbola
−5	0	0	−64	29	−68	0	0	Parabola
88	−6	−17	−79	97	−62	6,020	+	Ellipse
30	0	0	−53	84	−14	0	0	Parabola
1	19	−57	99	38	51	589	+	Ellipse
15	−39	35	98	−52	−40	−579	−	Hyperbola

Although the above given example indicates that transformations are necessary for knowledge discovery in certain scenarios, we usually do not have any clue on how such transformation can be found and when should them be applied. As [24] indicates, the best source for obtaining the correct transformation is usually the expert knowledge. Sometimes the best transformation can be discovered by brute force (see Sect. 3.5.4).

3.5.3 Non-polynomial Approximations of Transformations

Sometimes polynomial transformations, including the lineal and quadratic ones, are not enough to create new attributes able to facilitate the KDD task. In other words, each problem requires its own set of transformations and such transformations can adopt any form. For instance, let us consider several triangles in the plane described by the (X, Y) coordinates of their vertices as shown in Table 3.2. Considering only these attributes does not provide us any information about the relationship of the triangles, but by computing the length of the segments given by:

$$A = \sqrt{(X_1 - X_2)^2 + (Y_1 - Y_2)^2} \quad (3.21)$$

$$B = \sqrt{(X_2 - X_3)^2 + (Y_2 - Y_3)^2} \quad (3.22)$$

$$C = \sqrt{(X_1 - X_3)^2 + (Y_1 - Y_3)^2} \quad (3.23)$$

we can observe that all the segments are of the same length. Obtaining this conclusions from the original attributes was impossible and this example is useful to illustrate that some specific and non-polynomial attribute transformations are needed but they are also highly dependent of the problem domain. Selecting the appropriate transformation is not easy and expert knowledge is usually the best alternative to do so.

Table 3.2 Six triangles in the plane

X_1	Y_1	X_2	Y_2	X_3	Y_3	A	B	C
0.00	2.00	−2.00	0.00	2.00	0.00	2.83	2.83	4.00
−0.13	−2.00	2.00	−0.13	−2.00	0.13	2.83	2.83	4.00
−1.84	0.79	−0.79	−1.84	0.79	1.84	2.83	2.83	4.00
1.33	1.49	−1.49	1.33	1.49	−1.33	2.83	2.83	4.00
−2.00	−0.08	0.08	−2.00	−0.08	2.00	2.83	2.83	4.00
−1.99	−0.24	0.24	−1.99	−0.24	1.99	2.83	2.83	4.00

Adding the attributes A, B and C shows that they are all congruent

3.5.4 Polynomial Approximations of Transformations

In the last two sections we have observed that specific transformations may be needed to extract knowledge from a data set. However, help from an expert is not always available to dictate the form of the transformation and attributes to use. In [24] the author shows that when no knowledge is available, a transformation f can be approximated via a polynomial transformation using a brute search with one degree at a time.

Given a set of attributes $X_1, X_2, \ldots, X_n$, we want to compute a derived attribute Y from such a set of already existing features. We set the attribute Y as a function of the original attributes:

$$Y = f(X_1, X_2, \ldots, X_n). \tag{3.24}$$

Please note that f can be any kind of function. Due to the number of instances being finite, f can be expressed as a polynomial approximation. Each tuple $X_i = (X_1, X_2, \ldots, X_n)$ can be considered as a point in Euclidean space. Using the Weisstrass approximation, there is a polynomial function f that takes the value Y_i for each instance X_i.

There are as many polynomials verifying $Y = f(X)$ as we want, being different in their expression but with identical output for the point (i.e. the instances) given in the data set. As the number of instances in the data set increases, the approximations will be better. We can consider two different cases for our approximations: an intrinsic transformation which is not a polynomial and when it is intrinsically polynomial.

For the case where the intrinsic transformation is not polynomial, let us take as an example the first two columns of Table 3.2 representing two vertices of a triangle as the input tuples X, and the segment length between them as the attribute that we want to model Y. Adding more examples of two points and their distance, a polynomial approximation obtained using MATLAB is:

$$Y = 0.3002Y_1 - 1.1089X_2 + 0.8086Y_2. \tag{3.25}$$

This is the polynomial approximation of degree one. Adding columns $X_1 \cdot X_1$, $X_1 \cdot Y_1, \ldots, Y_2, \ldots, Y_2$ to the reference table, we can find the degree 2 polynomial approximation, obtained as follows:

$$\begin{aligned} Y = 0.5148 \cdot X_1 + 1.6402 \cdot Y_1 - 1.2406 \cdot X_2 - 0.9264 \cdot Y_2 - 0.6987 \cdot X_1 \cdot X_1 \\ -0.8897 \cdot X_1 \cdot Y_1 + 1.0401 \cdot X_1 \cdot X_2 + 1.2587 \cdot X_1 \cdot Y_2 \\ -0.4547 \cdot Y_1 \cdot Y_1 + 0.9598 \cdot Y_1 \cdot X_2 + 0.8365 \cdot Y_1 \cdot Y_2 \\ -0.0503 \cdot X_2 \cdot X_2 - 1.8903 \cdot X_2 \cdot Y_2 - 0.0983 \cdot Y_2 \cdot Y_2. \end{aligned} \tag{3.26}$$

We can extend this approximation to a degree 3 polynomial using a longer table with added attributes. However, the polynomials obtained are not unique, as they depend on the data set instances' values. As the size of the data set increases the polynomials will continue to vary and better approximate the new attribute Y.

On the other hand, when the intrinsic transformation is polynomial we need to add the cartesian product of the attributes needed for the polynomial degree approximation. As an example let us recall the example presented in Table 3.1. If we add more conic sections to the table and approximate the discriminant Z^* by a polynomial of second degree we will eventually obtain a similar expression as the number of examples in the data set grows. First, we use the original $A, \ldots, F$ attributes which approximated using MATLAB with a polynomial of degree one obtains:

$$Z^* = -4.5601A + 5.2601 + 3.6277C + 2.0358D - 6.4963E + 0.07279F. \tag{3.27}$$

If we want to correctly apply an approximation of second degree to our data set, we will have the original and the new $A \cdots A, A \cdots B, \ldots, F \cdots F$ attributes. With enough examples, we will retrieve the discriminant expression given by $Z^* = B^2 - AC$. As [24] indicates, the approximation obtained must be rounded to avoid the limitations of the computer digital precision and to retrieve the discriminant true expression.

3.5.5 Rank Transformations

A change in an attribute distribution can result in a change of the model performance, as we may reveal relationships that were obscured by the previous attribute distribution. The simplest transformation to accomplish this in numerical attributes is to replace the value of an attribute with its rank. The attribute will be transformed into a new attribute containing integer values ranging from 1 to m, being m the number of instances in the data set.

Afterwards, we can transform the ranks to normal scores representing their probabilities in the normal distribution by spreading these values on the gaussian curve using a simple transformation given by:

$$y = \Phi^{-1}\left(\frac{r_i - \frac{3}{8}}{m + \frac{1}{4}}\right), \tag{3.28}$$

being r_i the rank of the observation i and Φ the cumulative normal function.

This transformation is useful to obtain a new variable that is very likely to behave like to a normally distributed one. However, this transformation cannot be applied separately to the training and test partitions [30]. Therefore, this transformation is only recommended when the test and training data is the same.

3.5.6 Box-Cox Transformations

A big drawback when selecting the optimal transformation for an attribute is that we do not know in advance which transformation will be the best to improve the model performance. The Box-Cox transformation aims to transform a continuous variable into an almost normal distribution. As [30] indicates, this can be achieved by mapping the values using following the set of transformations:

$$y = \begin{cases} x^{\lambda-1}/\lambda, & \lambda \neq 0 \\ log(x), & \lambda = 0 \end{cases} \tag{3.29}$$

All linear, inverse, quadratic and similar transformations are special cases of the Box-Cox transformations. Please note that all the values of variable x in Eq. (3.29) must be positive. If we have negative values in the attribute we must add a parameter c to offset such negative values:

$$y = \begin{cases} (x+c)^{\lambda-1}/g\lambda, & \lambda \neq 0 \\ log(x+c)/g, & \lambda = 0 \end{cases} \tag{3.30}$$

The parameter g is used to scale the resulting values, and it is often considered as the geometric mean of the data. The value of λ is iteratively found by testing different values in the range from -3.0 to 3.0 in small steps until the resulting attribute is as close as possible to the normal distribution.

In [30] a likelihood function to be maximized depending on the value of λ is defined based on the work of Johnson and Wichern [19]. This function is computed as:

$$L(\lambda) = -\frac{n}{2} ln \left[\frac{1}{m} \sum_{j=1}^{m} (y_j - \overline{y})^2 \right] + (\lambda - 1) \sum_{j=1}^{m} ln x_j, \tag{3.31}$$

where y_j is the transformation of the value x_j using Eq. (3.29), and $\overline{y}$ is the mean of the transformed values.

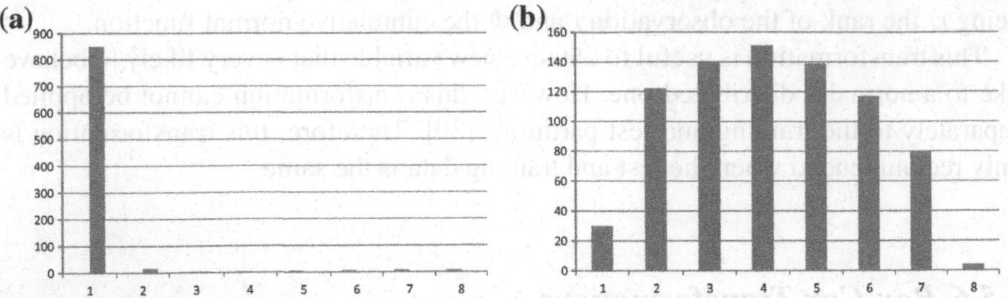

Fig. 3.1 Example of the histogram spreading made by a Box-Cox transformation: **a** before the transformation and **b** after the transformation

3.5.7 Spreading the Histogram

Spreading the histogram is a special case of Box-Cox transformations. As Box-Cox transforms the data to resemble a normal distribution, the histogram is thus spread as shown in Fig. 3.1.

When the user is not interested in converting the distribution to a normal one, but just spreading it, we can use two special cases of Box-Cox transformations [30]. Using the logarithm (with an offset if necessary) can be used to spread the right side of the histogram: $y = log(x)$. On the other hand, if we are interested in spreading the left side of the histogram we can simply use the power transformation $y = x^g$. However, as [30] shows, the power transformation may not be as appropriate as the Log transformation and it presents an important drawback: higher values of g may help to spread the histogram but they will also cause problems with the digital precision available.

3.5.8 Nominal to Binary Transformation

The presence of nominal attributes in the data set can be problematic, specially if the DM algorithm used cannot correctly handle them. This is the case of SVMs and ANNs. The first option is to transform the nominal variable to a numeric one, in which each nominal value is encoded by an integer, typically starting from 0 or 1 onwards. Although simple, this approach has two big drawbacks that discourage it:

- With this transformation we assume an ordering of the attribute values, as the integer values are ranked. However the original nominal values did not present any ranking among them.
- The integer values can be used in operations as numbers, whereas the nominal values cannot. This is even worse than the first point, as with this nominal to integer transformation we are establishing unequal differences between pairs of nominal values, which is not correct.

In order to avoid the aforementioned problems, a very typical transformation used for DM methods is to map each nominal attribute to a set of newly generated attributes. If N is the number of different values the nominal attribute has, we will substitute the nominal variable with a new set of binary attributes, each one representing one of the N possible values. For each instance, only one of the N newly created attributes will have a value of 1, while the rest will have the value of 0. The variable having the value 1 is the variable related to the original value that the old nominal attribute had. This transformation is also referred in the literature as 1-to-N transformation.

As [30] and [28] state, the new set of attributes are linearly dependent. That means that one of the attribute can be dismissed without loss of information as we can infer the value of one of the new attributes by knowing the values of the rest of them. A problem with this kind of transformation appears when the original nominal attribute has a large cardinality. In this case, the number of attributes generated will be large as well, resulting in a very sparse data set which will lead to numerical and performance problems.

3.5.9 Transformations via Data Reduction

In the previous sections, we have analyzed the processes to transform or create new attributes from the existing ones. However, when the data set is very large, performing complex analysis and DM can take a long computing time. Data reduction techniques are applied in these domains to reduce the size of the data set while trying to maintain the integrity and the information of the original data set as much as possible. In this way, mining on the reduced data set will be much more efficient and it will also resemble the results that would have been obtained using the original data set.

The main strategies to perform data reduction are Dimensionality Reduction (DR) techniques. They aim to reduce the number of attributes or instances available in the data set. Well known attribute reduction techniques are Wavelet transforms or Principal Component Analysis (PCA). Chapter 7 is devoted to attribute DR. Many techniques can be found for reducing the dimensionality in the number of instances, like the use of clustering techniques, parametric methods and so on. The reader will find a complete survey of IS techniques in Chap. 8. The use of binning and discretization techniques is also useful to reduce the dimensionality and complexity of the data set. They convert numerical attributes into nominal ones, thus drastically reducing the cardinality of the attributes involved. Chapter 9 presents a thorough presentation of these discretization techniques.

References

1. Agrawal, R., Srikant, R.: Searching with numbers. IEEE Trans. Knowl. Data Eng. **15**(4), 855–870 (2003)
2. Berry, M.J., Linoff, G.: Data Mining Techniques: For Marketing, Sales, and Customer Support. Wiley, New York (1997)

3. Cochinwala, M., Kurien, V., Lalk, G., Shasha, D.: Efficient data reconciliation. Inf. Sci. **137**(1–4), 1–15 (2001)
4. Cohen, W.W.: Integration of heterogeneous databases without common domains using queries based on textual similarity. In: Proceedings of the 1998 ACM SIGMOD International Conference on Management of Data. SIGMOD '98, pp. 201–212. New York (1998)
5. Dey, D., Sarkar, S., De, P.: Entity matching in heterogeneous databases: A distance based decision model. In: 31st Annual Hawaii International Conference on System Sciences (HICSS'98), pp. 305–313 (1998)
6. Do, H.H., Rahm, E.: Matching large schemas: approaches and evaluation. Inf. Syst. **32**(6), 857–885 (2007)
7. Doan, A., Domingos, P., Halevy, A.Y.: Reconciling schemas of disparate data sources: A machine-learning approach. In: Proceedings of the 2001 ACM SIGMOD International Conference on Management of Data, SIGMOD '01, pp. 509–520 (2001)
8. Doan, A., Domingos, P., Halevy, A.: Learning to match the schemas of data sources: a multi-strategy approach. Mach. Learn. **50**, 279–301 (2003)
9. Elmagarmid, A.K., Ipeirotis, P.G., Verykios, V.S.: Duplicate record detection: a survey. IEEE Trans. Knowl. Data Eng. **19**(1), 1–16 (2007)
10. Fellegi, I.P., Sunter, A.B.: A theory for record linkage. J. Am. Stat. Assoc. **64**, 1183–1210 (1969)
11. Gill, L.E.: OX-LINK: The Oxford medical record linkage system. In: Proceedings of the International Record Linkage Workshop and Exposition, pp. 15–33 (1997)
12. Gravano, L., Ipeirotis, P.G., Jagadish, H.V., Koudas, N., Muthukrishnan, S., Pietarinen, L., Srivastava, D.: Using q-grams in a DBMS for approximate string processing. IEEE Data Engineering Bull. **24**(4), 28–34 (2001)
13. Guha, S., Koudas, N., Marathe, A., Srivastava, D.: Merging the results of approximate match operations. In: Nascimento, M.A., Zsu, M.T., Kossmann, D., Miller, R.J., Blakeley, J.A., Schiefer, K.B. (eds.) VLDB. Morgan Kaufmann, San Francisco (2004)
14. Hall, M., Frank, E., Holmes, G., Pfahringer, B., Reutemann, P., Witten, I.H.: The WEKA data mining software: an update. SIGKDD Explor. Newsl. **11**(1), 10–18 (2009)
15. Han, J., Kamber, M., Pei, J.: Data mining: concepts and techniques. The Morgan Kaufmann Series in Data Management Systems, 2nd edn. Morgan Kaufmann, San Francisco (2006)
16. Hulse, J., Khoshgoftaar, T., Huang, H.: The pairwise attribute noise detection algorithm. Knowl. Inf. Syst. **11**(2), 171–190 (2007)
17. Jaro, M.A.: Unimatch: A record linkage system: User's manual. Technical report (1976)
18. Joachims, T.: Advances in kernel methods. In: Making Large-scale Support Vector Machine Learning Practical, pp. 169–184. MIT Press, Cambridge (1999)
19. Johnson, R.A., Wichern, D.W.: Applied Multivariate Statistical Analysis. Prentice-Hall, Englewood Cliffs (2001)
20. Kim, W., Choi, B.J., Hong, E.K., Kim, S.K., Lee, D.: A taxonomy of dirty data. Data Min. Knowl. Disc. **7**(1), 81–99 (2003)
21. Koudas, N., Marathe, A., Srivastava, D.: Flexible string matching against large databases in practice. In: Proceedings of the Thirtieth International Conference on Very Large Data Bases, VLDB '04, vol. 30, pp. 1078–1086. (2004)
22. Kukich, K.: Techniques for automatically correcting words in text. ACM Comput. Surv. **24**(4), 377–439 (1992)
23. Levenshtein, V.: Binary codes capable of correcting deletions. Insertions Reversals Sov. Phys. Doklady **163**, 845–848 (1965)
24. Lin, T.Y.: Attribute transformations for data mining I: theoretical explorations. Int. J. Intell. Syst. **17**(2), 213–222 (2002)
25. McCallum, A., Wellner, B.: Conditional models of identity uncertainty with application to noun coreference. Advances in Neural Information Processing Systems 17, pp. 905–912. MIT Press, Cambridge (2005)
26. Monge, A.E., Elkan, C.: The field matching problem: algorithms and applications. In: Simoudis, E., Han, J., Fayyad, U.M. (eds.) Proceedings of the Second International Conference on Knowledge Discovery and Data Mining (KDD-96), pp. 267–270. KDD, Portland, Oregon, USA (1996)

27. Philips, L.: Hanging on the metaphone. Comput. Lang. Mag. **7**(12), 39–44 (1990)
28. Pyle, D.: Data Preparation for Data Mining. Morgan Kaufmann, San Francisco (1999)
29. Ravikumar, P., Cohen, W.W.: A hierarchical graphical model for record linkage. In: Proceedings of the 20th Conference on Uncertainty in Artificial Intelligence, UAI '04, pp. 454–461 (2004)
30. Refaat, M.: Data Preparation for Data Mining Using SAS. Morgan Kaufmann, San Francisco (2007)
31. Ristad, E.S., Yianilos, P.N.: Learning string edit distance. IEEE Trans. Pattern Anal. Mach. Intell. **20**(5), 522–532 (1998)
32. Singla, P., Domingos, P.: Multi-relational record linkage. In: KDD-2004 Workshop on Multi-Relational Data Mining, pp. 31–48 (2004)
33. Verykios, V.S., Elmagarmid, A.K., Houstis, E.N.: Automating the approximate record-matching process. Inf. Sci. **126**(1–4), 83–98 (2000)
34. Winkler, W.E.: Improved decision rules in the Fellegi-Sunter model of record linkage. Technical report, Statistical Research Division, U.S. Census Bureau, Washington, DC (1993)
35. Zhang, S., Zhang, C., Yang, Q.: Data preparation for data mining. Appl. Artif. Intell. **17**(5–6), 375–381 (2003)

Chapter 4
Dealing with Missing Values

Abstract In this chapter the reader is introduced to the approaches used in the literature to tackle the presence of Missing Values (MVs). In real-life data, information is frequently lost in data mining, caused by the presence of missing values in attributes. Several schemes have been studied to overcome the drawbacks produced by missing values in data mining tasks; one of the most well known is based on preprocessing, formally known as imputation. After the introduction in Sect. 4.1, the chapter begins with the theoretical background which analyzes the underlying distribution of the missingness in Sect. 4.2. From this point on, the successive sections go from the simplest approaches in Sect. 4.3, to the most advanced proposals, focusing in the imputation of the MVs. The scope of such advanced methods includes the classic maximum likelihood procedures, like Expectation-Maximization or Multiple-Imputation (Sect. 4.4) and the latest Machine Learning based approaches which use algorithms for classification or regression in order to accomplish the imputation (Sect. 4.5). Finally a comparative experimental study will be carried out in Sect. 4.6.

4.1 Introduction

Many existing, industrial and research data sets contain MVs in their attribute values. Intuitively a MV is just a value for attribute that was not introduced or was lost in the recording process. There are various reasons for their existence, such as manual data entry procedures, equipment errors and incorrect measurements. The presence of such imperfections usually requires a preprocessing stage in which the data is prepared and cleaned [71], in order to be useful to and sufficiently clear for the knowledge extraction process. The simplest way of dealing with MVs is to discard the examples that contain them. However, this method is practical only when the data contains a relatively small number of examples with MVs and when analysis of the complete examples will not lead to serious bias during the inference [54].

MVs make performing data analysis difficult. The presence of MVs can also pose serious problems for researchers. In fact, inappropriate handling of the MVs in the analysis may introduce bias and can result in misleading conclusions being drawn

S. García et al., *Data Preprocessing in Data Mining*,
Intelligent Systems Reference Library 72, DOI 10.1007/978-3-319-10247-4_4

from a research study, and can also limit the generalizability of the research findings [96]. Three types of problems are usually associated with MVs in DM [5]:

1. loss of efficiency;
2. complications in handling and analyzing the data; and
3. bias resulting from differences between missing and complete data.

Recently some authors have tried to estimate how many MVs are needed to noticeably harm the prediction accuracy in classification [45].

Usually the treatment of MVs in DM can be handled in three different ways [27]:

- The first approach is to discard the examples with MVs in their attributes. Therefore deleting attributes with elevated levels of MVs is included in this category too.
- Another approach is the use of maximum likelihood procedures, where the parameters of a model for the complete portion of the data are estimated, and later used for imputation by means of sampling.
- Finally, the imputation of MVs is a class of procedures that aims to fill in the MVs with estimated ones. In most cases, a data set's attributes are not independent from each other. Thus, through the identification of relationships among attributes, MVs can be determined

We will focus our attention on the use of imputation methods. A fundamental advantage of this approach is that the MV treatment is independent of the learning algorithm used. For this reason, the user can select the most appropriate method for each situation faced. There is a broad family of imputation methods, from simple imputation techniques like mean substitution, KNN, etc.; to those which analyze the relationships between attributes such as: SVM-based, clustering-based, logistic regressions, maximum likelihood procedures and multiple imputation [6, 26].

The use of imputation methods for MVs is a task with a well established background. It is possible to track the first formal studies to several decades ago. The work of [54] laid the foundation of further work in this topic, specially in statistics. From their work, imputation techniques based on sampling from estimated data distributions followed, distinguishing between single imputation procedures (like Expectation-Maximization (EM) procedures [81]) and multiple imputation ones [82], the latter being more reliable and powerful but more difficult and restrictive to be applied.

These imputation procedures became very popular for quantitative data, and therefore they were easily adopted in other fields of knowledge, like bioinformatics [49, 62, 93], climatic science [85], medicine [94], etc. The imputation methods proposed in each field are adapted to the common characteristics of the data analyzed in it. With the popularization of the DM field, many studies in the treatment of MVs arose in this topic, particularly in the classification task. Some of the existent imputation procedures of other fields are adapted to be used in classification, for example adapting them to deal with qualitative data, while many specific approaches are proposed.

4.2 Assumptions and Missing Data Mechanisms

It is important to categorize the mechanisms which lead to the introduction of MVs [54]. The assumptions we make about the missingness mechanism and the MVs pattern of MVs can affect which treatment method could be correctly applied, if any.

When thinking about the missing data mechanism the probability distributions that lie beneath the registration of rectangular data sets should be taken into account, where the rows denote different registers, instances or cases, and the columns are the features or variables. A common assumption is that the instances are all independent and identically distributed (i.i.d.) draws of some multivariate probability distribution. This assumption is also made by Schafer in [82] where the schematic representation followed is depicted in Fig. 4.1.

X being the $n \times m$ rectangular matrix of data, we usually denote as x_i the ith row of X. If we consider the i.i.d. assumption, the probability function of the complete data can be written as follows:

$$P(X|\theta) = \prod_{i=1}^{n} f(x_i|\theta), \tag{4.1}$$

Attributes

Instances	1	2	3	4	5	...	m
1					?		
2			?				
3		?		?			
4							
5							
6						?	
7			?		?		
8							
9							
10			?			?	
11		?					
⋮				?			
n							?

Fig. 4.1 Data set with MVs denoted with a '?'

where f is the probability function for a single case and θ represents the parameters of the model that yield such a particular instance of data. The main problem is that the particular parameters' values θ for the given data are very rarely known. For this reason authors usually overcome this problem by considering distributions that are commonly found in nature and their properties are well known as well. The three distributions that standout among these are:

1. the multivariate normal distribution in the case of only real valued parameters;
2. the multinomial model for cross-classified categorical data (including loglinear models) when the data consists of nominal features; and
3. mixed models for combined normal and categorical features in the data [50, 55].

If we call X_{obs} the observed part of X and we denote the missing part as X_{mis} so that $X = (X_{obs}, X_{mis})$, we can provide a first intuitive definition of what *missing at random* (MAR) means. Informally talking, when the probability that an observation is missing may depend on X_{obs} but not on X_{mis} we can state that the missing data is missing at random.

In the case of MAR missing data mechanism, given a particular value or values for a set of features belonging to X_{obs}, the distribution of the rest of features is the same among the observed cases as it is among the missing cases. Following Schafer's example based on [79], let suppose that we dispose an $n \times p$ matrix called B of variables whose values are 1 or 0 when X elements are observed and missing respectively. The distribution of B should be related to X and to some unknown parameters ζ, so we dispose a probability model for B described by $P(B|X, \zeta)$. Having a MAR assumption means that this distribution does not depend on X_{mis}:

$$P(B|X_{obs}, X_{mis}, \zeta) = P(B|X_{obs}, \zeta). \tag{4.2}$$

Please be aware of MAR does not suggest that the missing data values constitute just another possible sample from the probability distribution. This condition is known as missing completely at random (MCAR). Actually MCAR is a special case of MAR in which the distribution of an example having a MV for an attribute does not depend on either the observed or the unobserved data. Following the previous notation, we can say that

$$P(B|X_{obs}, X_{mis}, \zeta) = P(B|\zeta). \tag{4.3}$$

Although there will generally be some loss of information, comparable results can be obtained with missing data by carrying out the same analysis that would have been performed with no MVs. In practice this means that, under MCAR, the analysis of only those units with complete data gives valid inferences.

Please note that MCAR is more restrictive than MAR. MAR requires only that the MVs behave like a random sample of all values in some particular subclasses defined by observed data. In such a way, MAR allows the probability of a missing datum to depend on the datum itself, but only indirectly through the observed values.

Recently a software package has been published in which the MCAR condition can be tested [43].

A third case arises when MAR does not apply as the MV depends on both the rest of observed values and the proper value itself. That is

$$P(B|X_{obs}, X_{mis}, \zeta) \tag{4.4}$$

is the actual probability estimation. This model is usually called not missing at random (NMAR) or missing not at random (MNAR) in the literature. This model of missingness is a challenge for the user as the only way to obtain an unbiased estimate is to model the missingness as well. This is a very complex task in which we should create a model accounting for the missing data that should be later incorporated to a more complex model used to estimate the MVs. However, even when we cannot account for the missingness model, the introduced bias may be still small enough. In [23] the reader can find an example of how to perform this.

4.3 Simple Approaches to Missing Data

In this section we introduce the most simplistic methods used to deal with MVs. As they are very simple, they usually do not take into account the missingness mechanism and they blindly perform the operation.

The most simple approach is to do not impute (DNI). As its name indicates, all the MVs remain unreplaced, so the DM algorithm must use their default MVs strategies if present. Often the objective is to verify whether imputation methods allow the classification methods to perform better than when using the original data sets. As a guideline, in [37] a previous study of imputation methods is presented. As an alternative for these learning methods that cannot deal with explicit MVs notation (as a special value for instance) another approach is to convert the MVs to a new value (encode them into a new numerical value), but such a simplistic method has been shown to lead to serious inference problems [82].

A very common approach in the specialized literature, even nowadays, is to apply case deletion or ignore missing (IM). Using this method, all instances with at least one MV are discarded from the data set. Although IM often results in a substantial decrease in the sample size available for the analysis, it does have important advantages. In particular, under the assumption that data is MCAR, it leads to unbiased parameter estimates. Unfortunately, even when the data are MCAR there is a loss in power using this approach, especially if we have to rule out a large number of subjects. And when the data is not MCAR, it biases the results. For example when low income individuals are less likely to report their income level, the resulting mean is biased in favor of higher incomes. The alternative approaches discussed below should be considered as a replacement for IM.

Often seen as a good choice, the substitution of the MVs for the global most common attribute value for nominal attributes, and global average value for numerical attributes (MC) [36] is widely used, specially when many instances in the data set contain MVs and to apply DNI would result in a very reduced and unrepresentative pre-processed data set. This method is very simple: for nominal attributes, the MV is replaced with the most common attribute value, and numerical values are replaced with the average of all values of the corresponding attribute.

A variant of MC is the concept most common attribute value for nominal attributes, and concept average value for numerical attributes (CMC) [36]. As stated in *MC*, the MV is replaced by the most repeated one if nominal or is the mean value if numerical, but considers only the instances with the same class as the reference instance.

Older and rarely used DM approaches may be put under this category. For example Hot deck imputation goes back over 50 years and was used quite successfully by the Census Bureau and others. It is referred from time to time [84] and thus it is interesting to describe it here partly for historical reasons and partly because it represents an approach of replacing data that is missing.

Hot deck has it origins in the surveys made in USA in the 40s and 50s, when most people felt impelled to participate in survey filling. As a consequence little data was missing and when any registers were effectively missing, a random complete case from the same survey was used to substitute the MVs. This process can be simulated nowadays by clustering over the complete data, and associating the instance with a cluster. Any complete example from the cluster can be used to fill in the MVs [6]. Cold deck is similar to hot deck, but the cases or instances used to fill in the MVs came from a different source. Traditionally this meant that the case used to fill the data was obtained from a different survey. Some authors have recently assessed the limitations imposed to the donors (the instances used to substitute the MVs) [44].

4.4 Maximum Likelihood Imputation Methods

At the same time Rubin et al. formalized the concept of missing data introduction mechanisms described in Sect. 4.2, they advised against use case deletion as a methodology (IM) to deal with the MVs. However, using MC or CMC techniques are not much better than replacing MVs with fixed values, as they completely ignore the mechanisms that yield the data values. In an ideal and rare case where the parameters of the data distribution θ were known, a sample from such a distribution conditioned to the other attributes' values or not depending of whether the MCAR, MAR or NMAR applies, would be a suitable imputed value for the missing one. The problem is that the parameters θ are rarely known and also very hard to estimate [38].

In a simple case such as flipping a coin, $P(heads) = \theta$ and $P(tails) = 1 - \theta$. Depending on the coin being rigged or not, the value of θ can vary and thus its value is unknown. Our only choice is to flip the coin several times, say n, obtaining h heads and $n - h$ tails. An estimation of θ would be $\widehat{\theta} = h/n$.

More formally, the likelihood of θ is obtained from a binomial distribution $P(\theta) = \binom{h}{n}\theta^h(1-\theta)^{n-h}$. Our $\widehat{\theta}$ can be proven to be the *maximum likelihood* estimate of θ.

So the next question arises: to solve a maximum likelihood type problem, can we analytically maximize the likelihood function? We have shown it can work with one dimensional Bernoulli problems like the coin toss, and that it also works with one dimensional Gaussian by finding the μ and σ parameters. To illustrate the latter case let us assume that we have the samples $1, 4, 7, 9$ obtained from a normal distribution and we want to estimate the population mean for the sake of simplicity, that is, in this simplistic case $\theta = \mu$. The maximum likelihood problem here is to choose a specific value of μ and compute $p(1) \cdot p(4) \cdot p(7) \cdot p(9)$. Intuitively one can say that this probability would be very small if we fix $\mu = 10$ and would be higher for $\mu = 4$ or $\mu = 5$. The value of μ that produces the maximum product of combined probabilities is what we call the maximum likelihood estimate of $\mu = \theta$. Again, in our case the maximum likelihood estimate would constitute the sample mean $\mu = 5.25$ and adding the variance to the problem can be solved again using the sample variance as the best estimator.

In real world data things are not that easy. We can have distribution that may not be well behaved or have too many parameters making the actual solution computationally too complex. Having a likelihood function made of a mixture of 100 100-dimensional Gaussians would yield 10,000 parameters and thus direct trial-error maximization is not feasible. The way to deal with such complexity is to introduce hidden variables in order to simplify the likelihood function and, in our case as well, to account for MVs. The observed variables are those that can be directly measured from the data, while hidden variables influence the data but are not trivial to measure. An example of an observed variable would be if it is sunny today, whereas the hidden variable can be $P(\textit{sunny today}|\textit{sunny yesterday})$.

Even simplifying with hidden variables does not allow us to reach the solution in a single step. The most common approach in these cases would be to use an iterative approach in which we obtain some parameter estimates, we use a regression technique to impute the values and repeat. However as the imputed values will depend on the estimated parameters θ, they will not add any useful information to the process and can be ignored. There are several techniques to obtain maximum likelihood estimators. The most well known and simplistic is the EM algorithm presented in the next section.

4.4.1 Expectation-Maximization (EM)

In a nutshell the EM algorithm estimates the parameters of a probability distribution. In our case this can be achieved from incomplete data. It iteratively maximizes the likelihood of the complete data X_{obs} considered as a function dependent of the parameters [20].

That is, we want to model dependent random variables as the observed variable a and the hidden variable b that generates a. We stated that a set of unknown parameters θ governs the probability distributions $P_\theta(a), P_\theta(b)$. As an iterative process, the EM algorithm consists of two steps that are repeated until convergence: the expectation (E-step) and the maximization (M-step) steps.

The E-step tries to compute the expectation of $logP_\theta(y, x)$:

$$Q(\theta, \theta') = \sum_y P_{\theta'}(b|a)logP_\theta(b, a), \tag{4.5}$$

where θ' are the new distribution parameters. Please note that we are using the *log*. The reason for this is that we need to multiply the probabilities of each observed value for an specific set of parameters. But multiplying several probabilities will soon yield a very small number and thus produce a loss of precision in a computer due to limited digital accuracy. A typical solution is then to use the *log* of these probabilities and to look for the maximum log likelihood. As the logs will be negative, we are looking for the set of parameters whose likelihood is as close to 0 as possible. In the M-step we try to find the θ that maximizes Q.

How can we find the θ that maximizes Q? Let us review conditional expectation where A and B are random variables drawn from distributions $P(a)$ and $P(b)$ to resolve the M-step. The conditional distribution is given by $P(b|a) = \frac{P(b,a)}{P(a)}$ and then $E[B] = \sum_b P(b)b$. For a function depending on B $h(B)$ the expectation is trivially obtained by $E[h(B)] = \sum_b P(b)h(b)$. For a particular value $A(A = a)$ the expectation is $E[h(B)|a] = \sum_b P(b|a)h(b)$.

Remember that we want to pick a θ that maximizes the log likelihood of the observed (a) and the unobserved (b) variables given an observed variable a and the previous parameters θ'. The conditional expectation of $logP_\theta(b, a)$ given a and θ' is

$$E[logP(b, a|\theta)|a, \theta'] = \sum_y P(b|a, \theta')logP(b, a|\theta) \tag{4.6}$$

$$= \sum_y P_{\theta'}(b|a)logP_\theta(b, a). \tag{4.7}$$

The key is that if $\sum_b P_{\theta'}(b|a)logP_\theta(b, a) > \sum_b P_{\theta'}(b|a)logP_{\theta'}(b, a)$ then $P_\theta(a) > P_{\theta'}(a)$. If we can improve the expectation of the log likelihood, EM is improving the model of the observed variable a.

In any real world problem, we do not have a single point but a series of attributes $x_1, \ldots, x_n$. Assuming i.i.d. we can sum over all points to compute the expectation:

$$Q(\theta, \theta') = \sum_{i=1}^{n} \sum_b P_{\theta'}(b|x_i)logP_\theta(b, x_i) \tag{4.8}$$

The EM algorithm is not perfect: it can be stuck in local maxima and also depends on an initial θ value. The latter is usually resolved by using a bootstrap process in order to choose a correct initial θ. Also the reader may have noticed that we have not talked about any imputation yet. The reason is EM is a meta algorithm that it is adapted to a particular application.

To use EM for imputation first we need to choose a plausible set of parameters, that is, we need to assume that the data follows a probability distribution, which is usually seen as a drawback of these kind of methods. The EM algorithm works better with probability distributions that are easy to maximize, as Gaussian mixture models. In [85] an approach of EM using multivariate Gaussian is proposed as using multivariate Gaussian data can be parameterized by the mean and the covariance matrix.

In each iteration of the EM algorithm for imputation the estimates of the mean μ and the covariance Σ are represented by a matrix and revised in three phases. These parameters are used to apply a regression over the MVs by using the complete data. In the first one in each instance with MVs the regression parameters B for the MVs are calculated from the current estimates of the mean and covariance matrix and the available complete data. Next the MVs are imputed with their conditional expectation values from the available complete ones and the estimated regression coefficients

$$x_{mis} = \mu_{mis} + (x_{obs} - \mu_{obs})B + e, \tag{4.9}$$

where the instance x of n attributes is separated into the observed values x_{obs} and the missing ones x_{mis}. The mean and covariance matrix are also separated in such a way. The residual $e \in \mathbb{R}^{1\times n_{mis}}$ is assumed to be a random vector with mean zero and unknown covariance matrix. These two phases would complete the E-step. Please note that for the iteration of the algorithm the imputation is not strictly needed as only the estimates of the mean and covariance matrix are, as well as the regression parameters. But our ultimate goal is to have our data set filled, so we use the latest regression parameters to create the best imputed values so far.

In the third phase the M-step is completed by re-estimating the mean a covariance matrix. The mean is taken as the sample mean of the completed data set and the covariance is the sample covariance matrix and the covariance matrices of the imputation errors as shown in [54]. That is:

$$\widehat{B} = \widehat{\Sigma}_{obs,obs}^{-1}\widehat{\Sigma}_{obs,mis}, \text{ and} \tag{4.10}$$

$$\widehat{C} = \widehat{\Sigma}_{mis,mis} - \widehat{\Sigma}_{mis,obs}\widehat{\Sigma}_{obs,obs}^{-1}\widehat{\Sigma}_{obs,mis} \tag{4.11}$$

The hat accent $\widehat{A}$ designates an estimate of a quantity A. After updating B and C the mean and covariance matrix must be updated with

$$\widehat{\mu}^{(t+1)} = \frac{1}{n}\sum_{i=1}^{n} X_i \tag{4.12}$$

and

$$\widehat{\Sigma}^{(t+1)} = \frac{1}{\tilde{n}} \sum_{i=1}^{n} \left[\widehat{S}_i^{(t)} - (\widehat{\mu}^{(t+1)}) \widehat{\mu}^{(t+1)} \right], \tag{4.13}$$

where, for each instance $x = X_i$, the conditional expectation $\widehat{S}_i^{(t)}$ of the cross-products is composed of three parts. The two parts that involve the available values in the instance,

$$E(x_{obs}^T x_{obs} | x_{obs}; \widehat{\mu}^{(t)}, \widehat{\Sigma}^{(t)}) = x_{obs}^T x_{obs} \tag{4.14}$$

and

$$E(x_{mis}^T x_{mis} | x_{obs}; \widehat{\mu}^{(t)}, \widehat{\Sigma}^{(t)}) = \widehat{x}_{mis}^T \widehat{x}_{mis} + \widehat{C}, \tag{4.15}$$

is the sum of the cross-product of the imputed values and the residual covariance matrix $\widehat{C} = Cov(x_{miss}, x_{mis} | x_{obs}; \widehat{\mu}^{(t)}, \widehat{\Sigma}^{(t)})$, the conditional covariance matrix of the imputation error. The normalization constant $\tilde{n}$ of the covariance matrix estimate [Eq. (4.13)] is the number of degrees of freedom of the sample covariance matrix of the completed data set.

The first estimation of the mean and covariance matrix needs to rely on a completely observed data set. One solution in [85] is to fill the data set with the initial estimates of the mean and covariance matrices. The process ends when the estimates of the mean and covariance matrix do not change over a predefined threshold. Please note that this EM approach is only well suited for numeric data sets, constituting a limitation for the application of EM, although an extension for mixed numerical and nominal attributes can be found in [82].

The EM algorithm is still valid nowadays, but is usually part of a system in which it helps to evolve some distributions like GTM neural networks in [95]. Still some research is being carried out for EM algorithms in which its limitations are being improved and also are applied to new fields like semi-supervised learning [97]. The most well known version of the EM for real valued data sets is the one introduced in [85] where the basic EM algorithm presented is extended with a regularization parameter.

4.4.2 Multiple Imputation

One big problem of the maximum likelihood methods like EM is that they tend to underestimate the inherent errors produced by the estimation process, formally standard errors. The Multiple Imputation (MI) approach was designed to take this into account to be a less biased imputation method, at the cost of being computationally expensive. MI is a Monte Carlo approach described very well by [80] in which we generate multiple imputed values from the observed data in a very similar way to the EM algorithm: it fills the incomplete data by repeatedly solving the observed-data. But a significative difference between the two methods is attained: while EM generates a single imputation in each step from the estimated parameters at each step, MI performs several imputations that yield several complete data sets.

This repeated imputation can be done thanks to the use of Markov Chain Monte Carlo methods, as the several imputations are obtained by introducing a random component, usually from a standard normal distribution. In a more advanced fashion, MI also considers that the parameters estimates are in fact sample estimates. Thus, the parameters are not directly estimated from the available data but, as the process continues, they are drawn from their Bayesian posterior distributions given the data at hand. These assumptions means that only in the case of MCAR or MAR missingness mechanisms hold MI should be applied.

As a result Eq. (4.9) can be applied with slight changes as the *e* term is now a sample from a standard normal distribution and is applied more than once to obtain several imputed values for a single MV. As indicated in the previous paragraph, MI has a Bayesian nature that forces the user to specify a prior distribution for the parameters θ of the model from which the *e* term is drawn. In practice [83] is stressed out that the results depend more on the election of the distribution for the data than the distribution for θ. Unlike the single imputation performed by EM where only one imputed value for each MV is created (and thus only one value of *e* is drawn), MI will create several versions of the data set, where the observed data X_{obs} is essentially the same, but the imputed values for X_{mis} will be different in each data set created. This process is usually known as data augmentation (DA) [91] as depicted in Fig. 4.2.

Surprisingly not many imputation steps are needed. Rubin claims in [80] that only 3–5 steps are usually needed. He states that the efficiency of the final estimation built upon *m* imputations is approximately:

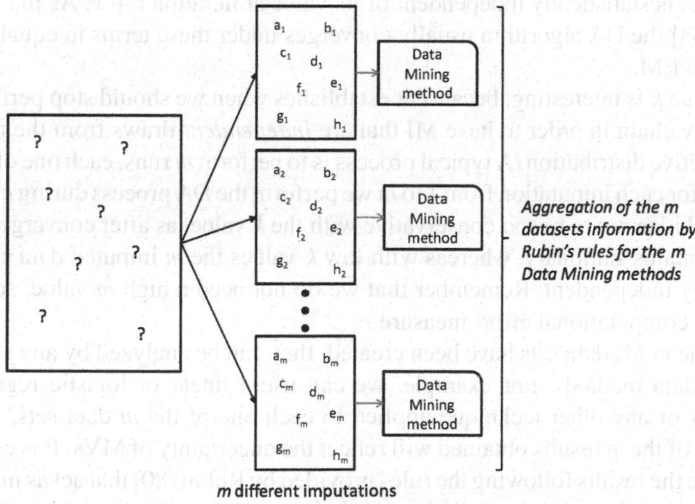

Fig. 4.2 Multiple imputation process by data augmentation. Every MV denoted by a '?' is replaced by several imputed and different values that will be used to continue the process later

$$\left(1+\frac{\gamma}{m}\right)^{-1},$$

where γ is the fraction of missing data in the data set. With a 30 % of MVs in each data set, which is a quite high amount, with 5 different final data sets a 94 % of efficiency will be achieved. Increasing the number to $m = 10$ slightly raises the efficiency to 97 %, which is a low gain paying the double computational effort.

To start we need an estimation of the mean and covariance matrices. A good approach is to take them from a solution provided from an EM algorithm once their values have stabilized at the end of its execution [83]. Then the DA process starts by alternately filling the MVs and then making inferences about the unknown parameters in a stochastic fashion. First DA creates an imputation using the available values of the parameters of the MVs, and then draws new parameter values from the Bayesian posterior distribution using the observed and missing data. Concatenating this process of simulating the MVs and the parameters is what creates a Markov chain that will converge at some point. The distribution of the parameters θ will stabilize to the posterior distribution averaged over the MVs, and the distribution of the MVs will stabilize to a predictive distribution: the proper distribution needed to drawn values for the MIs.

Large rates of MVs in the data sets will cause the convergence to be slow. However, the meaning of *convergence* is different to that used in EM. In EM the parameter estimates have converged when they no longer change from one iteration to the following over a threshold. In DA the distribution of the parameters do no change across iterations but the random parameter values actually continue changing, which makes the convergence of DA more difficult to assess than for EM. In [83] the authors propose to reinterpret convergence in DA in terms of lack of serial dependence: DA can be said to have converged by k cycles if the value of any parameter at iteration $t \in 1, 2, \ldots$ is statistically independent of its value at iteration $t + k$. As the authors show in [83] the DA algorithm usually converges under these terms in equal or less cycles than EM.

The value k is interesting, because it establishes when we should stop performing the Markov chain in order to have MI that are *independent* draws from the missing data predictive distribution. A typical process is to perform m runs, each one of length k. That is, for each imputation from 1 to m we perform the DA process during k cycles. It is a good idea not to be too conservative with the k value, as after convergence the process remains stationary, whereas with low k values the m imputed data sets will not be truly independent. Remember that we do not need a high m value, so k acts as the true computational effort measure.

Once the m MI data sets have been created, they can be analyzed by any standard complete-data methods. For example, we can use a linear or logistic regression, a classifier or any other technique applied to each one of the m data sets, and the variability of the m results obtained will reflect the uncertainty of MVs. It is common to combine the results following the rules provided by Rubin [80] that act as measures of ordinary sample variation to obtain a single inferential statement of the parameters of interest.

Rubin's rules to obtain an overall set of estimated coefficients and standard errors proceed as follows. Let $\widehat{R}$ denote the estimation of interest and U its estimated variance, R being either an estimated regression coefficient or a kernel parameter of a SVM, whatever applies. Once the MIs have been obtained, we will have $\widehat{R}_1, \widehat{R}_2, \ldots, \widehat{R}_m$ estimates and their respective variances $U_1, U_2, \ldots, U_m$. The overall estimate, occasionally called the MI estimate is given by

$$\overline{R} = \frac{1}{m}\sum_{i=1}^{m}\widehat{R}_i. \tag{4.16}$$

The variance for the estimate has two components: the variability within each data set and across data sets. The within imputation variance is simply the average of the estimated variances:

$$\overline{U} = \frac{1}{m}\sum_{i=1}^{m} U_i, \tag{4.17}$$

whereas the between imputation variance is the sample variance of the proper estimates:

$$B = \frac{1}{m-1}\sum_{i=1}^{m}(\widehat{R}_i - \overline{R})^2. \tag{4.18}$$

The total variance T is the corrected sum of these two components with a factor that accounts for the simulation error in $\widehat{R}$,

$$T = \widehat{U} + \left(1 + \frac{1}{m}\right)B. \tag{4.19}$$

The square root of T is the overall standard error associated to $\overline{R}$. In the case of no MVs being present in the original data set, all $\widehat{R}_1, \widehat{R}_2, \ldots, \widehat{R}_m$ would be the same, then $B = 0$ and $T = \overline{U}$. The magnitude of B with respect to $\overline{U}$ indicates how much information is contained in the missing portion of the data set relative to the observed part.

In [83] the authors elaborate more on the confidence intervals extracted from $\overline{R}$ and how to test the null hypothesis of $R = 0$ by comparing the ratio $\frac{\overline{R}}{\sqrt{T}}$ with a Student's t-distribution with degrees of freedom

$$df = (m-1)\left(1 + \frac{m\overline{U}}{(m+1)B}\right)^2, \tag{4.20}$$

in the case the readers would like to further their knowledge on how to use this hypothesis to check whether the number of MI m was large enough.

The MI algorithm has been widely used in many research fields. Focusing on DM methods to increase the robustness of MI [19], alleviate the parameter selection process [35] and improve Rubin's rules to aggregate models have been proposed [86]. New extensions to new problems like one-class [48] can be found, as well as hybridizations with innovative techniques such as Gray System Theory [92]. Implementing MI is not trivial and reputed implementations can be found in statistical packages as R [9] (see Chap. 10) and Stata [78].

4.4.3 Bayesian Principal Component Analysis (BPCA)

The MV estimation method based on BPCA [62] consists of three elementary processes. They are (1) principal component (PC) regression, (2) Bayesian estimation, and (3) an EM-like repetitive algorithm. In the following we describe each of these processes.

4.4.3.1 PC Regression

For the time being, we consider a situation where there is no MV. PCA represents the variation of D-dimensional example vectors y as a linear combination of principal axis vectors $w_l (1 \leq l \leq K)$ whose number is relatively small ($K < D$):

$$y = \sum_{l=1}^{K} x_l w_l + \epsilon \tag{4.21}$$

The linear coefficients $x_l (1 \leq l \leq K)$ are called factor scores. ϵ denotes the residual error. Using a specifically determined number K, PCA obtains x_l and w_l such that the sum of squared error $\| \epsilon \|^2$ over the whole data set Y is minimized.

When there is no MV, x_l and w_l are calculated as follows. A covariance matrix S for the example vectors $y_i (1 \leq i \leq N)$ is given by

$$S = \frac{1}{N} \sum_{i=1}^{N} (y_i - \mu)(y_i - \mu)^T, \tag{4.22}$$

where μ is the mean vector of y: $\mu = (1/N) \sum_{i=1}^{N} y_i$. T denotes the transpose of a vector or a matrix. For description convenience, Y is assumed to be row-wisely normalized by a preprocess, so that $\mu = 0$ holds. With this normalization, the result by PCA is identical to that by SVD.

Let $\lambda_1 \geq \lambda_2 \geq \cdots \geq \lambda_D$ and $u_1, u_2, \ldots, u_D$ denote the eigenvalues and the corresponding eigenvectors, respectively, of S. We also define the lth principal axis vector by $w_l = \sqrt{\lambda_l} u_l$ .With these notations, the lth factor score for an example vector

y is given by $x_l = (w_l/\lambda_l)^T y$. Now we assume the existence of MVs. In PC regression, the missing part y_{miss} in the expression vector y is estimated from the observed part y_{obs} by using the PCA result. Let w^l_{obs} and w^l_{miss} be parts of each principal axis w_l, corresponding to the observed and missing parts, respectively, in y. Similarly, let $W = (W_{obs}, W_{miss})$ where W_{obs} or W_{miss} denotes a matrix whose column vectors are $w^1_{obs}, \ldots, w^K_{obs}$ or $w^1_{miss}, \ldots, w^K_{miss}$, respectively.

Factor scores $x = (x_1, \ldots, x_K)$ for the example vector y are obtained by minimization of the residual error

$$err = \| y_{obs} - W_{obs}x \|^2 .$$

This is a well-known regression problem, and the least square solution is given by

$$x = (W^{obsT} W_{obs})^{-1} W^{obsT} y_{obs}.$$

Using x, the missing part is estimated as

$$y_{miss} = W_{miss}x \tag{4.23}$$

In the PC regression above, W should be known beforehand. Later, we will discuss the way to determine the parameter.

4.4.3.2 Bayesian Estimation

A parametric probabilistic model, which is called probabilistic PCA (PPCA), has been proposed recently. The probabilistic model is based on the assumption that the residual error ϵ and the factor scores $x_l (1 \leq l \leq K)$ in Equation (reflinearcomb) obey normal distributions:

$$p(x) = \mathcal{N}_K(x|0, I_K),$$
$$p(\epsilon) = \mathcal{N}_D(\epsilon|0, (1/\tau)I_D),$$

where $\mathcal{N}_K(x|\mu, \Sigma)$ denotes a K-dimensional normal distribution for x, whose mean and covariance are μ and Σ, respectively. I_K is a $(K \times K)$ identity matrix and τ is a scalar inverse variance of ϵ. In this PPCA model, a complete log-likelihood function is written as:

$$\begin{aligned} ln\, p(y, x|\theta) &= ln\, p(y, x|W, \mu, \tau) \\ &= -\frac{\tau}{2} \| y - Wx - \tau \|^2 - \frac{1}{2} \| x \|^2 + \frac{D}{2} ln\, \tau - \frac{K+D}{2} ln 2\Pi, \end{aligned}$$

where $\theta \equiv W, \mu, \tau$ is the parameter set. Since the maximum likelihood estimation of the PPCA is identical to PCA, PPCA is a natural extension of PCA to a probabilistic model.

We present here a Bayesian estimation method for PPCA from the authors. Bayesian estimation obtains the posterior distribution of θ and X, according to the Bayes' theorem:

$$p(\theta, X|Y) \propto p(Y, X|\theta)p(\theta). \tag{4.24}$$

$p(\theta)$ is called a prior distribution, which denotes a priori preference for parameter θ. The prior distribution is a part of the model and must be defined before estimation. We assume conjugate priors for τ and μ, and a hierarchical prior for W, namely, the prior for $W, p(W|\tau, \alpha)$, is parameterized by a hyperparameter $\alpha \in \mathbb{R}^K$.

$$p(\theta|\alpha) \equiv p(\mu, W, \tau|\alpha) = p(\mu|\tau)p(\tau)\prod_{j=1}^{K} p(w_j|\tau, \alpha_j),$$

$$p(\mu|tau) = \mathcal{N}(\mu|\overline{\mu}_0, (\gamma_{\mu 0}^{\tau})^{-1} I_m),$$
$$p(w_j|\tau, \alpha_j) = \mathcal{N}(w_j|0, (\alpha_j \tau)^{-1} I_m),$$
$$p(\tau) = \mathcal{G}(\tau|\overline{\tau}_0, \gamma_{\tau 0})$$

$\mathcal{G}(\tau|\overline{\tau}, \gamma_\tau)$ denotes a Gamma distribution with hyperparameters $\overline{\tau}$ and γ_τ:

$$\mathcal{G}(\tau|\overline{\tau}, \gamma_\tau) \equiv \frac{(\gamma_\tau \overline{\tau}^{-1})^{\gamma_\tau}}{\Gamma(\gamma_\tau)} \exp\left[-\gamma_\tau \overline{\tau}^{-1}\tau + (\gamma_\tau - 1) ln \tau\right]$$

where $\Gamma(\cdot)$ is a Gamma function.

The variables used in the above priors, $\gamma_{\mu 0}$, $\overline{\mu}_0$, $\gamma_{\tau 0}$ and $\overline{\tau}_0$ are deterministic hyperparameters that define the prior. Their actual values should be given before the estimation. We set $\gamma_{\mu 0} = \gamma_{\tau 0} = 10^{-10}$, $\overline{\mu}_0 = 0$ and $\overline{\tau}_0 = 1$, which corresponds to an almost non-informative prior.

Assuming the priors and given a whole data set $Y = y$, the type-II maximum likelihood hyperparameter α_{ML-II} and the posterior distribution of the parameter, $q(\theta) = p(\theta|Y, \alpha_{ML-II})$, are obtained by Bayesian estimation.

The hierarchical prior $p(W|\alpha, \tau)$, which is called an automatic relevance determination (ARD) prior, has an important role in BPCA. The jth principal axis w_j has a Gaussian prior, and its variance $1/(\alpha_j \tau)$ is controlled by a hyperparameter α_j which is determined by type-II maximum likelihood estimation from the data. When the Euclidian norm of the principal axis, $\| w_j \|$, is small relatively to the noise variance $1/\tau$, the hyperparameter α_j gets large and the principal axis w_j shrinks nearly to be 0. Thus, redundant principal axes are automatically suppressed.

4.4.3.3 EM-Like Repetitive Algorithm

If we know the true parameter θ_{true}, the posterior of the MVs is given by

$$q(Y_{miss}) = p(Y_{miss}|Y_{obs}, \theta_{true}),$$

which produces equivalent estimation to the PC regression. Here, $p(Y_{miss}|Y_{obs}, \theta_{true})$ is obtained by marginalizing the likelihood (4.24) with respect to the observed variables Y_{obs}. If we have the parameter posterior $q(\theta)$ instead of the true parameter, the posterior of the MVs is given by

$$q(Y_{miss}) = \int d\theta q(\theta) p(Y_{miss}|Y_{obs}, \theta),$$

which corresponds to the Bayesian PC regression. Since we do not know the true parameter naturally, we conduct the BPCA. Although the parameter posterior $q(\theta)$ can be easily obtained by the Bayesian estimation when a complete data set Y is available, we assume that only a part of Y, Y_{obs}, is observed and the rest Y_{miss} is missing. In that situation, it is required to obtain $q(\theta)$ and $q(Y_{miss})$ simultaneously.

We use a variational Bayes (VB) algorithm, in order to execute Bayesian estimation for both model parameter θ and MVs Y_{miss}. Although the VB algorithm resembles the EM algorithm that obtains maximum likelihood estimators for θ and Y_{miss}, it obtains the posterior distributions for θ and Y_{miss}, $q(\theta)$ and $q(Y_{miss})$, by a repetitive algorithm.

The VB algorithm is implemented as follows: (a) the posterior distribution of MVs, $q(Y_{miss})$, is initialized by imputing each of the MVs to instance-wise average; (b) the posterior distribution of the parameter θ, $q(\theta)$, is estimated using the observed data Y_{obs} and the current posterior distribution of MVs, $q(Y_{miss})$; (c) the posterior distribution of the MVs, $q(Y_{miss})$, is estimated using the current $q(\theta)$; (d) the hyperparameter α is updated using both of the current $q(\theta)$ and the current $q(Y_{miss})$; (e) repeat (b)–(d) until convergence.

The VB algorithm has been proved to converge to a locally optimal solution. Although the convergence to the global optimum is not guaranteed, the VB algorithm for BPCA almost always converges to a single solution. This is probably because the objective function of BPCA has a simple landscape. As a consequence of the VB algorithm, therefore, $q(\theta)$ and $q(Y_{miss})$ are expected to approach the global optimal posteriors.

Then, the MVs in the expression matrix are imputed to the expectation with respect to the estimated posterior distribution:

$$\widehat{Y}_{miss} = \int y_{miss} q(Y_{miss}) dY_{miss}. \tag{4.25}$$

4.5 Imputation of Missing Values. Machine Learning Based Methods

The imputation methods presented in Sect. 4.4 originated from statistics application and thus they model the relationship between the values by searching for the hidden distribution probabilities. In Artificial Intelligence modeling the unknown relationships between attributes and the inference of the implicit information contained in a sample data set has been done using ML models. Immediately many authors noticed that the same process that can be carried out to predict a continuous or a nominal value from a previous learning process in regression or classification can be applied to predict the MVs. The use of ML methods for imputation alleviates us from searching for the estimated underlying distribution of the data, but they are still subject to the MAR assumption in order to correctly apply them.

Batista [6] tested the classification accuracy of two popular classifiers (C4.5 and CN2) considering the proposal of KNN as an imputation (KNNI) method and MC. Both CN2 and C4.5 (like [37]) algorithms have their own MV estimation. From their study, KNNI results in good accuracy, but only when the attributes are not highly correlated to each other. Related to this work, [1] have investigated the effect of four methods that deal with MVs. As in [6], they use KNNI and two other imputation methods (MC and median imputation). They also use the KNN and Linear Discriminant Analysis classifiers. The results of their study show that no significantly harmful effect in accuracy is obtained from the imputation procedure. In addition to this, they state that the KNNI method is more robust with the increment of MVs in the data set in respect to the other compared methods.

The idea of using ML or Soft Computing techniques as imputation methods spread from this point on. Li et al. [53] uses a fuzzy clustering method: the Fuzzy K-Means (FKMI). They compare the FKMI with Mean substitution and KMI (K-Means imputation). Using a Root Mean Square Error error analysis, they state that the basic KMI algorithm outperforms the MC method. Experiments also show that the overall performance of the FKMI method is better than the basic KMI method, particularly when the percentage of MVs is high. Feng et al. [29] uses an SVM for filling in MVs (SVMI) but they do not compare this with any other imputation methods. Furthermore, they state that we should select enough complete examples without MVs as the training data set in this case.

In the following we proceed to describe the main details of the most used imputation methods based on ML techniques. We have tried to stay as close as possible to the original notation used by the authors so the interested reader can easily continue his or her exploration of details in the corresponding paper.

4.5.1 Imputation with K-Nearest Neighbor (KNNI)

Using this instance-based algorithm, every time an MV is found in a current instance, KNNI computes the KNN and a value from them is imputed. For nominal values, the most common value among all neighbors is taken, and for numerical values the average value is used. Therefore, a proximity measure between instances is needed for it to be defined. The Euclidean distance (it is a case of a L_p norm distance) is the most commonly used in the literature.

In order to estimate a MV y_{ih} in the ith example vector y_i by KNNI [6], we first select K examples whose attribute values are similar to y_i. Next, the MV is estimated as the average of the corresponding entries in the selected K expression vectors. When there are other MVs in y_i and/or y_j, their treatment requires some heuristics. The missing entry y_{ih} is estimated as average:

$$y_{\hat{ih}} = \frac{\sum_{j \in I_{Kih}} y_{jh}}{|I_{Kih}|}, \tag{4.26}$$

where I_{Kih} is now the index set of KNN examples of the ith example, and if y_{jh} is missing the jth attribute is excluded from I_{Kih}. Note that KNNI has no theoretical criteria for selecting the best K-value and the K-value has to be determined empirically.

4.5.2 Weighted Imputation with K-Nearest Neighbour (WKNNI)

The Weighted KNN method [93] selects the instances with similar values (in terms of distance) to incomplete instance, so it can impute as *KNNI* does. However, the estimated value now takes into account the different distances to the neighbors, using a weighted mean or the most repeated value according to a similarity measure. The similarity measure $s_i(y_j)$ between two examples y_i and y_j is defined by the Euclidian distance calculated over observed attributes in y_i. Next we define the measure as follows:

$$1/s_i = \sum_{h_i \in O_i \bigcap O_j} (y_{ih} - y_{jh})^2, \tag{4.27}$$

where $O_i = \{h|$ the hth component of$y_i$is observed$\}$.

The missing entry y_{ih} is estimated as average weighted by the similarity measure:

$$y_{\hat{ih}} = \frac{\sum_{j \in I_{Kih}} s_i(y_j) y_{jh}}{\sum_{j \in I_{Kih}} s_i(y_j)}, \tag{4.28}$$

where I_{Kih} is the index set of KNN examples of the ith example, and if y_{jh} is missing the jth attribute is excluded from I_{Kih}. Note that KNNI has no theoretical criteria for selecting the best K-value and the K-value has to be determined empirically.

4.5.3 K-means Clustering Imputation (KMI)

In K-means clustering [53], the intra-cluster dissimilarity is measured by the summation of distances between the objects and the centroid of the cluster they are assigned to. A cluster centroid represents the mean value of the objects in the cluster. Given a set of objects, the overall objective of clustering is to divide the data set into groups based on the similarity of objects, and to minimize the intra-cluster dissimilarity. KMI measures the intra-cluster dissimilarity by the addition of distances among the objects and the centroid of the cluster which they are assigned to. A cluster centroid represents the mean value of the objects in the cluster. Once the clusters have converged, the last process is to fill in all the non-reference attributes for each incomplete object based on the cluster information. Data objects that belong to the same cluster are taken to be nearest neighbors of each other, and KMI applies a nearest neighbor algorithm to replace MVs, in a similar way to KNNI.

Given a set of N objects $X = x_1, x_2, ldots, x_N$ where each object has S attributes, we use $x_{ij}(1 \leq i \leq N and 1 \leq j \leq S)$ to denote the value of attribute j in object x_i. Object x_i is called a *complete* object, if $\{x_{ij} \neq \phi | \forall 1 \leq j \leq S\}$, and an incomplete object, if $\{x_{ij} = \phi | \exists 1 \leq j \leq S\}$, and we say object x_i has a MV on attribute j. For any incomplete object x_i, we use $R = \{j | x_{ij} \neq \phi, 1 \leq j \leq S\}$ to denote the set of attributes whose values are available, and these attributes are called *reference* attributes. Our objective is to obtain the values of non-reference attributes for the incomplete objects. By K-means clustering method, we divide data set X into K clusters, and each cluster is represented by the centroid of the set of objects in the cluster. Let $V = v_1, \ldots, v_k$ be the set of K clusters, where $v_k (1 \leq k \leq K)$ represents the centroid of cluster k. Note that v_k is also a vector in a S-dimensional space. We use $d(v_k, x_i)$ to denote the distance between centroid v_k and object x_i.

KMI can be divided into three processes. First, randomly select K complete data objects as K centroids. Second, iteratively modify the partition to reduce the sum of the distances for each object from the centroid of the cluster to which the object belongs. The process terminates once the summation of distances is less than a user-specified threshold $\varepsilon = 100$, or no change on the centroids were made in last iteration. The last process is to fill in all the non-reference attributes for each incomplete object based on the cluster information. Data objects that belong to the same cluster are taken as nearest neighbors of each other, and we apply a nearest neighbor algorithm to replace missing data. We use as a distance measure the Euclidean distance.

4.5.4 Imputation with Fuzzy K-means Clustering (FKMI)

In fuzzy clustering, each data object has a membership function which describes the degree to which this data object belongs to a certain cluster. Now we want to extend the original K-means clustering method to a fuzzy version to impute missing data [1, 53]. The reason for applying the fuzzy approach is that fuzzy clustering provides a better description tool when the clusters are not well-separated, as is the case in missing data imputation. Moreover, the original K-means clustering may be trapped in a local minimum status if the initial points are not selected properly. However, continuous membership values in fuzzy clustering make the resulting algorithms less susceptible to get stuck in a local minimum situation.

In fuzzy clustering, each data object x_i has a membership function which describes the degree to which this data object belongs to certain cluster v_k. The membership function is defined in the next equation

$$U(v_k, x_i) = \frac{d(v_k, x_i)^{-2/(m-1)}}{\sum_{j=1}^{K} d(v_j, x_i)^{-2/(m-1)}} \tag{4.29}$$

where $m > 1$ is the fuzzifier, and $\sum_{j=1}^{K} U(v_j, x_i) = 1$ for any data object $x_i (1 \leq i \leq N)$. Now we can not simply compute the cluster centroids by the mean values. Instead, we need to consider the membership degree of each data object. Equation (4.30) provides the formula for cluster centroid computation:

$$v_k = \frac{\sum_{i=1}^{N} U(v_k, x_i) \times x_i}{\sum_{i=1}^{N} U(v_k, x_i)} \tag{4.30}$$

Since there are unavailable data in incomplete objects, we use only reference attributes to compute the cluster centroids.

The algorithm for missing data imputation with fuzzy K-means clustering method also has three processes. Note that in the initialization process, we pick K centroids which are evenly distributed to avoid local minimum situation. In the second process, we iteratively update membership functions and centroids until the overall distance meets the user-specified distance threshold ε. In this process, we cannot assign the data object to a concrete cluster represented by a cluster centroid (as did in the basic K-mean clustering algorithm), because each data object belongs to all K clusters with different membership degrees. Finally, we impute non-reference attributes for each incomplete object. We replace non-reference attributes for each incomplete data object x_i based on the information about membership degrees and the values of cluster centroids, as shown in next equation:

$$x_{i,j} = \sum_{k=1}^{K} U(x_i, v_k) \times v_{k,j}, \text{ for any non-reference attribute } j \notin R \tag{4.31}$$

4.5.5 Support Vector Machines Imputation (SVMI)

Support Vector Machines Imputation [29] is an SVM regression based algorithm to fill in MVs, i.e. set the decision attributes (output or classes) as the condition attributes (input attributes) and the condition attributes as the decision attributes, so SVM regression can be used to predict the missing condition attribute values. SVM regression estimation seeks to estimate functions

$$f(x) = (wx) + b, \qquad w, x \in \mathbb{R}^n, b \in \mathbb{R} \tag{4.32}$$

based on data

$$(x_1, y_1), \ldots, (x_l, y_l) \in \mathbb{R} \times \mathbb{R} \tag{4.33}$$

by minimizing the regularized risk functional

$$\| W \|^2 /2 + C \bullet R^{\varepsilon}_{emp} \tag{4.34}$$

where C is a constant determining the trade-off between minimizing the training error, or empirical risk

$$R^{\varepsilon}_{emp} = \frac{1}{l} \sum_{i=1}^{l} |y_i - f(x_i)|_{\varepsilon} \tag{4.35}$$

and the model complexity term $\| W \|^2$. Here, we use the so-called ε-insensitive loss function

$$|y - f(x)|_{\varepsilon} = \max\{0, |y - f(x)| - \varepsilon\} \tag{4.36}$$

The main insight of the statistical learning theory is that in order to obtain a small risk, one needs to control both training error and model complexity, i.e. explain the data with a simple model. The minimization of Eq. (4.36) is equivalent to the following constrained optimization problem [17]: minimize

$$\tau(w, \xi^{(*)}) = \frac{1}{2} \| w \|^2 + C \frac{1}{l} \sum_{i=1}^{l} (\xi_i + \xi_i^*) \tag{4.37}$$

subject to the following constraints

$$((w \bullet x_i) + b) - y_i \le \varepsilon + \xi_i \tag{4.38}$$

$$y_i - ((w \bullet x_i) + b) \le \varepsilon + \xi_i^* \tag{4.39}$$

$$\xi_i^{(*)} \ge 0, \quad \varepsilon \ge 0 \tag{4.40}$$

As mentioned above, at each point x_i we allow an error of magnitude ε. Errors above ε are captured by the slack variables ξ^* (see constraints 4.38 and 4.39). They are penalized in the objective function via the regularization parameter C chosen a priori.

In the ν-SVM the size of ε is not defined a priori but is itself a variable. Its value is traded off against model complexity and slack variables via a constant $\nu \in (0, 1]$ minimize

$$\tau(W, \xi^{(*)}, \varepsilon) = \frac{1}{2} \parallel W \parallel^2 + C \bullet (\nu\varepsilon + \frac{1}{l}\sum_{i=1}^{l}(\xi_i + \xi_i^*)) \quad (4.41)$$

subject to the constraints 4.38–4.40. Using Lagrange multipliers techniques, one can show [17] that the minimization of Eq. (4.37) under the constraints 4.38–4.40 results in a convex optimization problem with a global minimum. The same is true for the optimization problem 4.41 under the constraints 4.38–4.40. At the optimum, the regression estimate can be shown to take the form

$$f(x) = \sum_{i=1}^{l}(\alpha_i^* - \alpha_i)(x_i \bullet x) + b \quad (4.42)$$

In most cases, only a subset of the coefficients $(\alpha_i^* - \alpha_i)$ will be nonzero. The corresponding examples x_i are termed support vectors (SVs). The coefficients and the SVs, as well as the offset b; are computed by the ν-SVM algorithm. In order to move from linear (as in Eq. 4.42) to nonlinear functions the following generalization can be done: we map the input vectors x_i into a high-dimensional feature space Z through some chosen a priori nonlinear mapping $\Phi : X_i \rightarrow Z_i$. We then solve the optimization problem 4.41 in the feature space Z. In this case, the inner product of the input vectors $(x_i \bullet x)$ in Eq. (4.42) is replaced by the inner product of their icons in feature space Z, $(\Phi(x_i) \bullet \Phi(x))$. The calculation of the inner product in a high-dimensional space is computationally very expensive. Nevertheless, under general conditions (see [17] and references therein) these expensive calculations can be reduced significantly by using a suitable function k such that

$$(\Phi(x_i) \bullet \Phi(x)) = k(x_i \bullet x), \quad (4.43)$$

leading to nonlinear regressions functions of the form:

$$f(x) = \sum_{i=1}^{l}(\alpha_i^* - \alpha_i)k(x_i, x) + b \quad (4.44)$$

The nonlinear function k is called a kernel [17]. We mostly use a Gaussian kernel

$$k(x, y) \vDash \exp(- \parallel x - y \parallel^2 / (2\sigma_{kernel}^2)) \quad (4.45)$$

We can use SVM regression [29] to predict the missing condition attribute values. In order to do that, first we select the examples in which there are no missing attribute values. In the next step we set one of the condition attributes (input attribute), some of those values are missing, as the decision attribute (output attribute), and the decision attributes as the condition attributes by contraries. Finally, we use SVM regression to predict the decision attribute values.

4.5.6 Event Covering (EC)

Based on the work of Wong et al. [99], a mixed-mode probability model is approximated by a discrete one. First, we discretize the continuous components using a minimum loss of information criterion. Treating a mixed-mode feature n-tuple as a discrete-valued one, the authors propose a new statistical approach for synthesis of knowledge based on cluster analysis: (1) detect from data patterns which indicate statistical interdependency; (2) group the given data into clusters based on detected interdependency; and (3) interpret the underlying patterns for each of the clusters identified. The method of synthesis is based on author's *event–covering* approach. With the developed inference method, we are able to estimate the MVs in the data.

The cluster initiation process involves the analysis of the nearest neighbour distance distribution on a subset of samples, the selection of which is based on a mean probability criterion. Let $X = (X_1, X_2, \ldots, X_n)$ be a random n-tuple of related variables and $x = (x_1, x_2, \ldots, x_n)$ be its realization. Then a sample can be represented as x. Let S be an ensemble of observed samples represented as n-tuples. The nearest-neighbour distance of a sample x_i to a set of examples S is defined as:

$$D(x_i, S) = min_{x_j \in S_{x_i \neq x_j}} d(x_i, x_j) \tag{4.46}$$

where $d(x_i, x_j)$ is a distance measure. Since we are using discrete values, we have adopted the Hamming distance. Let C be a set of examples forming a simple cluster. We define the maximum within-cluster nearest-neighbour distance as

$$D_c^* = max_{x_i \in C} D(x_i, C) \tag{4.47}$$

D_c^* reflects an interesting characteristic of the cluster configuration: that is, the smaller the D_c^*, the denser the cluster.

Using a mean probability criterion to select a similar subset of examples, the isolated samples can be easily detected by observing the wide gaps in the nearest-neighbour distance space. The probability distribution from which the criterion is derived for the samples can be estimated using a second-order probability estimation. An estimation of $P(x)$ known as the *dependence tree product approximation* can be expressed as:

$$\widehat{P}(x) = \prod_{j=1}^{n} P(x_{mj}|x_{m_{k(j)}}), 0 < k(j) < 1 \tag{4.48}$$

where (1) the index set $m_1, m_2, \ldots, m_n$ is a permutation of the integer set $1, 2, \ldots, n$, (2) the ordered pairs $x_{mj}, x_{m_{k(j)}}$ are chosen so that they the set of branches of a spanning tree defined on X with their summed MI maximized, and (3) $P(x_{m1}|x_{m0}) = P(x_{m1})$. The probability defined above is known to be the best second-order approximation of the high-order probability distribution. Then corresponding to each x in the ensemble, a probability $P(x)$ can be estimated.

In general, it is more likely for samples of relatively high probability to form clusters. By introducing the mean probability below, samples can be divided into two subsets: those above the mean and those below. Samples above the mean will be considered first for cluster initiation.

Let $S = x$. The mean probability is defined as

$$\mu_s = \sum_{x \in S} P(x)/|S| \tag{4.49}$$

where $|S|$ is the number of samples in S. For more details in the probability estimation with *dependence tree product approximation*, please refer to [13].

When distance is considered for cluster initiation, we can use the following criteria in assigning a sample x to a cluster.

1. If there exists more than one cluster, say $C_k|k = 1, 2, \ldots$, such that $D(x, C_k) \leq D^*$ for all k, then all these clusters can be merged together.
2. If exactly one cluster C_k exists, such that $D(x, C_k) \leq D^*$, then x can be grouped into C_k.
3. If $D(x, C_K) > D^*$ for all clusters C_k, then x may not belong to any cluster.

To avoid including distance calculation of outliers, we use a simple method suggested in [99] which assigns D^* the maximum value of all nearest-neighbor distances in L provided there is a sample in L having a nearest-neighbor distance value of $D^* - 1$ (with the distance values rounded to the nearest integer value).

After finding the initial clusters along with their membership, the regrouping process is thus essentially an inference process for estimating the cluster label of a sample. Let $C = a_{c1}, a_{c2}, \ldots, a_{cq}$ be the set of labels for all possible clusters to which x can be assigned. For X_k in X, we can form a contingency table between X_k and C. Let a_{ks} and a_{cj} be possible outcomes of X_k and C respectively, and let $obs(a_{ks}$ and $obsa_{cj}$ be the respectively marginal frequencies of their observed occurrences. The expected relative frequency of (a_{ks}, a_{cj}) is expressed as:

$$exp(a_{ks}, a_{cj}) = \frac{obs(a_{ks}) \times obs(a_{cj})}{|S|} \tag{4.50}$$

Let $obs(a_{ks}, a_{cj})$ represent the actual observed frequency of (a_{ks}, a_{cj}) in S. The expression

$$D = \sum_{j=1}^{q} \frac{(obs_{ks} - exp(a_{ks}, a_{cj}))^2}{exp(a_{ks}, a_{cj})} \tag{4.51}$$

summing over the outcomes of C in the contingency table, possesses an asymptotic chi-squared property with $(q-1)$ degrees of freedom. D can then be used in a criterion for testing the statistical dependency between a_{ks}, and C at a presumed significant level as described below. For this purpose, we define a mapping

$$h_k^c(a_{ks}, C) = \begin{cases} 1, & \text{if } D > \chi^2(q-1); \\ 0, & \text{otherwise.} \end{cases} \tag{4.52}$$

where $\chi^2(q-1)$ is the tabulated chi-squared value. The subset of selected events of X_k, which has statistical interdependency with C,is defined as

$$E_k^c = \left\{a_{ks} | h_k^c(a_{ks}, C) = 1\right\} \tag{4.53}$$

We call E_k^c the covered event subset of X_k with respect to C. Likewise, the covered event subset E_c^k of C with respect to X_k can be defined. After finding the covered event subsets of E_c^k and E_k^c between a variable pair (C, X_k), information measures can be used to detect the statistical pattern of these subsets. An interdependence redundancy measure between X_k^c and C^k can be defined as

$$R(X_k^c, C^k) = \frac{I(X_k^c, C^k)}{H(X_k^c, C^k)} \tag{4.54}$$

where $I(X_k^c, C^k)$ is the expected MI and $H(X_k^c, C^k)$ is the Shannon's entropy defined respectively on X_k^c and C^k:

$$I(X_k^c, C^k) = \sum_{a_{cu} \in E_c^k} \sum_{a_{ks} \in E_k^c} P(a_{cu}, a_{ks}) \log \frac{P(a_{cu}, a_{ks})}{P(a_{cu})P(a_{ks})} \tag{4.55}$$

and

$$H(X_k^c, C^k) = -\sum_{a_{cu} \in E_c^k} \sum_{a_{ks} \in E_k^c} P(a_{cu}, a_{ks}) \log P(a_{cu}, a_{ks}). \tag{4.56}$$

The interdependence redundancy measure has a chi-squared distribution:

$$I(X_k^c, C^k) \frac{\chi^2_{df}}{2|S|H(x_k^c, C^k)} \tag{4.57}$$

where *df* is the corresponding degree of freedom having the value $(|E_c^k|-1)(|E_k^c|-1)$. A chi-squared test is then used to select interdependent variables in X at a presumed significant level.

The cluster regrouping process uses an information measure to regroup data iteratively. Wong et al. have proposed an information measure called *normalized surprisal* (NS) to indicate significance of joint information. Using this measure, the information conditioned by an observed event x_k is weighted according to $R(X_k^c, C^K)$, their measure of interdependency with the cluster label variable. Therefore, the higher the interdependency of a conditioning event, the more relevant the event is. NS measures the joint information of a hypothesized value based on the selected set of significant components. It is defined as

$$NS(a_{cj}|x'(a_{cj})) = \frac{I(a_{cj}|x'(a_{cj}))}{m\left(\sum_{k=1}^{m} R(X_k^c, C^k)\right)} \tag{4.58}$$

where $I(a_{cj}|x'(a_{cj}))$ is the summation of the weighted conditional information defined on the incomplete probability distribution scheme as

$$\begin{aligned} I(a_{cj}|x'(a_{cj})) &= \sum_{k=1}^{m} R(X_k^c, C^k) I(a_{cj}|x_k)) \\ &= \sum_{k=1}^{m} R(X_k^c, C^k) \left(-log \frac{P(a_{cj}|x_k)}{\sum_{a_{cu} \in E_c^k} P(a_{cu}|x_k)} \right) \end{aligned} \tag{4.59}$$

In rendering a meaningful calculation in the incomplete probability scheme formulation, x_k is selected if

$$\sum_{a_{cu} \in E_c^k} P(a_{cu}|x_k) > T \tag{4.60}$$

where $T \geq 0$ is a size threshold for meaningful estimation. NS can be used in a decision rule in the regrouping process. Let $C = \{a_{c1}, \ldots, a_{cq}\}$ be the set of possible cluster labels. We assign a_{cj} to x_e if

$$NS(a_{cj}|x'(a_{cj})) = \min_{a_{cu} \in C} NS(a_{cu}|x'(a_{cu})).$$

If no component is selected with respect to all hypothesized cluster labels, or if there is more than one label associated with the same minimum NS, then the sample is assigned a dummy label, indicating that the estimated cluster label is still uncertain. Also, zero probability may be encountered in the probability estimation, an unbiased probability based on *Entropy minimax*. In the regrouping algorithm, the cluster label for each sample is estimated iteratively until a stable set of label assignments is attained.

Once the clusters are stable, we take the examples with MVs. Now we use the distance $D(x_i, S) = \min_{x_j \in S_{x_i \neq x_j}} d(x_i, x_j)$ to find the nearest cluster C_i to such instance. From this cluster we compute the centroid x' such that

$$D(x', C_i) < D(x_i, C_i) \tag{4.61}$$

for all instances x_i of the cluster C_i. Once the centroid is obtained, the MV of the example is imputed with the value of the proper attribute of x_i.

4.5.7 Singular Value Decomposition Imputation (SVDI)

In this method, we employ singular value decomposition (4.62) to obtain a set of mutually orthogonal expression patterns that can be linearly combined to approximate the values of all attributes in the data set [93]. These patterns, which in this case are identical to the principle components of the data values' matrix, are further referred to as eigenvalues.

$$A_{m\times m} = U_{m\times m}\Sigma_{m\times n}V^T_{n\times n}. \tag{4.62}$$

Matrix V^T now contains eigenvalues, whose contribution to the expression in the eigenspace is quantified by corresponding eigenvalues on the diagonal of matrix Σ. We then identify the most significant eigenvalues by sorting the eigenvalues based on their corresponding eigenvalue. Although it has been shown that several significant eigenvalues are sufficient to describe most of the expression data, the exact fraction of eigenvalues best for estimation needs to be determined empirically.

Once k most significant eigenvalues from V^T are selected, we estimate a MV j in example i by first regressing this attribute value against the k eigenvalues and then use the coefficients of the regression to reconstruct j from a linear combination of the k eigenvalues. The jth value of example i and the jth values of the k eigenvalues are not used in determining these regression coefficients. It should be noted that SVD can only be performed on complete matrices; therefore we originally substitute row average for all MVs in matrix A, obtaining A'. We then utilize an Regularized EM method to arrive at the final estimate, as follows. Each MV in A' is estimated using the above algorithm, and then the procedure is repeated on the newly obtained matrix, until the total change in the matrix falls below the empirically determined (by the authors [93]) threshold of 0.01 (noted as *stagnation tolerance* in the *EM* algorithm). The other parameters of the *EM* algorithm are the same for both algorithms.

4.5.8 Local Least Squares Imputation (LLSI)

In this method proposed in [49] a target instance that has MVs is represented as a linear combination of similar instances. Rather than using all available instances in the data, only similar instances based on a similarity measure are used, and for that reason the method has the "local" connotation. There are two steps in the LLSI. The first step is to select k instances by the L_2-norm. The second step is regression and estimation, regardless of how the k instances are selected. A heuristic k parameter selection method is used by the authors.

Throughout the section, we will use $X \in \mathbb{R}^{m\times n}$ to denote a dataset with m attributes and n instances. Since LLSI was proposed for microarrays, it is assumed that $m \gg n$. In the data set X, a row $x_i^T \in \mathbb{R}^{1\times n}$ represents expressions of the ith instance in n examples:

$$X = \begin{pmatrix} x_1^T \\ \vdots \\ x_m^T \end{pmatrix} \in \mathbb{R}^{m\times n}$$

A MV in the lth location of the ith instance is denoted as α, i.e.

$$X(i, l) = \mathbf{x}_i(l) = \alpha$$

For simplicity we first assume assuming there is a MV in the first position of the first instance, i.e.

$$X(1, 1) = \mathbf{x}_1(1) = \alpha.$$

4.5.8.1 Selecting the Instances

To recover a MV α in the first location $\mathbf{x}_1(1)$ of $\mathbf{x}_1$ in $X \in \mathbb{R}^{m\times n}$, the KNN instance vectors for $\mathbf{x}_1$,

$$\mathbf{x}_{S_i}^T \in \mathbb{R}^{1\times n}, \quad 1 \leq i \leq k,$$

are found for LLSimpute based on the L_2-norm (LLSimpute). In this process of finding the similar instances, the first component of each instance is ignored due to the fact that $\mathbf{x}_1(1)$ is missing. The LLSimpute based on the Pearson's correlation coefficient to select the k instances can be consulted in [49].

4.5.8.2 Local Least Squares Imputation

As imputation can be performed regardless of how the k-instances are selected, we present only the imputation based on L_2-norm for simplicity. Based on these k-neighboring instance vectors, the matrix $A \in \mathbb{R}^{k\times(n-1)}$ and the two vectors $\mathbf{b} \in \mathbb{R}^{k\times 1}$ and $\mathbf{w} \in \mathbb{R}^{(n-1)\times 1}$ are formed. The k rows of the matrix A consist of the KNN instances $\mathbf{x}_{S_i}^T \in \mathbb{R}^{1\times n}$, $1 \leq i \leq k$, with their first values deleted, the elements of the vector b consists of the first components of the k vectors $\mathbf{x}_{S_i}^T$, and the elements of the vector $\mathbf{w}$ are the $n-1$ elements of the instance vector $\mathbf{x}_1$ whose missing first item is deleted. After the matrix A, and the vectors $\mathbf{b}$ and $\mathbf{w}$ are formed, the least squares problem is formulated as

$$\min_x ||A^T\mathbf{z} - \mathbf{w}||_2 \tag{4.63}$$

Then, the MV α is estimated as a linear combination of first values of instances

$$\alpha = \mathbf{b}^T\mathbf{z} = \mathbf{b}^T(A^T)^\dagger\mathbf{w}, \tag{4.64}$$

where $(A^T)^\dagger$ is the pseudoinverse of A^T.

For example, assume that the target instance $\mathbf{x}_1$ has a MV in the first position among a total of six examples. If the MV is to be estimated by the k similar instances, the matrix A, and vectors $\mathbf{b}$ and $\mathbf{w}$ are constructed as

$$\begin{pmatrix} x_1^T \\ \vdots \\ x_m^T \end{pmatrix} = \begin{pmatrix} \alpha & \mathbf{w}^T \\ \mathbf{b} & A \end{pmatrix}$$

$$= \begin{pmatrix} \alpha & \mathbf{w}_1 & \mathbf{w}_2 & \mathbf{w}_3 & \mathbf{w}_4 & \mathbf{w}_5 \\ \mathbf{b}_1 & A_{1,1} & A_{1,2} & A_{1,3} & A_{1,4} & A_{1,5} \\ \vdots & \vdots & \vdots & \vdots & \vdots & \vdots \\ \mathbf{b}_k & A_{k,1} & A_{k,2} & A_{k,3} & A_{k,4} & A_{k,5} \end{pmatrix}$$

where α is the MV and $\mathbf{x}_{S_1}^T, \ldots, \mathbf{x}_{S_k}^T$ are instances similar to $\mathbf{x}_1^T$. From the second to the last components of the neighbor instances, a_i^T, $1 \leq i \leq k$, form the ith row vector of the matrix A. The vector w of the known elements of target instance $\mathbf{x}_1$ can be represented as a linear combination

$$\mathbf{w} \simeq \mathbf{z}_1\mathbf{a}_1 + \mathbf{z}_2\mathbf{a}_2 + \cdots + \mathbf{z}_k\mathbf{a}_k$$

where $\mathbf{z}_i$ are the coefficients of the linear combination, found from the least squares formulation (4.63). Accordingly, the MV α in $\mathbf{x}_1$ can be estimated by

$$\alpha = \mathbf{b}^T x = \mathbf{b}_1\mathbf{z}_1 + \mathbf{b}_2\mathbf{z}_2 + \cdots + \mathbf{b}_k\mathbf{z}_k$$

Now, we deal with the case in which there is more than one MV in a instance vector. In this case, to recover the total of q MVs in any of the locations of the instance $\mathbf{x}_1$, first, the KNN instance vectors for x_1,

$$\mathbf{x}_{S_i}^T \in \mathbb{R}^{1\times n}, \quad 1 \leq i \leq k,$$

are found. In this process of finding the similar instances, the q components of each instance at the q locations of MVs in $\mathbf{x}_1$ are ignored. Then, based on these

k neighboring instance vectors, a matrix $A \in \mathbb{R}^{k\times(n-q)}$ a matrix $B \in \mathbb{R}^{k\times q}$ and a vector $\mathbf{w} \in \mathbb{R}^{(n-q)\times 1}$ are formed. The ith row vector $\mathbf{a}_i^T$ of the matrix A consists of the ith nearest neighbor instances $\mathbf{x}_{S_i}^T \in \mathbb{R}^{1\times n}$, $1 \leq i \leq k$, with its elements at the q missing locations of MVs of $\mathbf{x}_1$ excluded. Each column vector of the matrix B consists of the values of the jth location of the MVs ($1 \leq j \leq q$) of the k vectors $\mathbf{x}_{S_i}^T$. The elements of the vector $\mathbf{w}$ are the $n - q$ elements of the instance vector $\mathbf{x}$ whose missing items are deleted. After the matrices A and B and a vector $\mathbf{w}$ are formed, the least squares problem is formulated as

$$\min_x ||A^T \mathbf{z} - \mathbf{w}||_2 \tag{4.65}$$

Then, the vector $\mathbf{u} = (\alpha_1, \alpha_2, \ldots, \alpha_q)^T$ of q MVs can be estimated as

$$\mathbf{u} = \begin{pmatrix} \alpha_1 \\ \vdots \\ \alpha_q \end{pmatrix} = B^T \mathbf{z} = B^T (A^T)^\dagger \mathbf{w}, \tag{4.66}$$

where $(A^T)^\dagger$ is the pseudoinverse of A^T.

Table 4.1 Recent and most well-known imputation methods involving ML techniques

Clustering		Kernel methods	
MLP hybrid	[4]	Mixture-kernel-based iterative estimator	[105]
Rough fuzzy subspace clustering	[89]	*Nearest neighbors*	
LLS based	[47]	ICkNNI	[40]
Fuzzy c-means with SVR and Gas	[3]	Iterative KNNI	[101]
Biclustering based	[32]	CGImpute	[22]
KNN based	[46]	Boostrap for maximum likelihood	[72]
Hierarchical Clustering	[30]	kDMI	[75]
K2 clustering	[39]	*Ensembles*	
Weighted K-means	[65]	Random Forest	[42]
Gaussian mixture clustering	[63]	Decision forest	[76]
ANNs		Group Method of Data Handling (GMDH)	[104]
RBFN based	[90]	Boostrap	[56]
Wavelet ANNs	[64]	*Similarity and correlation*	
Multi layer perceptron	[88]	FIMUS	[77]
ANNs framework	[34]	*Parameter estimation for regression imputation*	
Self-organizing maps	[58]	EAs for covariance matrix estimation	[31]
Generative Topographic Mapping	[95]	Iterative mutual information imputation	[102]
Bayesian networks		CMVE	[87]
Dynamic bayesian networks	[11]	DMI (EM + decision trees)	[74]
Bayesian networks with weights	[60]	WLLSI	[12]

4.5.9 Recent Machine Learning Approaches to Missing Values Imputation

Although we have tried to provide an extensive introduction to the most used and basic imputation methods based on ML techniques, there is a great amount of journal publications showing their application and particularization to real world problems. We would like to give the reader a summarization of the latest and more important imputation methods presented at the current date of publication, both extensions of the introduced ones and completely novel ones in Table 4.1.

4.6 Experimental Comparative Analysis

In this section we aim to provide the reader with a general overview of the behavior and properties of all the imputation methods presented above. However, this is not an easy task. The main question is: what is a good imputation method?

As multiple imputation is a very resource consuming approach, we will focus on the single imputation methods described in this chapter.

4.6.1 Effect of the Imputation Methods in the Attributes' Relationships

From an unsupervised data point of view, those imputation methods able to generate values close to the true but unknown MV should be the best. This idea has been explored in the literature by means of using complete data sets and then artificially introducing MVs. Please note that such a mechanism will act as a MCAR MV generator mechanism, validating the use of imputation methods. Then, imputation methods are applied to the data and an estimation of how far is the estimation to the original (and known) value. Authors usually choose the mean square error (MSE) or root mean square error (RMSE) to quantify and compare the imputation methods over a set of data sets [6, 32, 41, 77].

On the other hand, other problems arise when we do not have the original values or the problem is supervised. In classification, for example, it is more demanding to impute values that will constitute an easier and more generalizable problem. As a consequence in this paradigm a good imputation method will enable the classifier to obtain better accuracy. This is harder to measure, as we are relating two different values: the MV itself and the class label assigned to the example. Neither MSE or RMSE can provide us with such kind of information.

One way to measure how good the imputation is for the supervised task is to use Wilson's Noise Ratio. This measure proposed by [98] observes the noise in the data set. For each instance of interest, the method looks for the KNN (using the

Euclidean distance), and uses the class labels of such neighbors in order to classify the considered instance. If the instance is not correctly classified, then the variable *noise* is increased by one unit. Therefore, the final noise ratio will be

$$\text{Wilson's Noise} = \frac{noise}{\#\text{instances in the data set}}$$

After imputing a data set with different imputation methods, we can measure how disturbing the imputation method is for the classification task. Thus by using Wilson's noise ratio we can observe which imputation methods reduce the impact of the MVs as a noise, and which methods produce noise when imputing.

Another approach is to use the MI (MI) which is considered to be a good indicator of relevance between two random variables [18]. Recently, the use of the MI measure in FS has become well-known and seen to be successful [51, 52, 66]. The use of the MuI measure for continuous attributes has been tackled by [51], allowing us to compute the Mui measure not only in nominal-valued data sets.

In our approach, we calculate the Mui between each input attribute and the class attribute, obtaining a set of values, one for each input attribute. In the next step we compute the ratio between each one of these values, considering the imputation of the data set with one imputation method in respect to the not imputed data set. The average of these ratios will show us if the imputation of the data set produces a gain in information:

$$\text{Avg. Mui Ratio} = \frac{\sum_{x_i \in X} \frac{Mui_\alpha(x_i)+1}{Mui(x_i)+1}}{|X|}$$

where X is the set of input attributes, $Mui_\alpha(i)$ represents the Mui value of the ith attribute in the imputed data set and $Mui(i)$ is the Mui value of the ith input attribute in the not imputed data set. We have also applied the Laplace correction, summing 1 to both numerator and denominator, as an Mui value of zero is possible for some input attributes.

The calculation of $Mui(x_i)$ depends on the type of attribute x_i. If the attribute x_i is nominal, the Mui between x_i and the class label Y is computed as follows:

$$Mui_{nominal}(x_i) = I(x_i; Y) = \sum_{z \in x_i} \sum_{y \in Y} p(z, y) log_2 \frac{p(z, y)}{p(z)p(y)}.$$

On the other hand, if the attribute x_i is numeric, we have used the Parzen window density estimate as shown in [51] considering a Gaussian window function:

$$Mui_{numeric}(x_i) = I(x_i; Y) = H(Y) - H(C|X);$$

where $H(Y)$ is the entropy of the class label

$$H(Y) = -\sum_{y \in Y} p(y) log_2 p(y);$$

and $H(C|X)$ is the conditional entropy

$$H(Y|x_i) = -\sum_{z \in x_i} \sum_{y \in Y} p(z, y) log_2 p(y|z).$$

Considering each sample has the same probability, applying the Bayesian rule and approximating $p(y|z)$ by the Parzen window we get:

$$\hat{H}(Y|x_i) = -\sum_{j=1}^{n} \frac{1}{n} \sum_{y=1}^{N} \hat{p}(y|z_j) log_2 \hat{p}(y|z_j)$$

where n is the number of instances in the data set, N is the total number of class labels and $\hat{p}(c|x)$ is

$$\hat{p}(y|z) = \frac{\sum_{i \in I_c} exp\left(-\frac{(z-z_i)\Sigma^{-1}(z-z_i)}{2h^2}\right)}{\sum_{k=1}^{N} \sum_{i \in I_k} exp\left(-\frac{(z-z_i)\Sigma^{-1}(z-z_i)}{2h^2}\right)}.$$

In this case, I_c is the set of indices of the training examples belonging to class c, and Σ is the covariance of the random variable $(z - z_i)$.

Let us consider all the single imputation methods presented in this chapter. For the sake of simplicity we will omit the Multiple Imputation approaches, as it will require us to select a probability model for all the data sets, which would be infeasible. In Table 4.2 we have summarized the Wilson's noise ratio values for 21 data sets with MVs from those presented in Sect. 2.1. We must point out that the results of Wilson's noise ratio and Mui are related to a given data set. Hence, the characteristics of the proper data appear to determine the values of this measure.

Looking at the results from Table 4.2 we can observe which imputation methods reduce the impact of the MVs as noise, and which methods produce noise when imputing. In addition the MI ratio allows us to relate the attributes to the imputation results. A value of the Mui ratio higher than 1 will indicate that the imputation is capable of relating more of the attributes individually to the class labels. A value lower than 1 will indicate that the imputation method is adversely affecting the relationship between the individual attributes and the class label.

If we consider the average Mui ratio in Table 4.2 we can observe that the average ratios are usually close to 1; that is, the use of imputation methods appears to harm the relationship between the class label and the input attribute little or not at all, even improving it in some cases. However, the MI considers only one attribute at a time and therefore the relationships between the input attributes are ignored. The imputation

Table 4.2 Wilson's noise ratio values and average MI values per data set

Data set	Imp. Method	% Wilson's Noise	Avg. Mui ratio	Data set	Imp. Method	% Wilson's Noise	Avg. Mui ratio	Data set	Imp. Method	% Wilson's Noise	Avg. Mui ratio
CLE	MC	50.0000	0.998195	HOV	MC	7.9208	0.961834	HEP	MC	17.3333	0.963765
	CMC	50.0000	0.998585		CMC	5.4455	1.105778		CMC	16.0000	0.990694
	KNNI	50.0000	0.998755		KNNI	7.4257	0.965069		KNNI	20.0000	0.978564
	WKNNI	50.0000	0.998795		WKNNI	7.4257	0.965069		WKNNI	20.0000	0.978343
	KMI	50.0000	0.998798		KMI	7.4257	0.961525		KMI	20.0000	0.980094
	FKMI	50.0000	0.998889		FKMI	7.9208	0.961834		FKMI	17.3333	0.963476
	SVMI	50.0000	0.998365		SVMI	6.9307	0.908067		SVMI	17.3333	1.006819
	EM	66.6667	0.998152		EM	11.8812	0.891668		EM	22.6667	0.974433
	SVDI	66.6667	0.997152		SVDI	8.9109	0.850361		SVDI	21.3333	0.967673
	BPCA	50.0000	0.998701		BPCA	6.9307	1.091675		BPCA	21.3333	0.994420
	LLSI	50.0000	0.998882		LLSI	**4.9505**	**1.122904**		LLSI	18.6667	0.995464
	EC	**33.3333**	**1.000148**		EC	7.4257	1.007843		EC	**16.0000**	**1.024019**
WIS	MC	18.7500	0.999004	WAT	MC	31.5068	0.959488	MUS	MC	**0.0000**	1.018382
	CMC	**12.5000**	0.999861		CMC	**21.2329**	0.967967		CMC	**0.0000**	1.018382
	KNNI	**12.5000**	0.999205		KNNI	27.3973	0.961601		KNNI	**0.0000**	0.981261
	WKNNI	**12.5000**	0.999205		WKNNI	27.3973	0.961574		WKNNI	**0.0000**	0.981261
	KMI	**12.5000**	0.999322		KMI	27.3973	0.961361		KMI	**0.0000**	1.018382
	FKMI	**12.5000**	0.998923		FKMI	31.5068	0.961590		FKMI	**0.0000**	1.018382
	SVMI	**12.5000**	0.999412		SVMI	23.9726	0.967356		SVMI	**0.0000**	0.981261
	EM	**12.5000**	0.990030		EM	46.5753	0.933846		EM	**0.0000**	**1.142177**
	SVDI	**12.5000**	0.987066		SVDI	49.3151	0.933040		SVDI	**0.0000**	1.137152
	BPCA	**12.5000**	0.998951		BPCA	26.0274	0.964255		BPCA	**0.0000**	0.987472

(continued)

Table 4.2 (continued)

Data set	Imp.	% Wilson's	Avg. Mui	Data set	Imp.	% Wilson's	Avg. Mui	Data set	Imp.	% Wilson's	Avg. Mui
	Method	Noise	ratio		Method	Noise	ratio		Method	Noise	ratio
	LLSI	**12.5000**	0.999580		LLSI	25.3425	0.964063		LLSI	**0.0000**	0.977275
	EC	**12.5000**	**1.000030**		EC	22.6027	**1.027369**		EC	**0.0000**	1.017366
CRX	MC	18.9189	1.000883	SPO	MC	27.2727	0.997675	POS	MC	**33.3333**	1.012293
	CMC	18.9189	1.000966		CMC	**22.7273**	**1.022247**		CMC	**33.3333**	1.012293
	KNNI	21.6216	0.998823		KNNI	27.2727	0.999041		KNNI	**33.3333**	1.012293
	WKNNI	21.6216	0.998870		WKNNI	27.2727	0.999041		WKNNI	**33.3333**	1.012293
	KMI	21.6216	1.001760		KMI	27.2727	0.998464		KMI	**33.3333**	1.012293
	FKMI	18.9189	1.000637		FKMI	27.2727	0.997675		FKMI	**33.3333**	1.012293
	SVMI	13.5135	0.981878		SVMI	27.2727	1.015835		SVMI	**33.3333**	1.012293
	EM	32.4324	0.985609		EM	36.3636	0.982325		EM	**33.3333**	1.012293
	SVDI	27.0270	0.976398		SVDI	31.8182	0.979187		SVDI	**33.3333**	1.014698
	BPCA	21.6216	0.999934		BPCA	27.2727	1.006236		BPCA	**33.3333**	1.012293
	LLSI	18.9189	1.001594		LLSI	27.2727	1.004821		LLSI	**33.3333**	**1.018007**
	EC	**13.5135**	**1.008718**		EC	27.2727	1.018620		EC	**33.3333**	0.997034
BRE	MC	55.5556	0.998709	BAN	MC	25.4753	1.012922	ECH	MC	40.0000	0.981673
	CMC	55.5556	0.998709		CMC	24.3346	1.070857		CMC	40.0000	0.995886
	KNNI	55.5556	0.992184		KNNI	23.1939	0.940369		KNNI	46.6667	0.997912
	WKNNI	55.5556	0.992184		WKNNI	22.8137	0.940469		WKNNI	44.4444	**0.998134**
	KMI	55.5556	0.998709		KMI	25.4753	1.016101		KMI	46.6667	0.967169
	FKMI	55.5556	0.998709		FKMI	24.3346	1.020989		FKMI	40.0000	0.983606
	SVMI	55.5556	0.998709		SVMI	**21.2928**	**1.542536**		SVMI	44.4444	0.987678
	EM	**44.4444**	**1.013758**		EM	26.2357	1.350315		EM	51.1111	0.967861
	SVDI	**44.4444**	0.999089		SVDI	22.4335	1.365572		SVDI	48.8889	0.935855

Table 4.2 (continued)

Data set	Imp.	% Wilson's	Avg. Mui	Data set	Imp.	% Wilson's	Avg. Mui	Data set	Imp.	% Wilson's	Avg. Mui
	Method	Noise	ratio		Method	Noise	ratio		Method	Noise	ratio
	BPCA	66.6667	1.000201		BPCA	23.9544	1.010596		BPCA	44.4444	0.972327
	LLSI	66.6667	1.000201		LLSI	24.7148	1.015033		LLSI	**37.7778**	0.988591
	EC	66.6667	1.001143		EC	23.5741	1.102328		EC	48.8889	0.970029
AUT	MC	45.6522	0.985610	HOC	MC	19.3906	0.848649	SOY	MC	**2.4390**	1.056652
	CMC	41.3043	0.991113		CMC	**10.2493**	**2.039992**		CMC	**2.4390**	1.123636
	KNNI	41.3043	0.986239		KNNI	20.2216	0.834734		KNNI	**2.4390**	1.115818
	WKNNI	41.3043	0.985953		WKNNI	19.1136	0.833982		WKNNI	**2.4390**	1.115818
	KMI	41.3043	0.985602		KMI	21.8837	0.821936		KMI	**2.4390**	1.056652
	FKMI	45.6522	0.984694		FKMI	20.4986	0.849141		FKMI	**2.4390**	1.056652
	SVMI	43.4783	0.991850		SVMI	20.2216	0.843456		SVMI	**2.4390**	**1.772589**
	EM	58.6957	0.970557		EM	21.0526	0.775773		EM	**2.4390**	1.099286
	SVDI	52.1739	0.968938		SVDI	21.0526	0.750930		SVDI	7.3171	1.065865
	BPCA	43.4783	0.986631		BPCA	19.3906	0.964587		BPCA	7.3171	1.121603
	LLSI	45.6522	0.985362		LLSI	20.4986	0.926068		LLSI	**2.4390**	1.159610
	EC	**30.4348**	**1.007652**		EC	20.7756	0.911543		EC	**2.4390**	1.222631
PRT	MC	71.0145	0.949896	AUD	MC	38.7387	0.990711	MAM	MC	21.3740	0.974436
	CMC	**60.8696**	**1.120006**		CMC	**32.8829**	1.032162		CMC	**13.7405**	1.029154
	KNNI	69.5652	0.976351		KNNI	38.7387	0.993246		KNNI	25.9542	0.965926
	WKNNI	69.5652	0.976351		WKNNI	38.7387	0.993246		WKNNI	25.9542	0.965926
	KMI	71.0145	0.949896		KMI	38.7387	1.000235		KMI	24.4275	0.966885
	FKMI	71.0145	0.949896		FKMI	38.7387	0.990711		FKMI	20.6107	0.974228
	SVMI	68.1159	1.038152		SVMI	37.8378	1.007958		SVMI	16.7939	**1.272993**

(continued)

Table 4.2 (continued)

Data set	Imp. Method	% Wilson's Noise	Avg. Mui ratio	Data set	Imp. Method	% Wilson's Noise	Avg. Mui ratio	Data set	Imp. Method	% Wilson's Noise	Avg. Mui ratio
	EM	88.4058	0.461600		EM	53.6036	1.129168		EM	20.6107	0.980865
	SVDI	91.7874	0.485682		SVDI	46.3964	1.065091		SVDI	27.4809	1.052790
	BPCA	71.4976	0.987598		BPCA	40.5405	1.156676		BPCA	25.1908	0.978209
	LLSI	69.5652	1.016230		LLSI	36.9369	1.061197		LLSI	26.7176	0.994349
	EC	66.1836	1.053185		EC	37.8378	**1.209608**		EC	18.3206	1.269505
DER	MC	**0.0000**	1.000581	LUN	MC	80.0000	0.996176	OZO	MC	4.8035	0.982873
	CMC	**0.0000**	**1.002406**		CMC	80.0000	1.008333		CMC	**3.6390**	**0.989156**
	KNNI	**0.0000**	0.999734		KNNI	80.0000	0.996176		KNNI	4.3668	0.982759
	WKNNI	**0.0000**	0.999734		WKNNI	80.0000	0.996176		WKNNI	4.5124	0.982721
	KMI	**0.0000**	1.000581		KMI	80.0000	0.996176		KMI	4.9491	0.982495
	FKMI	**0.0000**	1.000581		FKMI	80.0000	0.996176		FKMI	4.0757	0.982951
	SVMI	**0.0000**	1.001566		SVMI	80.0000	1.006028		SVMI	3.7846	0.988297
	EM	**0.0000**	1.000016		EM	**20.0000**	1.067844		EM	4.8035	0.979977
	SVDI	**0.0000**	0.999691		SVDI	40.0000	**1.076334**		SVDI	4.8035	0.979958
	BPCA	**0.0000**	0.999633		BPCA	80.0000	0.996447		BPCA	4.3668	0.983318
	LLSI	**0.0000**	0.999170		LLSI	80.0000	1.007612		LLSI	4.2213	0.983508
	EC	**0.0000**	1.000539		EC	80.0000	1.002385		EC	4.8035	0.944747

methods estimate the MVs using such relationships and can afford improvements in the performance of the classifiers. Hence the highest values of average Mui ratios could be related to those methods which can obtain better estimates for the MVs, and maintaining the relationship degree between the class labels and the isolated input attributes. It is interesting to note that when analyzing the Mui ratio, the values do not appear to be as highly data dependant as Wilson's noise ratio, as the values for all the data sets are more or less close to each other.

If we count the methods with the lowest Wilson's noise ratios in each data set in Table 4.2, we find that the CMC method is first, with 12 times being the lowest one, and the EC method is second with 9 times being the lowest one. If we count the methods with the highest MI ratio in each data set, the EC method has the highest ratio for 7 data sets and is therefore the first one. The CMC method has the highest ratio for 5 data sets and is the second one in this case. Immediately the next question arises: are these methods also the best for the performance of the learning methods applied afterwards? We try to answer this question in the following.

4.6.2 Best Imputation Methods for Classification Methods

Our aim is to use the same imputation results as data sets used in the previous Sect. 4.6.1 as the input for a series of well known classifiers in order to shed light on the question "which is the best imputation method?". Let us consider a wide range of classifiers grouped by their nature, as that will help us to limit the comparisons needed to be made. We have grouped them in three sub-categories. In Table 4.3 we summarize the classification methods we have used, organized in these three categories. The description of the former categories is as follows:

- The first group is the *Rule Induction Learning* category. This group refers to algorithms which infer rules using different strategies.
- The second group represents the *Black Box Methods*. It includes ANNs, SVMs and statistical learning.
- The third and last group corresponds to the *Lazy Learning* (LL) category. This group incorporates methods which do not create any model, but use the training data to perform the classification directly.

Some methods do not work with numerical attributes (CN2, AQ and Naïve-Bayes). In order to discretize the numerical values, we have used the well-known discretizer proposed by [28]. For the SVM methods (C-SVM, ν-SVM and SMO), we have applied the usual preprocessing in the literature to these methods [25]. This preprocessing consists of normalizing the numerical attributes to the [0, 1] range, and binarizing the nominal attributes. Some of the presented classification methods in the previous section have their own MVs treatment that will trigger when no imputation is made (DNI): C4.5 uses a probabilistic approach to handling MVs and CN2 applies the MC method by default in these cases. For ANNs [24] proposed to replace MVs with zero so as not to trigger the corresponding neuron which the MV is applied to.

Table 4.3 Classifiers used by categories

Method	Acronym	References
Rule Induction Learning		
C4.5	C4.5	[73]
Ripper	Ripper	[16]
CN2	CN2	[14]
AQ-15	AQ	[59]
PART	PART	[33]
Slipper	Slipper	[15]
Scalable Rule Induction Induction	SRI	[68]
Rule Induction Two In One	Ritio	[100]
Rule Extraction System version 6	Rule-6	[67]
Black Box Methods		
Multi-Layer Perceptron	MLP	[61]
C-SVM	C-SVM	[25]
ν-SVM	ν-SVM	[25]
Sequential Minimal Optimization	SMO	[70]
Radial Basis Function Network	RBFN	[8]
RBFN Decremental	RBFND	[8]
RBFN Incremental	RBFNI	[69]
Logistic	LOG	[10]
Naïve-Bayes	NB	[21]
Learning Vector Quantization	LVQ	[7]
Lazy Learning		
1-NN	1-NN	[57]
3-NN	3-NN	[57]
Locally Weighted Learning	LWL	[2]
Lazy Learning of Bayesian Rules	LBR	[103]

As shown here all the detailed accuracy values for each fold, data set, imputation method and classifier would be too long, we have used Wilcoxon's Signed Rank test to summarize them. For each classifier, we have compared every imputation method along with the rest in pairs. Every time the classifier obtains a better accuracy value for an imputation method than another one and the statistical test yield a $p-value < 0.1$ we count it as a win for the former imputation method. In another case it is a tie when $p-value > 0.1$.

In the case of rule induction learning in Table 4.4 we show the average ranking or each imputation method for every classifier belonging to this group. We can observe that, for the rule induction learning classifiers, the imputation methods FKMI, SVMI and EC perform best. The differences between these three methods in average rankings are low. Thus we can consider that these three imputation methods are the most suitable for this kind of classifier. They are well separated from the other

Table 4.4 Average ranks for the Rule Induction Learning methods

	C45	Ripper	PART	Slipper	AQ	CN2	SRI	Ritio	Rules-6	Avg.	Ranks
IM	5	8.5	1	4	6.5	10	6.5	6	5	5.83	4
EC	2.5	8.5	6.5	1	6.5	5.5	6.5	6	1	4.89	3
KNNI	9	2.5	6.5	11	11	5.5	11.5	11	11	8.78	11
WKNNI	11	2.5	6.5	7	6.5	1	11.5	6	11	7.00	8
KMI	5	2.5	6.5	3	6.5	5.5	9.5	12	7.5	6.44	6
FKMI	7.5	2.5	6.5	10	2	5.5	1	2	3	4.44	1
SVMI	1	5.5	6.5	7	1	5.5	6.5	6	2	4.56	2
EM	13	12	6.5	7	12	13	3	6	4	8.50	10
SVDI	11	11	6.5	12	10	12	9.5	10	11	10.33	12
BPCA	14	13	13	7	13	14	13	13	13	12.56	14
LLSI	11	5.5	6.5	7	6.5	11	3	6	7.5	7.11	9
MC	7.5	8.5	6.5	2	6.5	5.5	3	6	7.5	5.89	5
CMC	5	8.5	12	13	3	5.5	6.5	1	7.5	6.89	7
DNI	2.5	14	14	14	14	5.5	14	14	14	11.78	13

Table 4.5 Average ranks for the Black Box methods

	RBFN	RBFND	RBFNI	LOG	LVQ	MLP	NB	ν-SVM	C-SVM	SMO	Avg.	Ranks
IM	9	6.5	4.5	6	3.5	13	12	10	5.5	5.5	7.55	10
EC	1	1	1	3	7	8.5	10	13	1	2	4.75	1
KNNI	5	6.5	10.5	9	7	11	6.5	8	5.5	5.5	7.45	9
WKNNI	13	6.5	4.5	10	10	4.5	6.5	4.5	5.5	5.5	7.05	6
KMI	3.5	2	7	3	11	3	4.5	8	5.5	9	5.65	2
FKMI	12	6.5	10.5	3	1.5	4.5	11	4.5	5.5	3	6.20	3
SVMI	2	11.5	2.5	7.5	3.5	1.5	13	8	11	9	6.95	5
EM	3.5	6.5	13	12	12.5	10	4.5	4.5	10	11.5	8.80	11
SVDI	9	6.5	7	11	12.5	8.5	3	11.5	12	11.5	9.25	12
BPCA	14	14	14	13	7	14	2	2	13	13	10.60	14
LLSI	6	6.5	10.5	7.5	7	6.5	9	4.5	5.5	9	7.20	7
MC	9	6.5	10.5	3	7	6.5	8	11.5	5.5	5.5	7.30	8
CMC	9	13	2.5	3	1.5	1.5	14	14	5.5	1	6.50	4
DNI	9	11.5	7	14	14	12	1	1	14	14	9.75	13

imputation methods and we cannot choose the best method from among these three. On the other hand, BPCA and DNI are the worst methods.

In Table 4.5 we can observe the rankings associated with the methods belonging to the black-boxes modeling category. As can be appreciated, for black-boxes modelling the differences between imputation methods are even more evident. We can select the EC method as the best solution, as it has a difference of ranking of almost 1 with KMI, which stands as the second best. This difference increases when considering

Table 4.6 Average ranks for the Lazy Learning methods

	1-NN	3-NN	LBR	LWL	Avg.	Ranks
IM	5	11	5	8	7.25	7
EC	9.5	13	9	8	9.88	12
KNNI	2.5	5.5	9	8	6.25	4
WKNNI	4	5.5	9	8	6.63	5
KMI	12	5.5	9	2.5	7.25	8
FKMI	6	1.5	9	2.5	4.75	3
SVMI	9.5	9	3	8	7.38	9
EM	11	5.5	9	2.5	7.00	6
SVDI	13	12	1	12	9.50	11
BPCA	14	14	13	13	13.50	14
LLSI	7.5	5.5	9	8	7.50	10
MC	7.5	1.5	3	2.5	3.63	1
CMC	1	5.5	3	8	4.38	2
DNI	2.5	10	14	14	10.13	13

the third best, FKMI. No other family of classifiers present this gap in the rankings. Therefore, in this family of classification methods we could, with some confidence, establish the EC method as the best choice. The DNI and IM methods are among the worst. This means that for the black-boxes modelling methods the use of some kind of MV treatment is mandatory, whereas the EC method is the most suitable one. As with the RIL methods, the BPCA method is the worst choice, with the highest ranking.

Finally the results for the last LL group are presented in Table 4.6. For the LL models, the MC method is the best with the lowest average ranking. The CMC method, which is relatively similar to MC, also obtains a low rank very close to MC's. Only the FKMI method obtains a low enough rank to be compared with the MC and CMC methods. The rest of the imputation methods are far from these lowest ranks with almost two points of difference in the ranking. Again, the DNI and IM methods obtain high rankings. The DNI method is one of the worst, with only the BPCA method performing worse. As with the black-boxes modelling models, the imputation methods produce a significant improvement in the accuracy of these classification methods.

4.6.3 Interesting Comments

In this last Section we have carried out an experimental comparison among the imputation methods presented in this chapter. We have tried to obtain the best imputation choice by means of non-parametric statistical testing. The results obtained concur with previous studies:

- The imputation methods which fill in the MVs outperform the case deletion (IM method) and the lack of imputation (DNI method).
- There is no universal imputation method which performs best for all classifiers.

In Sect. 4.6.1 we have analyzed the influence of the imputation methods in the data in respect to two measures. These two measures are the *Wilson's noise ratio* and the *average MI difference*. The first one quantifies the noise induced by the imputation method in the instances which contain MVs. The second one examines the increment or decrement in the relationship of the isolated input attributes with respect to the class label. We have observed that the CMC and EC methods are the ones which introduce less noise and maintain the MI better.

According to the results in Sect. 4.6.2, the particular analysis of the MVs treatment methods conditioned to the classification methods' groups seems necessary. Thus, we can stress the recommended imputation algorithms to be used based on the classification method's type, as in the case of the *FKMI* imputation method for the Rule Induction Learning group, the *EC* method for the black-boxes modelling Models and the *MC* method for the Lazy Learning models. We can confirm the positive effect of the imputation methods and the classifiers' behavior, and the presence of more suitable imputation methods for some particular classifier categories than others.

References

1. Acuna, E., Rodriguez, C.: Classification, Clustering and Data Mining Applications. Springer, Berlin (2004)
2. Atkeson, C.G., Moore, A.W., Schaal, S.: Locally weighted learning. Artif. Intell. Rev. **11**, 11–73 (1997)
3. Aydilek, I.B., Arslan, A.: A hybrid method for imputation of missing values using optimized fuzzy c-means with support vector regression and a genetic algorithm. Inf. Sci. **233**, 25–35 (2013)
4. Azim, S., Aggarwal, S.: Hybrid model for data imputation: using fuzzy c-means and multi layer perceptron. In: Advance Computing Conference (IACC), 2014 IEEE International, pp. 1281–1285 (2014)
5. Barnard, J., Meng, X.: Applications of multiple imputation in medical studies: from aids to nhanes. Stat. Methods Med. Res. **8**(1), 17–36 (1999)
6. Batista, G., Monard, M.: An analysis of four missing data treatment methods for supervised learning. Appl. Artif. Intell. **17**(5), 519–533 (2003)
7. Bezdek, J., Kuncheva, L.: Nearest prototype classifier designs: an experimental study. Int. J. Intell. Syst. **16**(12), 1445–1473 (2001)
8. Broomhead, D., Lowe, D.: Multivariable functional interpolation and adaptive networks. Complex Systems **11**, 321–355 (1988)
9. van Buuren, S., Groothuis-Oudshoorn, K.: MICE: multivariate imputation by chained equations in r. J. Stat. Softw. **45**(3), 1–67 (2011)
10. le Cessie, S., van Houwelingen, J.: Ridge estimators in logistic regression. Appl. Stat. **41**(1), 191–201 (1992)
11. Chai, L., Mohamad, M., Deris, S., Chong, C., Choon, Y., Ibrahim, Z., Omatu, S.: Inferring gene regulatory networks from gene expression data by a dynamic bayesian network-based model. In: Omatu, S., De Paz Santana, J.F., González, S.R., Molina, J.M., Bernardos, A.M.,

Rodríguez, J.M.C. (eds.) Distributed Computing and Artificial Intelligence, Advances in Intelligent and Soft Computing, pp. 379–386. Springer, Berlin (2012)
12. Ching, W.K., Li, L., Tsing, N.K., Tai, C.W., Ng, T.W., Wong, A.S., Cheng, K.W.: A weighted local least squares imputation method for missing value estimation in microarray gene expression data. Int. J. Data Min. Bioinform. **4**(3), 331–347 (2010)
13. Chow, C., Liu, C.: Approximating discrete probability distributions with dependence trees. IEEE Trans. Inf. Theor. **14**(3), 462–467 (1968)
14. Clark, P., Niblett, T.: The CN2 induction algorithm. Machine Learning **3**(4), 261–283 (1989)
15. Cohen, W., Singer, Y.: A simple and fast and effective rule learner. In: Proceedings of the Sixteenth National Conference on Artificial Intelligence, pp. 335–342 (1999)
16. Cohen, W.W.: Fast effective rule induction. In: Proceedings of the Twelfth International Conference on Machine Learning (ICML), pp. 115–123 (1995).
17. Cortes, C., Vapnik, V.: Support vector networks. Machine Learning **20**, 273–297 (1995)
18. Cover, T.M., Thomas, J.A.: Elements of Information Theory, 2 edn. Wiley, New York (1991)
19. Daniel, R.M., Kenward, M.G.: A method for increasing the robustness of multiple imputation. Comput. Stat. Data Anal. **56**(6), 1624–1643 (2012)
20. Dempster, A., Laird, N., Rubin, D.: Maximum likelihood estimation from incomplete data via the EM algorithm (with discussion). J. Roy. Statist. Soc. Ser. B **39**, 1–38 (1977)
21. Domingos, P., Pazzani, M.: On the optimality of the simple bayesian classifier under zero-one loss. Machine Learning **29**, 103–137 (1997)
22. Dorri, F., Azmi, P., Dorri, F.: Missing value imputation in dna microarrays based on conjugate gradient method. Comp. Bio. Med. **42**(2), 222–227 (2012)
23. Dunning, T., Freedman, D.: Modeling section effects, Sage, pp. 225–231 (2008)
24. Ennett, C.M., Frize, M., Walker, C.R.: Influence of missing values on artificial neural network performance. Stud. Health Technol. Inform. **84**, 449–453 (2001)
25. Fan, R.E., Chen, P.H., Lin, C.J.: Working set selection using second order information for training support vector machines. J. Machine Learning Res. **6**, 1889–1918 (2005)
26. Farhangfar, A., Kurgan, L., Dy, J.: Impact of imputation of missing values on classification error for discrete data. Pattern Recognit. **41**(12), 3692–3705 (2008). http://dx.doi.org/10.1016/j.patcog.2008.05.019
27. Farhangfar, A., Kurgan, L.A., Pedrycz, W.: A novel framework for imputation of missing values in databases. IEEE Trans. Syst. Man Cybern. Part A **37**(5), 692–709 (2007)
28. Fayyad, U., Irani, K.: Multi-interval discretization of continuous-valued attributes for classification learning. In: 13th International Joint Conference on Uncertainly in Artificial Intelligence(IJCAI93), pp. 1022–1029 (1993)
29. Feng, H., Guoshun, C., Cheng, Y., Yang, B., Chen, Y.: A SVM regression based approach to filling in missing values. In: Khosla, R., Howlett, R.J., Jain, L.C. (eds.) KES (3), Lecture Notes in Computer Science, vol. 3683, pp. 581–587. Springer, Berlin (2005)
30. Feng, X., Wu, S., Liu, Y.: Imputing missing values for mixed numeric and categorical attributes based on incomplete data hierarchical clustering. In: Proceedings of the 5th International Conference on Knowledge Science, Engineering and Management, KSEM'11, pp. 414–424 (2011)
31. Figueroa García, J.C., Kalenatic, D., Lopez Bello, C.A.: Missing data imputation in multivariate data by evolutionary algorithms. Comput. Hum. Behav. **27**(5), 1468–1474 (2011)
32. de França, F.O., Coelho, G.P., Zuben, F.J.V.: Predicting missing values with biclustering: a coherence-based approach. Pattern Recognit. **46**(5), 1255–1266 (2013)
33. Frank, E., Witten, I.: Generating accurate rule sets without global optimization. In: Proceedings of the 15th International Conference on Machine Learning, pp. 144–151 (1998)
34. Gheyas, I.A., Smith, L.S.: A neural network-based framework for the reconstruction of incomplete data sets. Neurocomputing **73**(16–18), 3039–3065 (2010)
35. Gibert, K.: Mixed intelligent-multivariate missing imputation. Int. J. Comput. Math. **91**(1), 85–96 (2014)
36. Grzymala-Busse, J., Goodwin, L., Grzymala-Busse, W., Zheng, X.: Handling missing attribute values in preterm birth data sets. In: 10th International Conference of Rough Sets and Fuzzy Sets and Data Mining and Granular Computing(RSFDGrC05), pp. 342–351 (2005)

37. Grzymala-Busse, J.W., Hu, M.: A comparison of several approaches to missing attribute values in data mining. In: Ziarko, W., Yao, Y.Y. (eds.) Rough Sets and Current Trends in Computing, Lecture Notes in Computer Science, vol. 2005, pp. 378–385. Springer, Berlin (2000)
38. Howell, D.: The analysis of missing data. SAGE Publications Ltd, London (2007)
39. Hruschka Jr, E.R., Ebecken, N.F.F.: Missing values prediction with k2. Intell. Data Anal. **6**(6), 557–566 (2002)
40. Hulse, J.V., Khoshgoftaar, T.M.: Incomplete-case nearest neighbor imputation in software measurement data. Inf. Sci. **259**, 596–610 (2014)
41. Ingsrisawang, L., Potawee, D.: Multiple imputation for missing data in repeated measurements using MCMC and copulas, pp. 1606–1610 (2012)
42. Ishioka, T.: Imputation of missing values for unsupervised data using the proximity in random forests. In: eLmL 2013, The 5th International Conference on Mobile, Hybrid, and On-line Learning, pp. 30–36 (2013)
43. Jamshidian, M., Jalal, S., Jansen, C.: Missmech: an R package for testing homoscedasticity, multivariate normality, and missing completely at random (mcar). J. Stat. Softw. **56**(6), 1–31 (2014)
44. Joenssen, D.W., Bankhofer, U.: Hot deck methods for imputing missing data: the effects of limiting donor usage. In: Proceedings of the 8th International Conference on Machine Learning and Data Mining in Pattern Recognition, MLDM'12, pp. 63–75 (2012)
45. Juhola, M., Laurikkala, J.: Missing values: how many can they be to preserve classification reliability? Artif. Intell. Rev. **40**(3), 231–245 (2013)
46. Keerin, P., Kurutach, W., Boongoen, T.: Cluster-based knn missing value imputation for dna microarray data. In: Systems, Man, and Cybernetics (SMC), 2012 IEEE International Conference on, pp. 445–450. IEEE (2012)
47. Keerin, P., Kurutach, W., Boongoen, T.: An improvement of missing value imputation in dna microarray data using cluster-based lls method. In: Communications and Information Technologies (ISCIT), 2013 13th International Symposium on, pp. 559–564 (2013)
48. Khan, S.S., Hoey, J., Lizotte, D.J.: Bayesian multiple imputation approaches for one-class classification. In: Kosseim, L., Inkpen, D. (eds.) Advances in Artificial Intelligence - 25th Canadian Conference on Artificial Intelligence, Canadian AI 2012, Toronto, ON, Canada, Proceedings, pp. 331–336. 28–30 May 2012
49. Kim, H., Golub, G.H., Park, H.: Missing value estimation for dna microarray gene expression data: local least squares imputation. Bioinform. **21**(2), 187–198 (2005)
50. Krzanowski, W.: Multiple discriminant analysis in the presence of mixed continuous and categorical data. Comput. Math. Appl. **12**(2, Part A), 179–185 (1986)
51. Kwak, N., Choi, C.H.: Input feature selection by mutual information based on parzen window. IEEE Trans. Pattern Anal. Mach. Intell. **24**(12), 1667–1671 (2002)
52. Kwak, N., Choi, C.H.: Input feature selection for classification problems. IEEE Trans. Neural Networks **13**(1), 143–159 (2002)
53. Li, D., Deogun, J., Spaulding, W., Shuart, B.: Towards missing data imputation: a study of fuzzy k-means clustering method. In: 4th International Conference of Rough Sets and Current Trends in Computing (RSCTC04), pp. 573–579 (2004)
54. Little, R.J.A., Rubin, D.B.: Statistical Analysis with Missing Data, 1st edn. Wiley Series in Probability and Statistics, New York (1987)
55. Little, R.J.A., Schluchter, M.D.: Maximum likelihood estimation for mixed continuous and categorical data with missing values. Biometrika **72**, 497–512 (1985)
56. Lu, X., Si, J., Pan, L., Zhao, Y.: Imputation of missing data using ensemble algorithms. In: Fuzzy Systems and Knowledge Discovery (FSKD), 2011 8th International Conference on, vol. 2, pp. 1312–1315 (2011)
57. McLachlan, G.: Discriminant Analysis and Statistical Pattern Recognition. Wiley, New York(2004)
58. Merlin, P., Sorjamaa, A., Maillet, B., Lendasse, A.: X-SOM and L-SOM: a double classification approach for missing value imputation. Neurocomputing **73**(7–9), 1103–1108 (2010)

59. Michalksi, R., Mozetic, I., Lavrac, N.: The multipurpose incremental learning system AQ15 and its testing application to three medical domains. In: 5th INational Conference on Artificial Intelligence (AAAI86), pp. 1041–1045 (1986)
60. Miyakoshi, Y., Kato, S.: Missing value imputation method by using Bayesian network with weighted learning. IEEJ Trans. Electron. Inf. Syst. **132**, 299–305 (2012)
61. Moller, F.: A scaled conjugate gradient algorithm for fast supervised learning. Neural Networks **6**, 525–533 (1990)
62. Oba, S., aki Sato, M., Takemasa, I., Monden, M., ichi Matsubara, K., Ishii, S.: A bayesian missing value estimation method for gene expression profile data. Bioinform. **19**(16), 2088–2096 (2003)
63. Ouyang, M., Welsh, W.J., Georgopoulos, P.: Gaussian mixture clustering and imputation of microarray data. Bioinform. **20**(6), 917–923 (2004)
64. Panigrahi, L., Ranjan, R., Das, K., Mishra, D.: Removal and interpolation of missing values using wavelet neural network for heterogeneous data sets. In: Proceedings of the International Conference on Advances in Computing, Communications and Informatics, ICACCI '12, pp. 1004–1009 (2012)
65. Patil, B., Joshi, R., Toshniwal, D.: Missing value imputation based on k-mean clustering with weighted distance. In: Ranka, S., Banerjee, A., Biswas, K., Dua, S., Mishra, P., Moona, R., Poon, S.H., Wang, C.L. (eds.) Contemporary Computing, Communications in Computer and Information Science, vol. 94, pp. 600–609. Springer, Berlin (2010)
66. Peng, H., Long, F., Ding, C.: Feature selection based on mutual information: Criteria of max-dependency, max-relevance, and min-redundancy. IEEE Trans. Pattern Anal. Mach. Intell. 27(8), pp. 1226–1238 (2005)
67. Pham, D.T., Afify, A.A.: Rules-6: a simple rule induction algorithm for supporting decision making. In: Industrial Electronics Society, 2005. IECON 2005. 31st Annual Conference of IEEE, pp. 2184–2189 (2005)
68. Pham, D.T., Afify, A.A.: SRI: a scalable rule induction algorithm. Proc. Inst. Mech. Eng. [C]: J. Mech. Eng. Sci. **220**, 537–552 (2006)
69. Plat, J.: A resource allocating network for function interpolation. Neural Comput. **3**(2), 213–225 (1991)
70. Platt, J.C.: Fast training of support vector machines using sequential minimal optimization. In: Advances in Kernel Methods: Support Vector Learning, pp. 185–208. MIT Press, Cambridge (1999)
71. Pyle, D.: Data Preparation for Data Mining. Morgan Kaufmann Publishers Inc., San Francisco (1999)
72. Qin, Y., Zhang, S., Zhang, C.: Combining knn imputation and bootstrap calibrated empirical likelihood for incomplete data analysis. Int. J. Data Warehouse. Min. **6**(4), 61–73 (2010)
73. Quinlan, J.R.: C4.5: Programs for Machine Learning. Morgan Kaufmann Publishers, San Francisco (1993)
74. Rahman, G., Islam, Z.: A decision tree-based missing value imputation technique for data pre-processing. In: Proceedings of the 9th Australasian Data Mining Conference - Volume 121, AusDM '11, pp. 41–50 (2011)
75. Rahman, M., Islam, M.: KDMI: a novel method for missing values imputation using two levels of horizontal partitioning in a data set. In: Motoda, H., Wu, Z., Cao, L., Zaiane, O., Yao, M., Wang, W. (eds.) Advanced Data Mining and Applications. Lecture Notes in Computer Science, vol. 8347, pp. 250–263. Springer, Berlin (2013)
76. Rahman, M.G., Islam, M.Z.: Missing value imputation using decision trees and decision forests by splitting and merging records: two novel techniques. Know.-Based Syst. **53**, 51–65 (2013)
77. Rahman, M.G., Islam, M.Z.: Fimus: a framework for imputing missing values using co-appearance, correlation and similarity analysis. Know.-Based Syst. **56**, 311–327 (2014)
78. Royston, P., White, I.R.: Multiple imputation by chained equations (MICE): implementation in STATA. J. Stat. Softw. **45**(4), 1–20 (2011)
79. Rubin, D.B.: Inference and missing data. Biometrika **63**(3), 581–592 (1976)

80. Rubin, D.B.: Multiple Imputation for Nonresponse in Surveys. Wiley, New York (1987)
81. Safarinejadian, B., Menhaj, M., Karrari, M.: A distributed EM algorithm to estimate the parameters of a finite mixture of components. Knowl. Inf. Syst. **23**(3), 267–292 (2010)
82. Schafer, J.L.: Analysis of Incomplete Multivariate Data. Chapman & Hall, London (1997)
83. Schafer, J.L., Olsen, M.K.: Multiple imputation for multivariate missing-data problems: a data analyst's perspective. Multivar. Behav. Res. **33**(4), 545–571 (1998)
84. Scheuren, F.: Multiple imputation: how it began and continues. Am. Stat. **59**, 315–319 (2005)
85. Schneider, T.: Analysis of incomplete climate data: estimation of mean values and covariance matrices and imputation of missing values. J. Clim. **14**, 853–871 (2001)
86. Schomaker, M., Heumann, C.: Model selection and model averaging after multiple imputation. Comput. Stat. Data Anal. **71**, 758–770 (2014)
87. Sehgal, M.S.B., Gondal, I., Dooley, L.: Collateral missing value imputation: a new robust missing value estimation algorithm for microarray data. Bioinform. **21**(10), 2417–2423 (2005)
88. Silva-Ramírez, E.L., Pino-Mejías, R., López-Coello, M., Cubiles-de-la Vega, M.D.: Missing value imputation on missing completely at random data using multilayer perceptrons. Neural Networks **24**(1), 121–129 (2011)
89. Simński, K.: Rough fuzzy subspace clustering for data with missing values. Comput. Inform. **33**(1), 131–153 (2014)
90. Somasundaram, R., Nedunchezhian, R.: Radial basis function network dependent exclusive mutual interpolation for missing value imputation. J. Comput. Sci. **9**(3), 327–334 (2013)
91. Tanner, M.A., Wong, W.: The calculation of posterior distributions by data augmentation. J. Am. Stat. Assoc. **82**, 528–540 (1987)
92. Ting, J., Yu, B., Yu, D., Ma, S.: Missing data analyses: a hybrid multiple imputation algorithm using gray system theory and entropy based on clustering. Appl. Intell. **40**(2), 376–388 (2014)
93. Troyanskaya, O., Cantor, M., Sherlock, G., Brown, P., Hastie, T., Tibshirani, R., Botstein, D., Altman, R.B.: Missing value estimation methods for dna microarrays. Bioinform. **17**(6), 520–525 (2001)
94. Unnebrink, K., Windeler, J.: Intention-to-treat: methods for dealing with missing values in clinical trials of progressively deteriorating diseases. Stat. Med. **20**(24), 3931–3946 (2001)
95. Vellido, A.: Missing data imputation through GTM as a mixture of t-distributions. Neural Networks **19**(10), 1624–1635 (2006)
96. Wang, H., Wang, S.: Mining incomplete survey data through classification. Knowl. Inf. Syst. 24(2), 221–233 (2010)
97. Williams, D., Liao, X., Xue, Y., Carin, L., Krishnapuram, B.: On classification with incomplete data. IEEE Trans. Pattern Anal. Mach. Intell. 29(3), 427–436 (2007)
98. Wilson, D.: Asymptotic properties of nearest neighbor rules using edited data. IEEE Trans. Syst. Man Cybern. **2**(3), 408–421 (1972)
99. Wong, A.K.C., Chiu, D.K.Y.: Synthesizing statistical knowledge from incomplete mixed-mode data. IEEE Trans. Pattern Anal. Mach. Intell. **9**(6), 796–805 (1987)
100. Wu, X., Urpani, D.: Induction by attribute elimination. IEEE Trans. Knowl. Data Eng. **11**(5), 805–812 (1999)
101. Zhang, S.: Nearest neighbor selection for iteratively knn imputation. J. Syst. Softw. **85**(11), 2541–2552 (2012)
102. Zhang, S., Wu, X., Zhu, M.: Efficient missing data imputation for supervised learning. In: Cognitive Informatics (ICCI), 2010 9th IEEE International Conference on, pp. 672–679 (2010)
103. Zheng, Z., Webb, G.I.: Lazy learning of bayesian rules. Machine Learning **41**(1), 53–84 (2000)
104. Zhu, B., He, C., Liatsis, P.: A robust missing value imputation method for noisy data. Appl. Intell. **36**(1), 61–74 (2012)
105. Zhu, X., Zhang, S., Jin, Z., Zhang, Z., Xu, Z.: Missing value estimation for mixed-attribute data sets. IEEE Transactions on Knowl. Data Eng. **23**(1), 110–121 (2011)

Chapter 5
Dealing with Noisy Data

Abstract This chapter focuses on the noise imperfections of the data. The presence of noise in data is a common problem that produces several negative consequences in classification problems. Noise is an unavoidable problem, which affects the data collection and data preparation processes in Data Mining applications, where errors commonly occur. The performance of the models built under such circumstances will heavily depend on the quality of the training data, but also on the robustness against the noise of the model learner itself. Hence, problems containing noise are complex problems and accurate solutions are often difficult to achieve without using specialized techniques—particularly if they are noise-sensitive. Identifying the noise is a complex task that will be developed in Sect. 5.1. Once the noise has been identified, the different kinds of such an imperfection are described in Sect. 5.2. From this point on, the two main approaches carried out in the literature are described. On the first hand, modifying and cleaning the data is studied in Sect. 5.3, whereas designing noise robust Machine Learning algorithms is tackled in Sect. 5.4. An empirical comparison between the latest approaches in the specialized literature is made in Sect. 5.5.

5.1 Identifying Noise

Real-world data is never perfect and often suffers from corruptions that may harm interpretations of the data, models built and decisions made. In classification, noise can negatively affect the system performance in terms of classification accuracy, building time, size and interpretability of the classifier built [99, 100]. The presence of noise in the data may affect the intrinsic characteristics of a classification problem. Noise may create small clusters of instances of a particular class in parts of the instance space corresponding to another class, remove instances located in key areas within a concrete class or disrupt the boundaries of the classes and increase overlapping among them. These alterations corrupt the knowledge that can be extracted from the problem and spoil the classifiers built from that noisy data with respect to the original classifiers built from the clean data that represent the most accurate implicit knowledge of the problem [100].

S. García et al., *Data Preprocessing in Data Mining*,
Intelligent Systems Reference Library 72, DOI 10.1007/978-3-319-10247-4_5

Noise is specially relevant in supervised problems, where it alters the relationship between the informative features and the measure output. For this reason noise has been specially studied in classification and regression where noise hinders the knowledge extraction from the data and spoils the models obtained using that noisy data when they are compared to the models learned from clean data from the same problem, which represent the real implicit knowledge of the problem [100]. In this sense, *robustness* [39] is the capability of an algorithm to build models that are insensitive to data corruptions and suffer less from the impact of noise; that is, the more robust an algorithm is, the more similar the models built from clean and noisy data are. Thus, a classification algorithm is said to be more robust than another if the former builds classifiers which are less influenced by noise than the latter. Robustness is considered more important than performance results when dealing with noisy data, because it allows one to know a priori the expected behavior of a learning method against noise in cases where the characteristics of noise are unknown.

Several approaches have been studied in the literature to deal with noisy data and to obtain higher classification accuracies on test data. Among them, the most important are:

- *Robust learners* [8, 75] These are techniques characterized by being less influenced by noisy data. An example of a robust learner is the C4.5 algorithm [75]. C4.5 uses pruning strategies to reduce the chances reduce the possibility that trees overfit to noise in the training data [74]. However, if the noise level is relatively high, even a robust learner may have a poor performance.
- *Data polishing methods* [84] Their aim is to correct noisy instances prior to training a learner. This option is only viable when data sets are small because it is generally time consuming. Several works [84, 100] claim that complete or partial noise correction in training data, with test data still containing noise, improves test performance results in comparison with no preprocessing.
- *Noise filters* [11, 48, 89] identify noisy instances which can be eliminated from the training data. These are used with many learners that are sensitive to noisy data and require data preprocessing to address the problem.

Noise is not the only problem that supervised ML techniques have to deal with. Complex and nonlinear boundaries between classes are problems that may hinder the performance of classifiers and it often is hard to distinguish between such overlapping and the presence of noisy examples. This topic has attracted recent attention with the appearance of works that have indicated relevant issues related to the degradation of performance:

- *Presence of small disjuncts* [41, 43] (Fig. 5.1a) The minority class can be decomposed into many sub-clusters with very few examples in each one, being surrounded by majority class examples. This is a source of difficulty for most learning algorithms in detecting precisely enough those sub-concepts.
- *Overlapping between classes* [26, 27] (Fig. 5.1b) There are often some examples from different classes with very similar characteristics, in particular if they are located in the regions around decision boundaries between classes. These examples refer to overlapping regions of classes.

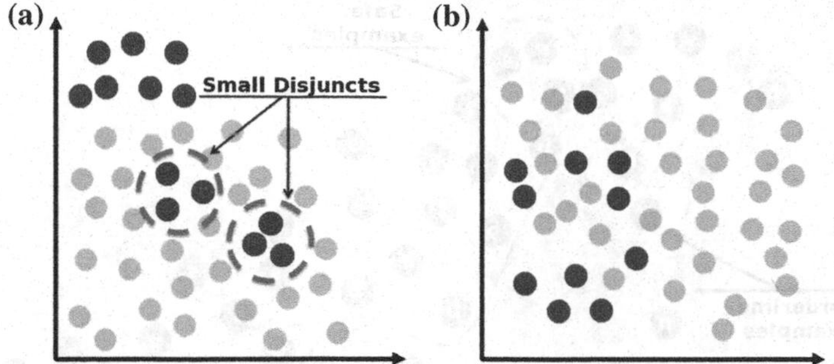

Fig. 5.1 Examples of the interaction between classes: **a** small disjuncts and **b** overlapping between classes

Closely related to the overlapping between classes, in [67] another interesting problem is pointed out: the higher or lower presence of examples located in the area surrounding class boundaries, which are called borderline examples. Researchers have found that misclassification often occurs near class boundaries where overlapping usually occurs as well and it is hard to find a feasible solution [25]. The authors in [67] showed that classifier performance degradation was strongly affected by the quantity of borderline examples and that the presence of other noisy examples located farther outside the overlapping region was also very difficult for re-sampling methods.

- *Safe examples* are placed in relatively homogeneous areas with respect to the class label.
- *Borderline examples* are located in the area surrounding class boundaries, where either the minority and majority classes overlap or these examples are very close to the difficult shape of the boundary—in this case, these examples are also difficult as a small amount of the attribute noise can move them to the wrong side of the decision boundary [52].
- *Noisy examples* are individuals from one class occurring in the safe areas of the other class. According to [52] they could be treated as examples affected by class label noise. Notice that the term *noisy examples* will be further used in this book in the wider sense of [100] where noisy examples are corrupted either in their attribute values or the class label.

The examples belonging to the two last groups often do not contribute to correct class prediction [46]. Therefore, one could ask a question whether removing them (all or the most difficult misclassification part) should improve classification performance. Thus, this book examines the usage of noise filters to achieve this goal, because they are widely used obtaining good results in classification, and in the application of techniques designed to deal with noisy examples.

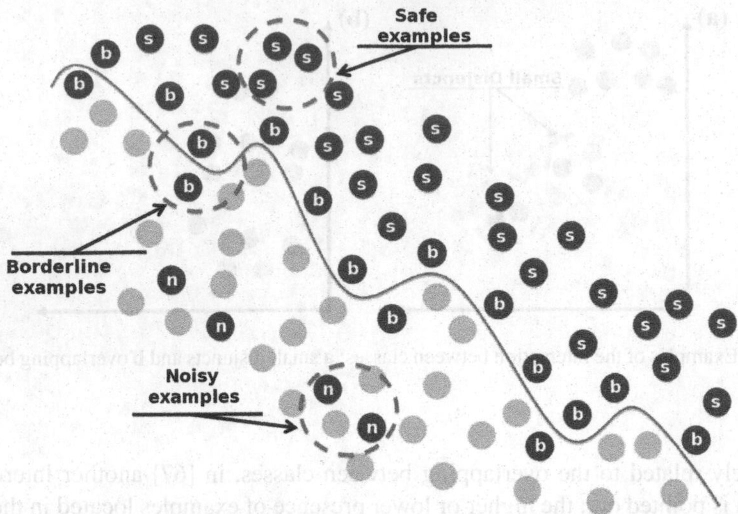

Fig. 5.2 The three types of examples considered in this book: safe examples (labeled as *s*), borderline examples (labeled as *b*) and noisy examples (labeled as *n*). The *continuous line* shows the decision boundary between the two classes

5.2 Types of Noise Data: Class Noise and Attribute Noise

A large number of components determine the quality of a data set [90]. Among them, the class labels and the attribute values directly influence the quality of a classification data set. The quality of the class labels refers to whether the class of each example is correctly assigned; otherwise, the quality of the attributes refers to their capability of properly characterizing the examples for classification purposes—obviously, if noise affects attribute values, this capability of characterization and therefore, the quality of the attributes, is reduced. Based on these two information sources, two types of noise can be distinguished in a given data set [12, 96]:

1. **Class noise** (also referred as *label noise*) It occurs when an example is incorrectly labeled. Class noise can be attributed to several causes, such as subjectivity during the labeling process, data entry errors, or inadequacy of the information used to label each example. Two types of class noise can be distinguished:
   - *Contradictory examples* There are duplicate examples in the data set having different class labels [31].
   - *Misclassifications* Examples are labeled with class labels different from their true label [102].
2. **Attribute noise** It refers to corruptions in the values of one or more attributes. Examples of attribute noise are: erroneous attribute values, missing or unknown attribute values, and incomplete attributes or "do not care" values.

In this book, class noise refers to misclassifications, whereas attribute noise refers to erroneous attribute values, because they are the most common in real-world data [100]. Furthermore, erroneous attribute values, unlike other types of attribute noise, such as MVs (which are easily detectable), have received less attention in the literature.

Treating class and attribute noise as corruptions of the class labels and attribute values, respectively, has been also considered in other works in the literature [69, 100]. For instance, in [100], the authors reached a series of interesting conclusions, showing that attribute noise is more harmful than class noise or that eliminating or correcting examples in data sets with class and attribute noise, respectively, may improve classifier performance. They also showed that attribute noise is more harmful in those attributes highly correlated with the class labels. In [69], the authors checked the robustness of methods from different paradigms, such as probabilistic classifiers, decision trees, instance based learners or SVMs, studying the possible causes of their behavior.

However, most of the works found in the literature are only focused on class noise. In [9], the problem of multi-class classification in the presence of labeling errors was studied. The authors proposed a generative multi-class classifier to learn with labeling errors, which extends the multi-class quadratic normal discriminant analysis by a model of the mislabeling process. They demonstrated the benefits of this approach in terms of parameter recovery as well as improved classification performance. In [32], the problems caused by labeling errors occurring far from the decision boundaries in Multi-class Gaussian Process Classifiers were studied. The authors proposed a Robust Multi-class Gaussian Process Classifier, introducing binary latent variables that indicate when an example is mislabeled. Similarly, the effect of mislabeled samples appearing in gene expression profiles was studied in [98]. A detection method for these samples was proposed, which takes advantage of the measuring effect of data perturbations based on the SVM regression model. They also proposed three algorithms based on this index to detect mislabeled samples. An important common characteristic of these works, also considered in this book, is that the suitability of the proposals was evaluated on both real-world and synthetic or noisy-modified real-world data sets, where the noise could be somehow quantified.

In order to model class and attribute noise, we consider four different synthetic noise schemes found in the literature, so that we can simulate the behavior of the classifiers in the presence of noise as presented in the next section.

5.2.1 Noise Introduction Mechanisms

Traditionally the label noise introduction mechanism has not attracted as much attention in its consequences as it has in the knowledge extracted from it. However, as the noise treatment is being embedded in the classifier design, the nature of noise becomes more and more important. Recently, the authors in Frenay and Verleysen [19] have adopted the statistical analysis for the MVs introduction described

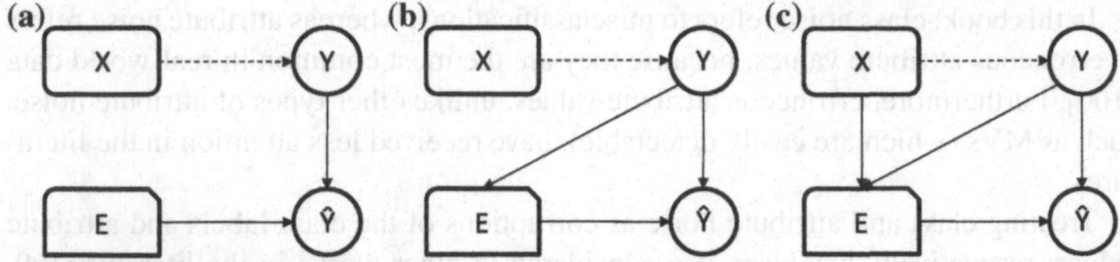

Fig. 5.3 Statistical taxonomy of label noise as described in [19]. **a** Noisy completely at random (NCAR), **b** Noisy at random (NAR), and **c** Noisy not at random (NNAR). X is the array of input attributes, Y is the true class label, $\hat{Y}$ is the actual class label and E indicates whether a labeling error occurred ($Y \neq \hat{Y}$). *Arrows* indicate statistical dependencies

in Sect. 4.2. That is, we will distinguish between three possible statistical models for label noise as depicted in Fig. 5.3. In the three subfigures of Fig. 5.3 the dashed arrow points out a the implicit relation between the input features and the output that is desired to be modeled by the classifier. In the most simplistic case in which the noise procedure is not dependent of either the true value of the class Y or the input attribute values X, the label noise is called noise completely at random or NCAR as shown in Fig. 5.3a. In [7] the observed label is different from the true class with a probability $p_n = P(E = 1)$, that is also called the error rate or noise rate. In binary classification problems, the labeling error in NCAR is applied symmetrically to both class labels and when $p_n = 0.5$ the labels will no longer provide useful information. In multiclass problems when the error caused by noise (i.e. $E = 1$) appears the class label is changed by any other different one available. In the case in which the selection of the erroneous class label is made by a uniform probability distribution, the noise model is known as *uniform label/class noise*.

Things get more complicated in the noise at random (NAR) model. Although the noise is independent of the inputs X, the true value of the class make it more or less prone to be noisy. This asymmetric labeling error can be produced by the different cost of extracting the true class, as for example in medical case-control studies, financial score assets and so on. Since the wrong class label is subject to a particular true class label, the labeling probabilities can be defined as:

$$P(\hat{Y} = \hat{y}|Y = y) = \sum_{e \in 0,1} P(\hat{Y} = \hat{y}|E = e, Y = y)P(E = e|Y = y). \qquad (5.1)$$

Of course this probability definition span over all the class labels and the possibly erroneous class that the could take. As shown in [70] this conforms a transition matrix γ where each position γ_{ij} shows the probability of $P(\hat{Y} = c_i|Y = c_j)$ for the possible class labels c_i and c_j. Some examples can be examined with detail in [19]. The NCAR model is a special case of the NAR label noise model in which the probability of each position γ_{ij} denotes the independency between $\hat{Y}$ and Y: $\gamma_{ij} = P(\hat{Y}_i, Y = c_j)$.

Apart from the uniform class noise, the NAR label noise has widely studied in the literature. An example is the pairwise label noise, where two selected class labels are chosen to be labeled with the other with certain probability. In this pairwise label noise (or pairwise class noise) only two positions of the γ matrix are nonzero outside of the diagonal. Another problem derived from the NAR noise model is that it is not trivial to decide whether the class labels are useful or not.

The third and last noise model is the noisy not at random (NNAR), where the input attributes somehow affect the probability of the class label being erroneous as shown in Fig. 5.3c. An example of this illustrated by Klebanov [49] where evidence is given that difficult samples are randomly labeled.It also occurs that those examples similar to existing ones are labeled by experts in a biased way, having more probability of being mislabeled the more similar they are. NNAR model is the more general case of class noise [59] where the error E depends on both X and Y and it is the only model able to characterize mislabelings in the class borders or due to poor sampling density. As shown in [19] the probability of error is much more complex than in the two previous cases as it has to take into account the density function of the input over the input feature space $\mathcal{X}$ when continuous:

$$p_n = P(E = 1) = \sum_{c_i \in C} \times \int_{x \in \mathcal{X}} P(X = x | Y = y) P(E = 1 | X = x, Y = y) dx. \tag{5.2}$$

As a consequence the perfect identification and estimation of the NNAR noise is almost impossible, relying in approximating it from the expert knowledge of the problem and the domain.

In the case of attribute noise, the modelization described above can be extended and adapted. In this case, we can distinguish three possibilities as well:

- When the noise appearance does not depend either on the rest of the input features' values or the class label the NCAR noise model applies. This type of noise can occur when distortions in the measures appear at random, for example in faulty hand data insertion or network errors that do not depend in the data content itself.
- When the attribute noise depends on the true value x_i but not on the rest of input values $x_1, \ldots, x_{i-1}, x_{i+1}, \ldots, x_n$ or the observed class label y the NAR model is applicable. An illustrative example is when the different temperatures affect their registration in climatic data in a different way depending on the proper temperature value.
- In the last case the noise probability will depend on the value of the feature x_i but also on the rest of the input feature values $x_1, \ldots, x_{i-1}, x_{i+1}, \ldots, x_n$. This is a very complex situation in which the value is altered when the rest of features present a particular combination of values, as in medical diagnosis when some test results are filled by an expert prediction without conducting the test due to high costs.

For the sake of brevity we will not develop the probability error equations here as their expressions would vary depending on the nature of the input feature, being different from real valued ones with respect to nominal attributes. However we must point out that in attribute noise the probability dependencies are not the only important aspect to be considered. The probability distribution of the noise is also fundamental.

For numerical data, the noisy datum $\hat{x}_i$ may be a slight variation of the true value x_i or a completely random value. The density function of the noise values is very rarely known. Simple examples of the first type of noise would be perturbations caused by a normal distribution with the mean centered in the true value and with a fixed variance. The second type of noise is usually estimated by assigning an uniform probability to all the possible values of the input feature's range. This procedure is also typical with nominal data, where no preference of one value is taken. Again note that the distribution of the noise is not the same as the probability of its appearance discussed above: first the noise must be introduced with a certain probability (following the NCAR, NAR or NNAR models) and then the noise value is stated or analyzed to follow the aforementioned density functions.

5.2.2 Simulating the Noise of Real-World Data Sets

Checking the effect of noisy data on the performance of classifier learning algorithms is necessary to improve their reliability and has motivated the study of how to generate and introduce noise into the data. Noise generation can be characterized by three main characteristics [100]:

1. **The place where the noise is introduced** Noise may affect the input attributes or the output class, impairing the learning process and the resulting model.
2. **The noise distribution** The way in which the noise is present can be, for example, uniform [84, 104] or Gaussian [100, 102].
3. **The magnitude of generated noise values** The extent to which the noise affects the data set can be relative to each data value of each attribute, or relative to the minimum, maximum and standard deviation for each attribute [100, 102, 104].

In contrast to other studies in the literature, this book aims to clearly explain how noise is defined and generated, and also to properly justify the choice of the noise introduction schemes. Furthermore, the noise generation software has been incorporated into the KEEL tool (see Chap. 10) for its free usage. The two types of noise considered in this work, class and attribute noise, have been modeled using four different noise schemes; in such a way that, the presence of these types of noise will allow one to simulate the behavior of the classifiers in these two scenarios:

1. **Class noise** usually occurs on the boundaries of the classes, where the examples may have similar characteristics—although it can occur in any other area of the domain. In this book, class noise is introduced using an *uniform class noise scheme* [84] (randomly corrupting the class labels of the examples) and a *pairwise class noise scheme* [100, 102] (labeling examples of the majority class with the second majority class). Considering these two schemes, noise affecting any class label and noise affecting only the two majority classes is simulated respectively.

Whereas the former can be used to simulate a NCAR noise model, the latter is useful to produce a particular NAR noise model.

2. **Attribute noise** can proceed from several sources, such as transmission constraints, faults in sensor devices, irregularities in sampling and transcription errors [85]. The erroneous attribute values can be totally unpredictable, i.e., random, or imply a low variation with respect to the correct value. We use the *uniform attribute noise scheme* [100, 104] and the *Gaussian attribute noise scheme* in order to simulate each one of the possibilities, respectively. We introduce attribute noise in accordance with the hypothesis that interactions between attributes are weak [100]; as a consequence, the noise introduced into each attribute has a low correlation with the noise introduced into the rest.

Robustness is the capability of an algorithm to build models that are insensitive to data corruptions and suffer less from the impact of noise [39]. Thus, a classification algorithm is said to be more robust than another if the former builds classifiers which are less influenced by noise than the latter, i.e., more robust. In order to analyze the degree of robustness of the classifiers in the presence of noise, we will compare the performance of the classifiers learned with the original (without induced noise) data set with the performance of the classifiers learned using the noisy data set. Therefore, those classifiers learned from noisy data sets being more similar (in terms of results) to the noise free classifiers will be the most robust ones.

5.3 Noise Filtering at Data Level

Noise filters are preprocessing mechanisms to detect and eliminate noisy instances in the training set. The result of noise elimination in preprocessing is a reduced training set which is used as an input to a classification algorithm. The separation of noise detection and learning has the advantage that noisy instances do not influence the classifier building design [24].

Noise filters are generally oriented to detect and eliminate instances with class noise from the training data. Elimination of such instances has been shown to be advantageous [23]. However, the elimination of instances with attribute noise seems counterproductive [74, 100] since instances with attribute noise still contain valuable information in other attributes which can help to build the classifier. It is also hard to distinguish between noisy examples and true exceptions, and henceforth many techniques have been proposed to deal with noisy data sets with different degrees of success.

We will consider three noise filters designed to deal with mislabeled instances as they are the most common and the most recent: the *Ensemble Filter* [11], the *Cross-Validated Committees Filter* [89] and the *Iterative-Partitioning Filter* [48]. It should be noted that these three methods are ensemble-based and vote-based filters. A motivation for using ensembles for filtering is pointed out in [11]: when it is assumed that some instances in the data have been mislabeled and that the label errors are independent of the particular model being fitted to the data, collecting information from different models will provide a better method for detecting mislabeled instances than collecting information from a single model.

The implementations of these three noise filters can be found in KEEL (see Chap. 10). Their descriptions can be found in the following subsections. In all descriptions we use D_T to refer to the training set, D_N to refer to the noisy data identified in the training set (initially, $D_N = \emptyset$) and Γ is the number of folds in which the training data is partitioned by the noise filter.

The three noise filters presented below use a voting scheme to determine which instances to eliminate from the training set. There are two possible schemes to determine which instances to remove: consensus and majority schemes. The consensus scheme removes an instance if it is misclassified by all the classifiers, while the majority scheme removes an instance if it is misclassified by more than half of the classifiers. Consensus filters are characterized by being conservative in rejecting good data at the expense of retaining bad data. Majority filters are better at detecting bad data at the expense of rejecting good data.

5.3.1 Ensemble Filter

The *Ensemble Filter* (EF) [11] is a well-known filter in the literature. It attempts to achieve an improvement in the quality of the training data as a preprocessing step in classification, by detecting and eliminating mislabeled instances. It uses a set of learning algorithms to create classifiers in several subsets of the training data that serve as noise filters for the training set.

The identification of potentially noisy instances is carried out by performing an Γ-FCV on the training data with μ classification algorithms, called filter algorithms. In the developed experimentation for this book we have utilized the three filter algorithms used by the authors in [11], which are C4.5, 1-NN and LDA [63]. The complete process carried out by EF is described below:

- Split the training data set D_T into Γ equal sized subsets.
- For each one of the μ filter algorithms:
  - For each of these Γ parts, the filter algorithm is trained on the other $\Gamma - 1$ parts. This results in Γ different classifiers.
  - These Γ resulting classifiers are then used to tag each instance in the excluded part as either correct or mislabeled, by comparing the training label with that assigned by the classifier.
- At the end of the above process, each instance in the training data has been tagged by each filter algorithm.
- Add to D_N the noisy instances identified in D_T using a voting scheme, taking into account the correctness of the labels obtained in the previous step by the μ filter algorithms. We use a consensus vote scheme in this case.
- Remove the noisy instances from the training set: $D_T \leftarrow D_T \backslash D_N$.

5.3.2 Cross-Validated Committees Filter

The *Cross-Validated Committees Filter* (CVCF) [89] uses ensemble methods in order to preprocess the training set to identify and remove mislabeled instances in classification data sets. CVCF is mainly based on performing an Γ-FCV to split the full training data and on building classifiers using decision trees in each training subset. The authors of CVCF place special emphasis on using ensembles of decision trees such as C4.5 because they think that this kind of algorithm works well as a filter for noisy data.

The basic steps of CVCF are the following:

- Split the training data set D_T into Γ equal sized subsets.
- For each of these Γ parts, a base learning algorithm is trained on the other $\Gamma - 1$ parts. This results in Γ different classifiers. We use C4.5 as base learning algorithm in our experimentation as recommended by the authors.
- These Γ resulting classifiers are then used to tag each instance in the training set D_T as either correct or mislabeled, by comparing the training label with that assigned by the classifier.
- Add to D_N the noisy instances identified in D_T using a voting scheme (the majority scheme in our experimentation), taking into account the correctness of the labels obtained in the previous step by the Γ classifier built.
- Remove the noisy instances from the training set: $D_T \leftarrow D_T \backslash D_N$.

5.3.3 Iterative-Partitioning Filter

The *Iterative-Partitioning Filter* (IPF) [48] is a preprocessing technique based on the *Partitioning Filter* [102]. It is employed to identify and eliminate mislabeled instances in large data sets. Most noise filters assume that data sets are relatively small and capable of being learned after only one time, but this is not always true and partitioning procedures may be necessary.

IPF removes noisy instances in multiple iterations until a stopping criterion is reached. The iterative process stops if, for a number of consecutive iterations s, the number of identified noisy instances in each of these iterations is less than a percentage p of the size of the original training data set. Initially, we have a set of noisy instances $D_N = \emptyset$ and a set of good data $D_G = \emptyset$. The basic steps of each iteration are:

- Split the training data set D_T into Γ equal sized subsets. Each of these is small enough to be processed by an induction algorithm once.
- For each of these Γ parts, a base learning algorithm is trained on this part. This results in Γ different classifiers. We use C4.5 as the base learning algorithm in our experimentation as recommended by the authors.

- These Γ resulting classifiers, are then used to tag each instance in the training set D_T as either correct or mislabeled, by comparing the training label with that assigned by the classifier.
- Add to D_N the noisy instances identified in D_T using a voting scheme, taking into account the correctness of the labels obtained in the previous step by the Γ classifier built. For the IPF filter we use the majority vote scheme.
- Add to D_G a percentage y of the good data in D_T. This step is useful when we deal with large data sets because it helps to reduce them faster. We do not eliminate good data with the IPF method in our experimentation (we set $y = 0$, so D_G is always empty) and nor do we lose generality.
- Remove the noisy instances and the good data from the training set: $D_T \leftarrow D_T \backslash \{D_N \cup D_G\}$.

At the end of the iterative process, the filtered data is formed by the remaining instances of D_T and the good data of D_G; that is, $D_T \cup D_G$.

A particularity of the voting schemes in IPF is that a noisy instance should also be misclassified by the model which was induced in the subset containing that instance as an additional condition. Moreover, by varying the required number of filtering iterations, the level of conservativeness of the filter can also be varied in both schemes, consensus and majority.

5.3.4 More Filtering Methods

Apart from the three aforementioned filtering methods, we can find many more in the specialized literature. We try to provide a helpful summary of the most recent and well-known ones in the following Table 5.1. For the sake of brevity, we will not carry out a deep description of these methods as done in the previous sections. A recent categorization of the different filtering procedures made by Frenay and Verleysen [19] will be followed as it matches our descriptions well.

5.4 Robust Learners Against Noise

Filtering the data has also one major drawback: some instances will be dropped from the data sets, even if they are valuable. Instead of filtering the data set or modifying the learning algorithm, we can use other approaches to diminish the effect of noise in the learned model. In the case of labeled data, one powerful approach is to train not a single classifier but several ones, taking advantage of their particular strengths. In this section we provide a brief insight into classifiers that are known to be robust to noise to a certain degree, even when the noise is not treated or cleansed. As said in Sect. 5.1 C4.5 has been considered as a paradigmatic robust learner against noise. However, it is also true that classical decision trees have been considered

Table 5.1 Filtering approaches by category as of [19]

Detection based on thresholding of a measure		Partition filtering for large data sets	
Measure: classification confidence	[82]	For large and distributed data sets	[102, 103]
Least complex correct hypothesis	[24]	*Model influence*	
Classifier predictions based		LOOPC	[57]
Cost sensitive learning based	[101]	Single perceptron perturbation	[33]
SVM based	[86]	*Nearest neighbor based*	
ANN based	[42]	CNN	[30]
Multi classifier system	[65]	BBNR	[15]
C4.5	[44]	IB3	[3]
Nearest instances to a candidate	[78, 79]	Tomek links	[88]
Voting filtering		PRISM	[81]
Ensembles	[10, 11]	DROP	[93]
Bagging	[89]	*Graph connections based*	
ORBoost	[45]	Grabiel graphs	[18]
Edge analysis	[92]	Neighborhood graphs	[66]

sensitive to class noise as well [74]. This instability has make them very suitable for ensemble methods. As a countermeasure for this lack of stability some strategies can be used. The first one is to carefully select an appropriate splitting criteria measure. In [2] several measures are compared to minimize the impact of label noise in the constructed trees, empirically showing that the imprecise info-gain measure is able to improve the accuracy and reduce the tree growing size produced by the noise.

Another approach typically described as useful to deal with noise in decision trees is the use of pruning. Pruning tries to stop the overfitting caused by the overspecialization over the isolated (and usually noisy) examples. The work of [1] eventually shows that the usage of pruning helps to reduce the effect and impact of the noise in the modeled trees. C4.5 is the most famous decision tree and it includes this pruning strategy by default, and can be easily adapted to split under the desired criteria.

We have seen that the usage of ensembles is a good strategy to create accurate and robust filters. Whether an ensemble of classifiers is robust or not against noise can be also asked.

Many ensemble approaches exist and their noise robustness has been tested. An ensemble is a system where the base learners are all of the same type built to be as varied as possible. The two most classic approaches bagging and boosting were compared in [16] showing that bagging obtains better performance than boosting when label noise is present. The reason shown in [1] indicates that boosting (or the particular implementation made by AdaBoost) increase the weights of noisy instances too much, making the model construction inefficient and imprecise, whereas mislabeled instances favour the variability of the base classifiers in bagging [19]. As AdaBoost is not the only boosting algorithm, other implementations as LogitBoost and BrownBoost have been checked as more robust to class noise [64]. When the base classifiers are different we talk of Multiple Classifier Systems (MCSs). They are thus a generalization of the classic ensembles and they should offer better improvements in noisy environments. They are tackled in Sect. 5.4.1.

We can separate the labeled instances in several "bags" or groups, each one containing only those instances belonging to the same class. This type of decomposition is well suited for those classifiers that can only work with binary classification problems, but has also been suggested that this decomposition can help to diminish the effects of noise. This decomposition is expected to decrease the overlapping between the classes and to limit the effect of noisy instances to their respective bags by simplifying the problem and thus alleviating the effect of the noise if the whole data set were considered.

5.4.1 Multiple Classifier Systems for Classification Tasks

Given a set of problems, finding the best overall classification algorithm is sometimes difficult because some classifiers may excel in some cases and perform poorly in others. Moreover, even though the optimal match between a learning method and a problem is usually sought, this match is generally difficult to achieve and perfect solutions are rarely found for complex problems [34, 36]. This is one reason for using Multi-Classifier Systems [34, 36, 72], since it is not necessary to choose a specific learning method. All of them might be used, taking advantage of the strengths of each method, while avoiding its weaknesses. Furthermore, there are other motivations to combine several classifiers [34]:

- To avoid the choice of some arbitrary but important initial conditions, e.g. those involving the parameters of the learning method.
- To introduce some randomness to the training process in order to obtain different alternatives that can be combined to improve the results obtained by the individual classifiers.
- To use complementary classification methods to improve dynamic adaptation and flexibility.

Several works have claimed that simultaneously using classifiers of different types, complementing each other, improves classification performance on difficult problems, such as satellite image classification [60], fingerprint recognition [68] and foreign exchange market prediction [73]. Multiple Classifier Systems [34, 36, 72, 94] are presented as a powerful solution to these difficult classification problems, because they build several classifiers from the same training data and therefore allow the simultaneous usage of several feature descriptors and inference procedures. An important issue when using MCSs is the way of creating *diversity* among the classifiers [54], which is necessary to create discrepancies among their decisions and hence, to take advantage of their combination.

MCSs have been traditionally associated with the capability of working accurately with problems involving noisy data [36]. The main reason supporting this hypothesis could be the same as one of the main motivations for combining classifiers: the improvement of the generalization capability (due to the complementarity of each classifier), which is a key question in noisy environments, since it might allow one to avoid the overfitting of the new characteristics introduced by the noisy examples [84]. Most of the works studying MCSs and noisy data are focused on techniques like bagging and boosting [16, 47, 56], which introduce diversity considering different samples of the set of training examples and use only one baseline classifier. For example, in [16] the suitability of randomization, bagging and boosting to improve the performance of C4.5 was studied. The authors reached the conclusion that with a low noise level, boosting is usually more accurate than bagging and randomization. However, bagging outperforms the other methods when the noise level increases. Similar conclusions were obtained in the paper of Maclin and Opitz [56]. Other works [47] compare the performance of boosting and bagging techniques dealing with imbalanced and noisy data, reaching also the conclusion that bagging methods generally outperforms boosting ones. Nevertheless, explicit studies about the adequacy of MCSs (different from bagging and boosting, that is, those introducing diversity using different base classifiers) to deal with noisy data have not been carried out yet. Furthermore, most of the existing works are focused on a concrete type of noise and on a concrete combination rule. On the other hand, when data is suffering from noise, a proper study on how the robustness of each single method influences the robustness of the MCS is necessary, but this fact is usually overlooked in the literature.

There are several strategies to use more than one classifier for a single classification task [36]:

- *Dynamic classifier selection* This is based on the fact that one classifier may outperform all others using a global performance measure but it may not be the best in all parts of the domain. Therefore, these types of methods divide the input domain into several parts and aim to select the classifier with the best performance in that part.
- *Multi-stage organization* This builds the classifiers iteratively. At each iteration, a group of classifiers operates in parallel and their decisions are then combined. A dynamic selector decides which classifiers are to be activated at each stage based on the classification performances of each classifier in previous stages.
- *Sequential approach* A classifier is used first and the other ones are used only if the first does not yield a decision with sufficient confidence.
- *Parallel approach* All available classifiers are used for the same input example in parallel. The outputs from each classifier are then combined to obtain the final prediction.

Although the first three approaches have been explored to a certain extent, the majority of classifier combination research focuses on the fourth approach, due to its simplicity and the fact that it enables one to take advantage of the factors presented in the previous section. For these reasons, this book focus on the fourth approach.

5.4.1.1 Decisions Combination in Multiple Classifiers Systems

As has been previously mentioned, parallel approaches need a posterior phase of combination after the evaluation of a given example by all the classifiers. Many decisions combination proposals can be found in the literature, such as the intersection of decision regions [29], voting methods [62], prediction by top choice combinations [91], use of the Dempster–Shafer theory [58, 97] or ranking methods [36]. In concrete, we will study the following four combination methods for the MCSs built with heterogeneous classifiers:

1. **Majority vote (MAJ)** [62] This is a simple but powerful approach, where each classifier gives a vote to the predicted class and the one with the most votes is chosen as the output.
2. **Weighted majority vote (W-MAJ)** [80] Similarly to MAJ, each classifier gives a vote for the predicted class, but in this case, the vote is weighted depending on the competence (accuracy) of the classifier in the training phase.
3. **Naïve Bayes** [87] This method assumes that the base classifiers are mutually independent. Hence, the predicted class is the one that obtains the highest posterior probability. In order to compute these probabilities, the confusion matrix of each classifier is considered.
4. **Behavior-Knowledge Space (BKS)** [38] This is a multinomial method that indexes a cell in a look-up table for each possible combination of classifiers outputs. A cell is labeled with the class to which the majority of the instances in that cell belong to. A new instance is classified by the corresponding cell label; in case the cell is not labeled or there is a tie, the output is given by MAJ.

We always use the same training data set to train all the base classifiers and to compute the parameters of the aggregation methods, as is recommended in [53]. Using a separate set of examples to obtain such parameters can imply some important training data to be ignored and this fact is generally translated into a loss of accuracy of the final MCS built.

In MCSs built with heterogeneous classifiers, all of them may not return a confidence value. Even though each classifier can be individually modified to return a confidence value for its predictions, such confidences will come from different computations depending on the classifier adapted and their combination could become meaningless. Nevertheless, in MCSs built with the same type of classifier, this fact does not occur and it is possible to combine their confidences since these are homogeneous among all the base classifiers [53]. Therefore, in the case of bagging, given that the same classifier is used to train all the base classifiers, the confidence of the prediction can be used to compute a weight and, in turn, these weights can be used in a weighted voting combination scheme.

5.4.2 *Addressing Multi-class Classification Problems by Decomposition*

Usually, the more classes in a problem, the more complex it is. In multi-class learning, the generated classifier must be able to separate the data into more than a pair of classes, which increases the chances of incorrect classifications (in a two-class balanced problem, the probability of a correct random classification is 1/2, whereas in a multi-class problem it is 1/M). Furthermore, in problems affected by noise, the boundaries, the separability of the classes and therefore, the prediction capabilities of the classifiers may be severely hindered.

When dealing with multi-class problems, several works [6, 50] have demonstrated that decomposing the original problem into several binary subproblems is an easy, yet accurate way to reduce their complexity. These techniques are referred to as binary decomposition strategies [55]. The most studied schemes in the literature are: *One-vs-One* (OVO) [50], which trains a classifier to distinguish between each pair of classes, and *One-vs-All* (OVA) [6], which trains a classifier to distinguish each class from all other classes. Both strategies can be encoded within the Error Correcting Output Codes framework [5, 17]. However, none of these works provide any theoretical nor empirical results supporting the common assumption that assumes a better behavior against noise of decomposition techniques compared to not using decomposition. Neither do they show what type of noise is better handled by decomposition techniques.

Consequently, we can consider the usage of the OVO strategy, which generally out-stands over OVA [21, 37, 76, 83], and check its suitability with noisy training data. It should be mentioned that, in real situations, the existence of noise in the data sets is usually unknown-therefore, neither the type nor the quantity of noise in the data set can be known or supposed *a priori*. Hence, tools which are able to manage the presence of noise in the data sets, despite its type or quantity (or unexistence), are of great interest. If the OVO strategy (which is a simple yet effective methodology when clean data sets are considered) is also able to properly (better than the baseline non-OVO version) handle the noise, its usage could be recommended in spite of the presence of noise and without taking into account its type. Furthermore, this strategy can be used with any of the existing classifiers which are able to deal with two-class problems. Therefore, the problems of algorithm level modifications and preprocessing techniques could be avoided; and if desired, they could also be combined.

5.4.2.1 Decomposition Strategies for Multi-class Problems

Several motivations for the usage of binary decomposition strategies in multi-class classification problems can be found in the literature [20, 21, 37, 76]:

- The separation of the classes becomes easier (less complex), since less classes are considered in each subproblem [20, 61]. For example, in [51], the classes in a

digit recognition problem were shown to be linearly separable when considered in pairs, becoming a simpler alternative than learning a unique non-linear classifier over all classes simultaneously.
- Classification algorithms, whose extension to multi-class problems is not easy, can address multi-class problems using decomposition techniques [20].
- In [71], the advantages of using decomposition were pointed out when the classification errors for different classes have distinct costs. The binarization allows the binary classifiers generated to impose preferences for some of the classes.
- Decomposition allows one to easily parallelize the classifier learning, since the binary subproblems are independent and can be solved with different processors.

Dividing a problem into several new subproblems, which are then independently solved, implies the need of a second phase where the outputs of each problem need to be aggregated. Therefore, decomposition includes two steps:

1. *Problem division*. The problem is decomposed into several binary subproblems which are solved by independent binary classifiers, called *base classifiers* [20]. Different decomposition strategies can be found in the literature [55]. The most common one is OVO [50].
2. *Combination of the outputs*. [21] The different outputs of the binary classifiers must be aggregated in order to output the final class prediction. In [21], an exhaustive study comparing different methods to combine the outputs of the base classifiers in the OVO and OVA strategies is developed. Among these combination methods, the Weighted Voting [40] and the approaches in the framework of probability estimates [95] are highlighted.

This book focuses the OVO decomposition strategy due to the several advantages shown in the literature with respect to OVA [20, 21, 37, 76]:

- OVO creates simpler borders between classes than OVA.
- OVO generally obtains a higher classification accuracy and a shorter training time than OVA because the new subproblems are easier and smaller.
- OVA has more of a tendency to create imbalanced data sets which can be counterproductive [22, 83].
- The application of the OVO strategy is widely extended and most of the software tools considering binarization techniques use it as default [4, 13, 28].

5.4.2.2 One-vs-One Decomposition Scheme

The OVO decomposition strategy consists of dividing a classification problem with M classes into $M(M-1)/2$ binary subproblems. A classifier is trained for each new subproblem only considering the examples from the training data corresponding to the pair of classes (λ_i, λ_j) with $i < j$ considered.

When a new instance is going to be classified, it is presented to all the the binary classifiers. This way, each classifier discriminating between classes λ_i and λ_j provides a confidence degree $r_{ij} \in [0, 1]$ in favor of the former class (and hence, r_{ji} is computed by $1 - r_{ij}$). These outputs are represented by a score matrix R:

$$R = \begin{pmatrix} - & r_{12} & \cdots & r_{1M} \\ r_{21} & - & \cdots & r_{2M} \\ \vdots & & & \vdots \\ r_{M1} & r_{M2} & \cdots & - \end{pmatrix} \tag{5.3}$$

The final output is derived from the score matrix by different aggregation models. The most commonly used and simplest combination, also considered in the experiments of this book, is the application of a voting strategy:

$$Class = \arg\max_{i=1,\ldots,M} \sum_{1 \le j \ne i \le M} s_{ij} \tag{5.4}$$

where s_{ij} is 1 if $r_{ij} > r_{ji}$ and 0 otherwise. Therefore, the class with the largest number of votes will be predicted. This strategy has proved to be competitive with different classifiers obtaining similar results in comparison with more complex strategies [21].

5.5 Empirical Analysis of Noise Filters and Robust Strategies

In this section we want to illustrate the advantages of the noise approaches described above.

5.5.1 Noise Introduction

In the data sets we are going to use (taken from Chap. 2), as in most of the real-world data sets, the initial amount and type of noise present is unknown. Therefore, no assumptions about the base noise type and level can be made. For this reason, these data sets are considered to be noise free, in the sense that no recognizable noise has been introduced. In order to control the amount of noise in each data set and check how it affects the classifiers, noise is introduced into each data set in a supervised manner. Four different noise schemes proposed in the literature, as explained in Sect. 5.2, are used in order to introduce a noise level x% into each data set:

1. **Introduction of class noise.**
    - **Uniform class noise** [84] x% of the examples are corrupted. The class labels of these examples are randomly replaced by another one from the M classes.
    - **Pairwise class noise** [100, 102] Let X be the majority class and Y the second majority class, an example with the label X has a probability of $x/100$ of being incorrectly labeled as Y.

2. **Introduction of attribute noise**
   - **Uniform attribute noise** [100, 104] x% of the values of each attribute in the data set are corrupted. To corrupt each attribute A_i, x% of the examples in the data set are chosen, and their A_i value is assigned a random value from the domain $\mathbb{D}_i$ of the attribute A_i. An uniform distribution is used either for numerical or nominal attributes.
   - **Gaussian attribute noise** This scheme is similar to the uniform attribute noise, but in this case, the A_i values are corrupted, adding a random value to them following Gaussian distribution of $mean = 0$ and *standard deviation* = (*max-min*)/5, being *max* and *min* the limits of the attribute domain ($\mathbb{D}_i$). Nominal attributes are treated as in the case of the uniform attribute noise.

In order to create a noisy data set from the original, the noise is introduced into the training partitions as follows:

1. A level of noise $x\%$, of either class noise (uniform or pairwise) or attribute noise (uniform or Gaussian), is introduced into a copy of the full original data set.
2. Both data sets, the original and the noisy copy, are partitioned into 5 equal folds, that is, with the same examples in each one.
3. The training partitions are built from the noisy copy, whereas the test partitions are formed from examples from the base data set, that is, the noise free data set.

We introduce noise, either class or attribute noise, only into the training sets since we want to focus on the effects of noise on the training process. This will be carried out observing how the classifiers built from different noisy training data for a particular data set behave, considering the accuracy of those classifiers, with the same clean test data. Thus, the accuracy of the classifier built over the original training set without additional noise acts as a reference value that can be directly compared with the accuracy of each classifier obtained with the different noisy training data. Corrupting the test sets also affects the accuracy obtained by the classifiers and therefore, our conclusions will not only be limited to the effects of noise on the training process.

The accuracy estimation of the classifiers in a data set is obtained by means of 5 runs of a stratified 5-FCV. Hence, a total of 25 runs per data set, noise type and level are averaged. 5 partitions are used because, if each partition has a large number of examples, the noise effects will be more notable, facilitating their analysis.

The robustness of each method is estimated with the *relative loss of accuracy* (RLA) (Eq. 5.5), which is used to measure the percentage of variation of the accuracy of the classifiers at a concrete noise level with respect to the original case with no additional noise:

$$RLA_{x\%} = \frac{Acc_{0\%} - Acc_{x\%}}{Acc_{0\%}}, \tag{5.5}$$

where $RLA_{x\%}$ is the relative loss of accuracy at a noise level $x\%$, $Acc_{0\%}$ is the test accuracy in the original case, that is, with 0% of induced noise, and $Acc_{x\%}$ is the test accuracy with a noise level $x\%$.

5.5.2 Noise Filters for Class Noise

The usage of filtering is claimed to be useful in the presence of noise. This section tries to show whether this claim is true or not and to what extent. As a simple but representative case of study, we show the results of applying noise filters based on detecting and eliminating mislabeled training instances. We want to illustrate how applying filters is a good strategy to obtain better results in the presence of even low amounts of noise. As filters are mainly designed for class noise, we will focus on the two types of class noise described in this chapter: the uniform class noise and the pairwise class noise.

Three popular classifiers will be used to obtain the accuracy values that are C4.5, Ripper and a SVM. Their selection is not made at random: SVMs are known to be very accurate but also sensitive to noise. Ripper is a rule learning algorithm able to perform averagely well, but as we saw in Sect. 5.4 rule learners are also sensitive to noise when they are not designed to cope with it. The third classifier is C4.5 using the pruning strategy, that it is known for diminishing the effects of noise in the final tree. Table 5.2 shows the average results for the three noise filters for each kind of class noise studied. The amount of noise ranges from 5 to 20 %, enough to show the differences between no filtering (labeled as "None") and the noise filters. The results shown are the average over all the data sets considered in order to ease the reading.

The evolution of the results and their tendencies can be better depicted by using a graphical representation. Figure 5.4a shows the performance of SVM from an amount of 0 % of controlled pairwise noise to the final 20 % introduced. The accuracy can be seen to drop from an initial amount of 90–85 % by only corrupting 20 % of the class labels. The degradation is even worse in the case of uniform class noise depicted in Fig. 5.4b, as all the class labels can be affected. The evolution of not using any

Table 5.2 Filtering of class noise over three classic classifiers

		Pairwise class noise					Uniform random class noise				
		0%	5%	10%	15%	20%	0%	5%	10%	15%	20%
	None	90.02	88.51	86.97	86.14	84.86	90.02	87.82	86.43	85.18	83.20
SVM	EF	90.49	89.96	89.07	88.33	87.40	90.49	89.66	88.78	87.78	86.77
	CVCF	90.56	89.86	88.94	88.28	87.76	90.48	89.56	88.72	87.92	86.54
	IPF	90.70	90.13	89.37	88.85	88.27	90.58	89.79	88.97	88.48	87.37
	None	82.46	81.15	80.35	79.39	78.49	82.46	79.81	78.55	76.98	75.68
Ripper	EF	83.36	82.87	82.72	82.43	81.53	83.46	83.03	82.87	82.30	81.66
	CVCF	83.17	82.93	82.64	82.03	81.68	83.17	82.59	82.19	81.69	80.45
	IPF	83.74	83.59	83.33	82.72	82.44	83.74	83.61	82.94	82.94	82.48
	None	83.93	83.66	82.81	82.25	81.41	83.93	82.97	82.38	81.69	80.28
C4.5	EF	84.18	84.07	83.70	83.20	82.36	84.16	83.96	83.53	83.38	82.66
	CVCF	84.15	83.92	83.24	82.54	82.13	84.15	83.61	83.00	82.84	81.61
	IPF	84.44	84.33	83.92	83.38	82.53	84.44	83.89	83.84	83.50	82.72

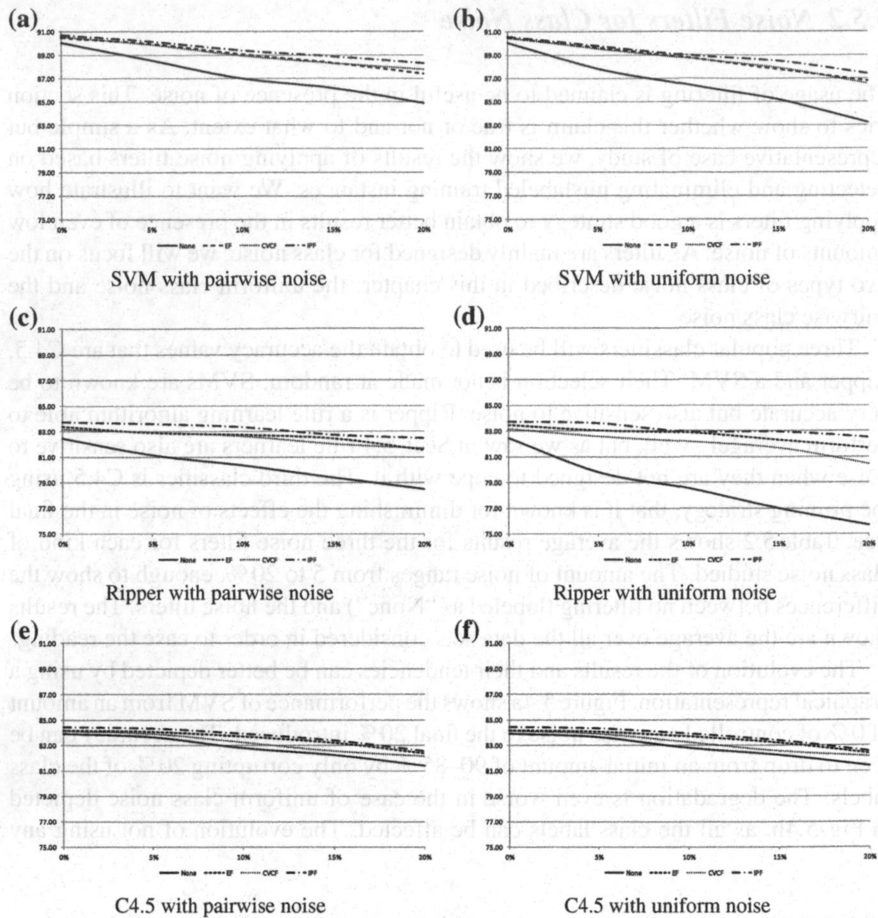

Fig. 5.4 Accuracy over different amounts and types of noise. The different filters used are named by their acronyms. "None" denotes the absence of any filtering. **a** SVM with pairwise noise **b** SVM with uniform noise **c** Ripper with pairwise noise **d** Ripper with uniform noise **e** C4.5 with pairwise noise **f** C4.5 with uniform noise

noise filter denoted by "None" is remarkably different from the lines that illustrate the usage of any noise filter. The IPF filter is slightly better than the other due to its greater sophistication, but in overall the use of filters is highly recommended. Even in the case of 0 % of controlled noise, the noise already present is also cleansed, allowing the filtering to improve even in this base case. Please note that the divergence appears even in the 5 % case, showing that noise filtering is worth trying in low noise frameworks.

Ripper obtains a lower overall accuracy than SVM, but the conclusions are akin: the usage of noise filters is highly recommended as can be seen in Fig. 5.4c, d. It is remarkable that not applying filtering for Ripper causes a fast drop in performance, indicating that the rule base modeled is being largely affected by the noise. Thanks to the use of the noise filter the inclusion of misleading rules is controlled, resulting in a smoother drop in performance, even slower than that for SVM.

The last case is also very interesting. Being that C4.5 is more robust against noise than SVM and Ripper, the accuracy drop over the increment of noise is lower. However the use of noise filters is still recommended as they improve both the initial case 0 % and the rest of levels. The greater differences between not filtering and the use of any filter are found in uniform class noise (Fig. 5.4f). As we indicated when describing the SVM case, uniform class noise is more disruptive but the use of filtering for C4.5 make its performance comparable to the case of pairwise noise (Fig. 5.4e).

Although not depicted here, the size of C4.5 trees, Ripper rule base size and the number of support vectors of SVM is lower with the usage of noise filters when the noise amount increases, resulting in a shorter time when evaluating examples for classification. This is specially critical for SVM, whose evaluation times dramatically increase with the increment of selected support vectors.

5.5.3 *Noise Filtering Efficacy Prediction by Data Complexity Measures*

In the previous Sect. 5.5.2 we have seen that the application of noise filters are beneficial in most cases, especially when higher amounts of noise are present in the data. However, applying a filter is not "free" in terms of computing time and information loss. Indiscriminate application of noise filtering may be interpreted as the outcome of the aforementioned example study, but it would be interesting to study the noise filters' behavior further and to obtain hints about whether filtering is useful or not depending on the data case.

In an ideal case, only the examples that are completely wrong would be erased from the data set. The truth is both correct examples and examples containing valuable information may be removed, as the filters are ML techniques with their inherent limitations. This fact implies that these techniques do not always provide an improvement in performance. The success of these methods depends on several circumstances, such as the kind and nature of the data errors, the quantity of noise removed or the capabilities of the classifier to deal with the loss of useful information related to the filtering. Therefore, the efficacy of noise filters, i.e., whether their use causes an improvement in classifier performance, depends on the noise-robustness and the generalization capabilities of the classifier used, but it also strongly depends on the characteristics of the data.

Describing the characteristics of the data is not an easy task, as specifying what "difficult" means is usually not straightforward or it simply does not depend on a single factor. Data complexity measures are a recent proposal to represent characteristics of the data that are considered difficult in classification tasks, e.g. the overlapping

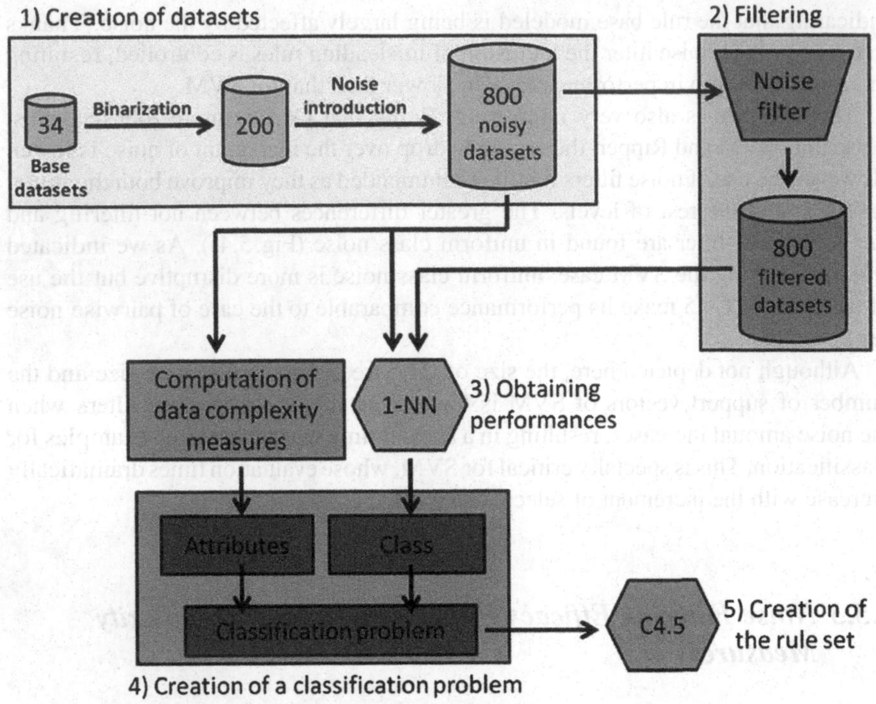

Fig. 5.5 Using C4.5 to build a rule set to predict noise filtering efficacy

among classes, their separability or the linearity of the decision boundaries. The most commonly used data complexity set of measures are those gathered together by Ho and Basu [35]. They consist of 12 metrics designed for binary classification problems that numerically estimate the difficulty of 12 different aspects of the data. For some measures lower/higher values mean a more difficult problem regarding to such a characteristic. Having a numeric description of the difficult aspects of the data opens a new question: can we predict which characteristics are related with noise and will they be successfully corrected by noise filters?

This prediction can help, for example, to determine an appropriate noise filter for a concrete noisy data set such a filter providing a signicant advantage in terms of the results or to design new noise filters which select more or less aggressive filtering strategies considering the characteristics of the data. Choosing a noise-sensitive learner facilitates the checking of when a filter removes the appropriate noisy examples in contrast to a robust learner-the performance of classiers built by the former is more sensitive to noisy examples retained in the data set after the ltering process.

A way to formulate rules that describe when it is appropriate to filter the data follows the scheme depicted in Fig. 5.5. From an initial set of 34 data sets from those described in Chap. 2, a large amount of two-class data sets are obtained by binarization along their data complexity measures. Thus the filtering efficacy is compared by using 1-NN as a classifier to obtain the accuracy of filtering versus not filtering. This comparison is achieved by using a Wilcoxon Signed Rank test. If the statistical test yields differences favouring the filtering, the two-class data set is labeled as appropriate for filtering, and not favorable in other case. As a result for each binary data set we will have 12 data complexity measures and a label describing whether the data set is eligible for filtering or not. A simple way to summarize this information into a rule set is to use a decision tree (C4.5) using the 12 data complexity values as the input features, and the appropriateness label as the class.

An important appreciation about the scheme presented in Fig. 5.5 is that for every label noise filter we want to consider, we will obtain a different set of rules. For the sake of simplicity we will limit this illustrative study to our selected filters—EF, CVCF and IPF—in Sect. 5.3.

How accurate is the set of rules when predicting the suitability of label noise filters? Using a 10-FCV over the data set obtained in the fourth step in Fig. 5.5, the training and test accuracy of C4.5 for each filter is summarized in Table 5.3.

The test accuracy above 80 % in all cases indicates that the description obtained by C4.5 is precise enough.

Using a decision tree is also interesting not only due to the generated rule set, but also because we can check which data complexity measures (that is, the input attributes) are selected first, and thus are considered as more important and discriminant by C4.5. Averaging the rank of selection of each data complexity measure over the 10 folds, Table 5.4 shows which complexity measures are the most dis-

Table 5.3 C4.5 accuracy in training and test for the ruleset describing the adequacy of label noise filters

Noise filter	% Acc. training	% Acc. Test
EF	0.9948	0.8176
CVCF	0.9966	0.8353
IPF	0.9973	0.8670

Table 5.4 Average rank of each data complexity measure selected by C4.5 (the lower the better)

Metric	EF	CVCF	IPF	Mean
F1	5.90	4.80	4.50	**5.07**
F2	1.00	1.00	1.00	**1.00**
F3	10.10	3.40	3.30	**5.60**
N1	9.10	9.90	7.10	8.70
N2	3.30	2.00	3.00	**2.77**
N3	7.80	8.50	9.50	8.60
N4	9.90	9.70	10.50	10.03
L1	7.90	10.00	6.00	7.97
L2	9.30	6.80	10.00	8.70
L3	4.60	8.70	5.90	6.40
T1	5.20	6.80	11.00	7.67

criminating, and thus more interesting, for C4.5 to discern when a noise filter will behave well or badly. Based on these rankings it is easy to observe that F2, N2, F1 and F3 are the predominant measures in the order of choice. Please remember that behind these acronyms, the data complexity measures aim to describe one particular source of difficulty for any classification problem. Following the order from the most important of these four outstanding measures to the least, the volume of overlap region (F2) is key to describe the effectiveness of a class noise filter. The less any attribute is overlapped, the better the filter is able to decide if the instance is noisy. It is complemented with the ratio of average intra/inter class distance as defined by the nearest neighbor rule. When the examples sharing the same class are closer than the examples of other classes the filtering is effective for 1-NN. This measure is expected to change if another classifier is chosen to build the classification problem. F1 and F3 are also measures of individual attribute overlapping as F2, but they are less important in general.

If the discriminant abilities of these complexity measures are as good as their ranks indicate, using only these few measures we can expect to obtain a better and more concise description of what a easy-to-filter problem is. In order to avoid the study of all the existing combinations of the five metrics, the following experimentation is mainly focused on the measures F2, N2 and F3, the most discriminative ones since the order results can be considered more important than the percentage results. The incorporation of F1 into this set is also studied. The prediction capability of the measure F2 alone, since is the most discriminative one, is also shown. All these results are presented in Table 5.5.

The use of the measure F2 alone to predict the noise filtering efficacy with good performance can be discarded, since its results are not good enough compared with the cases where more than one measure is considered. This fact reflects that the use of single measures does not provide enough information to achieve a good filtering efficacy prediction result. Therefore, it is necessary to combine several measures which examine different aspects of the data. Adding the rest of selected measures provides comparable results to those shown in Table 5.3 yet limits the complexity of the rule set obtained.

The work carried out in this section is studied further in [77], showing how a rule set obtained for one filter can be applied to other filters, how these rule sets are validated with unseen data sets and even increasing the number of filters involved.

Table 5.5 Performance results of C4.5 predicting the noise filtering efficacy (measures used: F2, N2, F3, and F1)

	F2		F2-N2-F3-F1		F2-N2-F3	
Noise Filter	Training	Test	Training	Test	Training	Test
CVCF	1.0000	0.5198	0.9983	0.7943	0.9977	0.8152
EF	1.0000	0.7579	0.9991	0.8101	0.9997	0.8421
IPF	1.0000	0.7393	0.9989	0.8119	0.9985	0.7725
Mean	1.0000	0.6723	0.9988	0.8054	0.9986	0.8099

5.5.4 Multiple Classifier Systems with Noise

We will dispose of three well-known classifiers to build the MCS used in this illustrative section. SVM [14], C4.5 [75] and KNN [63] are chosen based on their good performance in a large number of real-world problems. Moreover, they were selected because these methods have a highly differentiated and well known noise-robustness, which is important in order to properly evaluate the performance of MCSs in the presence of noise. Considering thec previous classifiers (SVM, C4.5 and KNN), a MCS composed by 3 individual classifiers (SVM, C4.5 and 1-NN) is built. Therefore, the MCSs built with heterogeneous classifiers (MCS3-1) will contain a noise-robust algorithm (C4.5), a noise-sensitive method (SVM) and a local distance dependent method with a low tolerance to noise (1-NN).

5.5.4.1 First Scenario: Data Sets with Class Noise

Table 5.6 shows the performance (top part of the table) and robustness (bottom part of table) results of each classification algorithm at each noise level on data sets with class noise. Each one of these parts in the table (performance and robustness parts) is divided into another two parts: one with the results of the uniform class noise

Table 5.6 Performance and robustness results on data sets with class noise

			Results				p-values MCS3-1 vs.		
		x%	SVM	C4.5	1-NN	MCS3-1	SVM	C4.5	1-NN
Performance	Uniform	0%	83.25	82.96	81.42	85.42	5.20E-03	1.80E-03	7.10E-04
		10%	79.58	82.08	76.28	83	1.10E-04	3.90E-01	1.30E-07
		20%	76.55	79.97	71.22	80.09	5.80E-05	9.5E-01*	1.00E-07
		30%	73.82	77.9	65.88	77.1	2.80E-04	3.5E-01*	6.10E-08
		40%	70.69	74.51	61	73.2	7.20E-03	2.0E-01*	7.00E-08
		50%	67.07	69.22	55.55	67.64	5.40E-01	1.4E-01*	1.10E-07
	Pairwise	0%	83.25	82.96	81.42	85.42	5.20E-03	1.80E-03	7.10E-04
		10%	80.74	82.17	77.73	83.95	1.20E-04	5.80E-02	1.20E-07
		20%	79.11	80.87	74.25	82.21	2.00E-04	3.50E-01	8.20E-08
		30%	76.64	78.81	70.46	79.52	2.10E-03	8.20E-01	5.60E-08
		40%	73.13	74.83	66.58	75.25	3.80E-02	8.0E-01*	4.50E-08
		50%	65.92	60.29	63.06	64.46	2.6E-02*	2.60E-05	1.10E-01
RLA	Uniform	10%	4.44	1.1	6.16	2.91	7.20E-03	1.5E-06*	1.00E-07
		20%	8.16	3.78	12.1	6.4	1.50E-02	3.1E-05*	1.20E-06
		30%	11.38	6.36	18.71	9.95	8.80E-02	4.6E-05*	1.80E-07
		40%	15.08	10.54	24.15	14.54	6.60E-01	8.2E-05*	1.50E-06
		50%	19.47	17	30.58	21.08	1.9E-01*	1.3E-04*	1.30E-05
	Pairwise	10%	2.97	1	4.2	1.73	6.60E-03	2.0E-04*	6.70E-06
		20%	4.86	2.66	8.21	3.86	7.20E-02	1.7E-03*	1.00E-05
		30%	7.81	5.33	12.75	7.11	3.80E-01	2.7E-03*	6.30E-06
		40%	12.01	10.19	17.2	12.08	5.50E-01	7.8E-03*	4.40E-05
		50%	20.3	26.7	21.18	24.13	1.4E-04*	2.60E-03	1.1E-01*

and another with the results of the pairwise class noise. A star '∗' next to a p-value indicates that the corresponding single algorithm obtains more ranks than the MCS in Wilcoxon's test comparing the individual classifier and the MCS. Note that the robustness can only be measured if the noise level is higher than 0 %, so the robustness results are presented from a noise level of 5 % and higher.

From the raw results we can extract some interesting conclusions. If we consider the performance results with uniform class noise we can observe that MCS3-k is statistically better than SVM, but in the case of C4.5 statistical differences are only found at the lowest noise level. For the rest of the noise levels, MCS3-1 is statistically equivalent to C4.5. Statistical differences are found between MCS3-1 and 1-NN for all the noise levels, indicating that MCS are specially suitable when taking noise sensitive classifiers into account.

In the case of pairwise class noise the conclusions are very similar. MCS3-1 statistically outperforms its individual components when the noise level is below 45 %, whereas it only performs statistically worse than SVM when the noise level reaches 50 % (regardless of the value of 1). MCS3-1 obtains more ranks than C4.5 in most of the cases; moreover, it is statistically better than C4.5 when the noise level is below 15 %. Again MCS3-1 statistically outperforms 1-NN regardless of the level of noise.

In uniform class noise MCS3-1 is significantly more robust than SVM up to a noise level of 30 %. Both are equivalent from 35 % onwards—even though MCS3-1 obtains more ranks at 35–40 % and SVM at 45–50 %. The robustness of C4.5 excels with respect to MCS3-1, observing the differences found. MCS3-1 is statistically better than 1-NN. The Robustness results with pairwise class noise present some differences with respect to uniform class noise. MCS3-1 statistically overcomes SVM up to a 20 % noise level, they are equivalent up to 45 % and MCS3-1 is outperformed by SVM at 50 %. C4.5 is statistically more robust than MCS3-1 (except in highly affected data sets, 45–50 %) and The superiority of MCS3-1 against 1-NN is notable, as it is statistically better at all noise levels.

It is remarkable that the uniform scheme is the most disruptive class noise for the majority of the classifiers. The higher disruptiveness of the uniform class noise in MCSs built with heterogeneous classifiers can be attributed to two main reasons: (i) this type of noise affects all the output domain, that is, all the classes, to the same extent, whereas the pairwise scheme only affects the two majority classes; (ii) a noise level x% with the uniform scheme implies that exactly x% of the examples in the data sets contain noise, whereas with the pairwise scheme, the number of noisy examples for the same noise level x% depends on the number of examples of the majority class N_{maj}; as a consequence, the global noise level in the whole data set is usually lower—more specifically, the number of noisy examples can be computed as $(x \cdot N_{maj})/100$.

With the performance results in uniform class noise MCS3-1 generally outperforms its single classifier components. MCS3-1 is better than SVM and 1-NN, whereas it only performs statistically better than C4.5 at the lowest noise levels. In pairwise class noise MCS3-1 improves SVM up to a 40 % noise level, it is better than C4.5 at the lowest noise levels—these noise levels are lower than those of the uniform class noise—and also outperforms 1-NN. Therefore, the behavior of MCS3-1 with respect to their individual components is better in the uniform scheme than in the pairwise one.

5.5.4.2 Second Scenario: Data Sets with Attribute Noise

Table 5.7 shows the performance and robustness results of each classification algorithm at each noise level on data sets with attribute noise.

At first glance we can appreciate that the results on data sets with uniform attribute noise are much worse than those on data sets with Gaussian noise for all the classifiers, including MCSs. Hence, the most disruptive attribute noise is the uniform scheme. As the uniform attribute noise is the most disruptive noise scheme, MCS3-1 outperforms SVM and 1-NN. However, with respect to C4.5, MCS3-1 is significantly better only at the lowest noise levels (up to 10–15 %), and is equivalent at the rest of the noise levels. With gaussian attribute noise MCS3-1 is only better than 1-NN and SVM, and better than C4.5 at the lowest noise levels (up to 25 %).

The robustness of the MCS3-1 with uniform attribute noise does not outperform that of its individual classifiers, as it is statistically equivalent to SVM and sometimes worse than C4.5. Regarding 1-NN, MCS3-1 performs better than 1-NN. When

Table 5.7 Performance and robustness results on data sets with attribute noise

			Results				p-values MCS3-1 vs.		
		x%	SVM	C4.5	1-NN	MCS3-1	SVM	C4.5	1-NN
Performance	Uniform	0%	83.25	82.96	81.42	85.42	5.20E-03	1.80E-03	7.10E-04
		10%	81.78	81.58	78.52	83.33	3.90E-03	8.30E-02	8.10E-06
		20%	78.75	79.98	75.73	80.64	4.40E-03	0.62	5.20E-06
		30%	76.09	77.64	72.58	77.97	2.10E-03	9.4E-01*	1.00E-05
		40%	72.75	75.19	69.58	74.84	9.90E-03	7.6E-01*	1.30E-06
		50%	69.46	72.12	66.59	71.36	1.20E-02	3.1E-01*	2.70E-05
	Gaussian	0%	83.25	82.96	81.42	85.42	5.20E-03	1.80E-03	7.10E-04
		10%	82.83	82.15	80.08	84.49	3.40E-03	4.80E-03	4.00E-05
		20%	81.62	81.16	78.52	83.33	4.40E-03	2.10E-02	1.00E-05
		30%	80.48	80.25	76.74	81.85	1.10E-02	2.00E-01	1.00E-05
		40%	78.88	78.84	74.82	80.34	1.60E-02	0.4	1.00E-05
		50%	77.26	77.01	73.31	78.49	0.022	3.40E-01	3.00E-05
RLA	Uniform	10%	1.38	1.68	3.63	2.4	6.7E-01*	6.6E-02*	1.30E-02
		20%	5.06	3.6	6.78	5.54	6.10E-01	5.2E-03*	1.30E-02
		30%	8.35	6.62	10.73	8.85	7.80E-01	2.6E-03*	1.20E-02
		40%	12.13	9.6	14.29	12.44	7.8E-01*	1.1E-02*	2.00E-03
		50%	16.28	13.46	17.7	16.68	8.5E-01*	3.4E-03*	2.00E-02
	Gaussian	10%	0.25	0.94	1.6	1.03	1.70E-01	2.4E-01*	2.60E-01
		20%	1.74	2.1	3.36	2.36	3.30E-01	1.2E-01*	1.00E-01
		30%	3.14	3.25	5.64	4.14	7.3E-01*	1.3E-02*	7.20E-02
		40%	5.23	5.02	7.91	5.93	9.40E-01	1.5E-02*	2.60E-03
		50%	7.28	7.29	9.51	8.14	0.98	6.0E-02*	0.051

focusing in gaussian noise the robustness results are better than those of the uniform noise. The main difference in this case is that MCS3-1 and MCS5 are not statistically worse than C4.5.

5.5.4.3 Conclusions

The results obtained have shown that the MCSs studied do not always significantly improve the performance of their single classification algorithms when dealing with noisy data, although they do in the majority of cases (if the individual components are not heavily affected by noise). The improvement depends on many factors, such as the type and level of noise. Moreover, the performance of the MCSs built with heterogeneous classifiers depends on the performance of their single classifiers, so it is recommended that one studies the behavior of each single classifier before building the MCS. Generally, the MCSs studied are more suitable for class noise than for attribute noise. Particularly, they perform better with the most disruptive class noise scheme (the uniform one) and with the least disruptive attribute noise scheme (the gaussian one).

The robustness results show that the studied MCS built with heterogeneous classifiers will not be more robust than the most robust among their single classification algorithms. In fact, the robustness can always be shown as an average of the robustness of the individual methods. The higher the robustness of the individual classifiers are, the higher the robustness of the MCS is.

5.5.5 Analysis of the OVO Decomposition with Noise

In this section, the performance and robustness of the classification algorithms using the OVO decomposition with respect to its baseline results when dealing with data suffering from noise are analyzed. In order to investigate whether the decomposition is able to reduce the effect of noise or not, a large number of data sets are created introducing different levels and types of noise, as suggested in the literature. Several well-known classification algorithms, with or without decomposition, are trained with them in order to check when decomposition is advantageous. The results obtained show that methods using the One-vs-One strategy lead to better performances and more robust classifiers when dealing with noisy data, especially with the most disruptive noise schemes. Section 5.5.5.1 is devoted to the study of the class noise scheme, whereas Sect. 5.5.5.2 analyzes the attribute noise case.

5.5.5.1 First Scenario: Data Sets with Class Noise

Table 5.8 shows the test accuracy and RLA results for each classification algorithm at each noise level along with the associated p-values between the OVO and the

Table 5.8 Test accuracy, RLA results and p-values on data sets with class noise. Cases where the baseline classifiers obtain more ranks than the OVO version in the Wilcoxon's test are indicated with a star (*)

			Uniform random class noise						Pairwise class noise					
			C4.5		Ripper		5-NN		C4.5		Ripper		5-NN	
			Base	OVO	Base	OVO	Base	OVO	Base	OVO	Base	OVO	Base	OVO
Test accuracy	Results	0%	81.66	82.7	77.92	82.15	82.1	83.45	81.66	82.7	77.92	82.15	82.1	83.45
		10%	80.5	81.71	71.3	79.86	81.01	82.56	80.94	81.86	75.94	80.71	81.42	82.82
		20%	78.13	80.27	66.71	77.35	79.55	81.36	79.82	81.03	74.77	79.62	79.41	81.01
		30%	75.22	78.87	62.91	74.98	77.21	79.82	78.49	79.26	73.38	78.05	75.29	76.81
		40%	71.1	76.88	58.32	72.12	73.82	76.83	76.17	76.91	71.6	76.19	69.89	71.65
		50%	64.18	73.71	53.79	67.56	68.04	72.73	63.63	63.52	67.11	65.78	64.02	65.52
	p-values	0%	-		-		-		0.007		0.0002		0.093	
		10%	0.0124		0.0001		0.0036		0.0033		0.0003		0.0137	
		20%	0.0028		0.0001		0.0017		0.0017		0.0002		0.0022	
		30%	0.0002		0.0001		0.0013		0.009		0.0001		0.01	
		40%	0.0002		0.0001		0.0111		0.0276		0.0003		0.004	
		50%	0.0001		0.0001		0.0008		0.5016(*)		0.0930(*)		0.0057	
RLA value	Results	0%	-		-		-		-		-		-	
		10%	1.56	1.28	9.35	2.79	1.46	1.12	0.91	1.01	2.38	1.72	0.89	0.82
		20%	4.63	3.15	15.44	5.86	3.48	2.67	2.39	2.13	3.52	3.03	3.29	2.93
		30%	8.36	4.9	20.74	8.82	6.4	4.53	4.16	4.49	5.13	4.84	8.05	7.81
		40%	13.46	7.41	26.72	12.35	10.3	8.11	7.01	7.38	7.25	7.12	14.37	13.68
		50%	21.87	11.29	32.74	18.1	17.47	13.12	21.03	22.28	12.22	18.75	21.06	20.67
	p-values	0%	-		-		-		-		-		-	
		10%	0.5257		0.0001		0.1354		0.8721(*)		0.3317		0.3507	
		20%	0.0304		0.0001		0.0479		0.0859		0.4781		0.0674	
		30%	0.0006		0.0001		0.0124		0.6813		0.6542		0.3507	
		40%	0.0001		0.0001		0.0333		0.6274		0.6274(*)		0.0793	
		50%	0.0001		0.0001		0.0015		0.0400(*)		0.0001(*)		0.062	

non-OVO version from the Wilcoxon's test. The few exceptions where the baseline classifiers obtain more ranks than the OVO version in the Wilcoxon's test are indicated with a star next to the p-value.

For random class noise the test accuracy of the methods using OVO is higher in all the noise levels. Moreover, the low p-values show that this advantage in favor of OVO is significant. The RLA values of the methods using OVO are lower than those of the baseline methods at all noise levels. These differences are also statistically significant as reflected by the low p-values. Only at some very low noise levels—5 % and 10 % for C4.5 and 5 % for 5-NN - the results between the OVO and the non-OVO version are statistically equivalent, but notice that the OVO decomposition does not hinder the results, simply the loss is not lower.

These results also show that OVO achieves more accurate predictions when dealing with **pairwise class noise**, however, it is not so advantageous with C4.5 or RIPPER as with 5-NN in terms of robustness when noise only affects one class. For example, the behavior of RIPPER with this noise scheme can be related to the hierarchical way in which the rules are learned: it starts learning rules of the class with the lowest number of examples and continues learning those classes with more examples. When introducing this type of noise, RIPPER might change its training order, but the remaining part of the majority class can still be properly learned, since it now has more priority. Moreover, the original second majority class, now with noisy examples, will probably be the last one to be learned and it would depend on how the rest of the classes have been learned. Decomposing the problem with OVO, a considerable number of classifiers will have a notable quantity of noise—those of the majority and the second majority classes—and hence, the tendency to predict the original majority class decreases—when the noise level is high, it strongly affects the accuracy, since the majority has more influence on it.

In contrast with the rest of noise schemes, with pairwise noise scheme, all the data sets have different real percentages of noisy examples at the same noise level of $x\%$. This is because each data set has a different number of examples of the majority class, and thus a noise level of $x\%$ does not affect all the data sets in the same way. In this case, the percentage of noisy examples with a noise level of $x\%$ is computed as $(x \cdot N_{maj})/100$, where N_{maj} is the percentage of examples of the majority class.

5.5.5.2 Second Scenario: Data Sets with Attribute Noise

In this section, the performance and robustness of the classification algorithms using OVO in comparison to its non-OVO version when dealing with data with attribute noise are analyzed. The test accuracy, RLA results and p-values of each classification algorithm at each noise level are shown in Table 5.9.

In the case of uniform attribute noise it can be pointed out that the test accuracy of the methods using OVO is always statistically better at all the noise levels. The RLA values of the methods using OVO are lower than those of the baseline methods at all noise levels—except in the case of C4.5 with a 5 % of noise level. Regarding the p-values, a clear tendency is observed, the p-value decreases when the noise level increases with all the algorithms. With all methods—C4.5, RIPPER and 5-NN—the p-values of the RLA results at the lowest noise levels (up to 20–25 %) show that the robustness of OVO and non-OVO methods is statistically equivalent. From that point on, the OVO versions statistically outperform the non-OVO ones. Therefore, the usage of OVO is clearly advantageous in terms of accuracy and robustness when

Table 5.9 Test accuracy, RLA results and p-values on data sets with attribute noise. Cases where the baseline classifiers obtain more ranks than the OVO version in the Wilcoxon's test are indicated with a star (*)

			Uniform random attribute noise						Gaussian attribute noise					
			C4.5		Ripper		5-NN		C4.5		Ripper		5-NN	
			Base	OVO	Base	OVO	Base	OVO	Base	OVO	Base	OVO	Base	OVO
Test accuracy	Results	0%	81.66	82.7	77.92	82.15	82.1	83.45	81.66	82.7	77.92	82.15	82.1	83.45
		10%	80.31	81.65	76.08	80.85	79.81	81.34	80.93	81.67	76.53	81.12	80.91	82.52
		20%	78.71	80.27	73.95	79.15	77.63	79.38	79.77	81.11	75.35	80.06	80.16	81.74
		30%	76.01	78.25	71.25	77.06	74.68	76.46	79.03	80.4	74.46	78.93	78.84	80.77
		40%	73.58	76.19	68.66	74.56	71.29	73.65	77.36	79.51	72.94	78.1	77.53	79.11
		50%	70.49	73.51	65.5	71.66	67.72	70.07	75.29	78.03	71.57	76.27	76.02	77.72
	p-values	0%	0.007		0.0002		0.093		0.007		0.0002		0.093	
		10%	0.0169		0.0003		0.091		0.1262		0.0004		0.0064	
		20%	0.0057		0.0003		0.0015		0.0048		0.0002		0.0036	
		30%	0.0043		0.0001		0.0112		0.0051		0.0003		0.0025	
		40%	0.0032		0.0001		0.0006		0.0019		0.0003		0.1262	
		50%	0.0036		0.0007		0.0011		0.0004		0.0008		0.0251	
RLA value	Results	0%	-		-		-		-		-		-	
		10%	1.82	1.32	2.56	1.62	3.03	2.62	0.92	1.27	1.9	1.26	1.68	1.13
		20%	3.88	3.11	5.46	3.77	5.72	5.03	2.4	1.99	3.49	2.59	2.51	2.09
		30%	7.54	5.77	9.2	6.42	9.57	8.76	3.42	2.91	4.66	3.95	4.34	3.32
		40%	10.64	8.25	12.81	9.64	13.73	12.08	5.67	4.03	6.74	4.96	6.03	5.38
		50%	14.74	11.74	17.21	13.33	18.14	16.55	8.37	5.87	8.5	7.35	7.82	7.16
	p-values	0%	-		-		-		-		-		-	
		10%	0.4781		0.5755		1.0000(*)		0.0766(*)		0.8519(*)		0.4115	
		20%	0.2471		0.1454		0.1354		0.8405		0.9108(*)		0.3905	
		30%	0.0304		0.0438		0.1672		0.6542		0.2627(*)		0.2627	
		40%	0.0569		0.0036		0.0111		0.1169		0.3905		0.9405	
		50%	0.0152		0.0064		0.0228		0.009		0.6542(*)		0.218	

noise affects the attributes in a random and uniform way. This behavior is particularly notable with the highest noise levels, where the effects of noise are expected to be more detrimental.

On the other hand, analyzing the gaussian attribute noise results in the test accuracy of the methods using OVO being better at all the noise levels. The low p-values show that this advantage, also in favor of OVO, is statistically significant. With respect to the RLA results the p-values show a clear decreasing tendency when the noise level increases in all the algorithms. In the case of C4.5, OVO is statistically better from a 35 % noise level onwards. RIPPER and 5-NN are statistically equivalent at all noise levels—although 5-NN with OVO obtains higher Wilcoxon's ranks.

Hence, the OVO approach is also suitable considering the accuracy achieved with this type of attribute noise. The robustness results are similar between the OVO and non-OVO versions with RIPPER and 5-NN. However, for C4.5 there are statistical differences in favor of OVO at the highest noise levels. It is important to note that in some cases, particularly in the comparisons involving RIPPER, some RLA results show that OVO is better than the non-OVO version in average but the latter obtains more ranks in the statistical test—even though these differences are not significant. This is due to the extreme results of some individual data sets, such as *led7digit* or *flare*, in which the RLA results of the non-OVO version are much worse than those of the OVO version. Anyway, we should notice that average results themselves are not meaningful and the corresponding non-parametric statistical analysis must be carried out in order to extract meaningful conclusions, which reflects the real differences between algorithms.

5.5.5.3 Conclusions

The results obtained have shown that the OVO decomposition improves the baseline classifiers in terms of accuracy when data is corrupted by noise in all the noise schemes shown in this chapter. The robustness results are particularly notable with the more disruptive noise schemes—the uniform random class noise scheme and the uniform random attribute noise scheme—where a larger amount of noisy examples and with higher corruptions are available, which produce greater differences (with statistical significance).

In conclusion, we must emphasize that one usually does not know the type and level of noise present in the data of the problem that is going to be addressed. Decomposing a problem suffering from noise with OVO has shown a better accuracy, higher robustness and homogeneity in all the classification algorithms tested. For this reason, the use of the OVO decomposition strategy in noisy environments can be recommended as an easy-to-applicate, yet powerful tool to overcome the negative effects of noise in multi-class problems.

References

1. Abellán, J., Masegosa, A.R.: Bagging decision trees on data sets with classification noise. In: Link S., Prade H. (eds.) FoIKS, Lecture Notes in Computer Science, vol. 5956, pp. 248–265. Springer, Heidelberg (2009)

2. Abellán, J., Masegosa, A.R.: Bagging schemes on the presence of class noise in classification. Expert Syst. Appl. **39**(8), 6827–6837 (2012)
3. Aha, D.W., Kibler, D.: Noise-tolerant instance-based learning algorithms. In: Proceedings of the 11th International Joint Conference on Artificial Intelligence, Vol. 1, IJCAI'89, pp. 794–799. Morgan Kaufmann Publishers Inc. (1989)
4. Alcalá-Fdez, J., Sánchez, L., García, S., del Jesus, M., Ventura, S., Garrell, J., Otero, J., Romero, C., Bacardit, J., Rivas, V., Fernández, J., Herrera, F.: KEEL: a software tool to assess evolutionary algorithms for data mining problems. Soft Comput. Fus. Found. Methodol. Appl. **13**, 307–318 (2009)
5. Allwein, E.L., Schapire, R.E., Singer, Y.: Reducing multiclass to binary: a unifying approach for margin classifiers. J. Mach. Learn. Res. **1**, 113–141 (2000)
6. Anand, R., Mehrotra, K., Mohan, C.K., Ranka, S.: Efficient classification for multiclass problems using modular neural networks. IEEE Trans. Neural Netw. **6**(1), 117–124 (1995)
7. Angluin, D., Laird, P.: Learning from noisy examples. Mach. Learn. **2**(4), 343–370 (1988)
8. Bonissone, P., Cadenas, J.M., Carmen Garrido, M., Díaz-Valladares, A.: A fuzzy random forest. Int. J. Approx. Reason. **51**(7), 729–747 (2010)
9. Bootkrajang, J., Kaban, A.: Multi-class classification in the presence of labelling errors. In: ESANN 2011, 19th European Symposium on Artificial Neural Networks, Bruges, Belgium, 27–29 April 2011, Proceedings, ESANN (2011)
10. Brodley, C.E., Friedl, M.A.: Identifying and eliminating mislabeled training instances. In: Clancey W.J., Weld D.S. (eds.) AAAI/IAAI, Vol. 1, pp. 799–805 (1996)
11. Brodley, C.E., Friedl, M.A.: Identifying mislabeled training data. J. Artif. Intell. Res. **11**, 131–167 (1999)
12. Catal, C., Alan, O., Balkan, K.: Class noise detection based on software metrics and ROC curves. Inf. Sci. **181**(21), 4867–4877 (2011)
13. Chang, C.C., Lin, C.J.: A library for support vector machines. ACM Trans. Intell. Syst. Technol. **2**(3), 1–27 (2011)
14. Cortes, C., Vapnik, V.: Support vector networks. Mach. Learn. **20**, 273–297 (1995)
15. Delany, S.J., Cunningham, P.: An analysis of case-base editing in a spam filtering system. In: Funk P., González-Calero P.A. (eds.) ECCBR, pp. 128–141 (2004)
16. Dietterich, T.G.: An experimental comparison of three methods for constructing ensembles of decision trees: bagging, boosting, and randomization. Mach. Learn. **40**(2), 139–157 (2000)
17. Dietterich, T.G., Bakiri, G.: Solving multiclass learning problems via error-correcting output codes. J. Artif. Intell. Res. **2**(1), 263–286 (1995)
18. Du, W., Urahama, K.: Error-correcting semi-supervised pattern recognition with mode filter on graphs. J. Adv. Comput. Intell. Intell. Inform. **15**(9), 1262–1268 (2011)
19. Frenay, B., Verleysen, M.: Classification in the presence of label noise: a survey. Neural Netw. Learn. Syst. IEEE Trans. **25**(5), 845–869 (2014)
20. Fürnkranz, J.: Round robin classification. J. Mach. Learn. Res. **2**, 721–747 (2002)
21. Galar, M., Fernández, A., Barrenechea, E., Bustince, H., Herrera, F.: An overview of ensemble methods for binary classifiers in multi-class problems: experimental study on one-vs-one and one-vs-all schemes. Pattern Recognit. **44**(8), 1761–1776 (2011)
22. Galar, M., Fernández, A., Tartas, E.B., Sola, H.B., Herrera, F.: A review on ensembles for the class imbalance problem: bagging-, boosting-, and hybrid-based approaches. IEEE Trans. Syst. Man Cybern. Part C **42**(4), 463–484 (2012)
23. Gamberger, D., Boskovic, R., Lavrac, N., Groselj, C.: Experiments with noise filtering in a medical domain. In: Proceedings of the Sixteenth International conference on machine learning, pp. 143–151. Morgan Kaufmann Publishers (1999)
24. Gamberger, D., Lavrac, N., Dzeroski, S.: Noise detection and elimination in data preprocessing: experiments in medical domains. Appl. Artif. Intell. **14**, 205–223 (2000)
25. García, V., Alejo, R., Sánchez, J., Sotoca, J., Mollineda, R.: Combined effects of class imbalance and class overlap on instance-based classification. In: Corchado, E., Yin, H., Botti, V., Fyfe, C. (eds.) Intelligent Data Engineering and Automated Learning IDEAL 2006. Lecture Notes in Computer Science, vol. 4224, pp. 371–378. Springer, Berlin (2006)

26. García, V., Mollineda, R., Sánchez, J.: On the k-NN performance in a challenging scenario of imbalance and overlapping. Pattern Anal. Appl. **11**(3–4), 269–280 (2008)
27. García, V., Sánchez, J., Mollineda, R.: An empirical study of the behavior of classifiers on imbalanced and overlapped data sets. In: Rueda, L., Mery, D., Kittler, J. (eds.) CIARP 2007. LNCS, vol. 4756, pp. 397–406. Springer, Heidelberg (2007)
28. Hall, M., Frank, E., Holmes, G., Pfahringer, B., Reutemann, P., Witten, I.H.: The WEKA data mining software: an update. SIGKDD Explor. Newsl. **11**(1), 10–18 (2009)
29. Haralick, R.M.: The table look-up rule. Commun. Stat. Theory Methods A **5**(12), 1163–1191 (1976)
30. Hart, P.E.: The condensed nearest neighbor rule. IEEE Trans. Inf. Theory **14**, 515–516 (1968)
31. Hernández, M.A., Stolfo, S.J.: Real-world data is dirty: data cleansing and the merge/purge problem. Data Min. Knowl. Discov. **2**, 9–37 (1998)
32. Hernández-Lobato, D., Hernández-Lobato, J.M., Dupont, P.: Robust multi-class gaussian process classification. In: Shawe-Taylor J., Zemel R.S., Bartlett P.L., Pereira F.C.N., Weinberger K.Q. (eds.) Advances in Neural Information Processing Systems 24: 25th Annual Conference on Neural Information Processing Systems 2011. Proceedings of a meeting held 12–14 December 2011, Granada, Spain, NIPS, pp. 280–288 (2011)
33. Heskes, T.: The use of being stubborn and introspective. In: Ritter, H., Cruse, H., Dean, J. (eds.) Prerational Intelligence: Adaptive Behavior and Intelligent Systems Without Symbols and Logic, pp. 725–741. Kluwer, Dordrecht (2001)
34. Ho, T.K.: Multiple classifier combination: lessons and next steps. In: Kandel, Bunke E. (eds.) Hybrid Methods in Pattern Recognition, pp. 171-198. World Scientific, New York (2002)
35. Ho, T.K., Basu, M.: Complexity measures of supervised classification problems. IEEE Trans. Pattern Anal. Mach. Intell. **24**(3), 289–300 (2002)
36. Ho, T.K., Hull, J.J., Srihari, S.N.: Decision combination in multiple classifier systems. IEEE Trans. Pattern Anal. Mach. Intell. **16**(1), 66–75 (1994)
37. Hsu, C.W., Lin, C.J.: A comparison of methods for multiclass support vector machines. IEEE Trans. Neural Netw. **13**(2), 415–425 (2002)
38. Huang, Y.S., Suen, C.Y.: A method of combining multiple experts for the recognition of unconstrained handwritten numerals. IEEE Trans. Pattern Anal. Mach. Intell. **17**, 90–93 (1995)
39. Huber, P.J.: Robust Statistics. Wiley, New York (1981)
40. Hüllermeier, E., Vanderlooy, S.: Combining predictions in pairwise classification: an optimal adaptive voting strategy and its relation to weighted voting. Pattern Recognit. **43**(1), 128–142 (2010)
41. Japkowicz, N.: Class imbalance: are we focusing on the right issue? In: II Workshop on learning from imbalanced data sets, ICML, pp. 17–23 (2003)
42. Jeatrakul, P., Wong, K., Fung, C.: Data cleaning for classification using misclassification analysis. J. Adv. Comput. Intell. Intell. Inform. **14**(3), 297–302 (2010)
43. Jo, T., Japkowicz, N.: Class Imbalances versus small disjuncts. SIGKDD Explor. **6**(1), 40–49 (2004)
44. John, G.H.: Robust decision trees: removing outliers from databases. In: Fayyad, U.M., Uthurusamy, R. (eds.) Proceedings of the First International Conference on Knowledge Discovery and Data Mining (KDD-95), pp. 174–179. Montreal, Canada, August (1995)
45. Karmaker, A., Kwek, S.: A boosting approach to remove class label noise. Int. J. Hybrid Intell. Syst. **3**(3), 169–177 (2006)
46. Kermanidis, K.L.: The effect of borderline examples on language learning. J. Exp. Theor. Artif. Intell. **21**, 19–42 (2009)
47. Khoshgoftaar, T., Van Hulse, J., Napolitano, A.: Comparing boosting and bagging techniques with noisy and imbalanced data. IEEE Trans. Syst. Man Cybern. Part A Syst. Hum. **41**(3), 552–568 (2011)
48. Khoshgoftaar, T.M., Rebours, P.: Improving software quality prediction by noise filtering techniques. J. Comput. Sci. Technol. **22**, 387–396 (2007)
49. Klebanov, B.B., Beigman, E.: Some empirical evidence for annotation noise in a benchmarked dataset. In: Human Language Technologies: The 2010 Annual Conference of the

North American Chapter of the Association for Computational Linguistics, HLT '10, pp. 438–446. Association for Computational Linguistics (2010)
50. Knerr, S., Personnaz, L., Dreyfus, G.: Single-layer learning revisited: a stepwise procedure for building and training a neural network. In: Fogelman Soulié F., Hérault J. (eds.) Neurocomputing: Algorithms, Architectures and Applications, pp. 41–50. Springer, Heidelberg (1990)
51. Knerr, S., Personnaz, L., Dreyfus, G., Member, S.: Handwritten digit recognition by neural networks with single-layer training. IEEE Trans. Neural Netw. **3**, 962–968 (1992)
52. Kubat, M., Matwin, S.: Addresing the curse of imbalanced training sets: one-side selection. In: Proceedings of the 14th International Conference on Machine Learning, pp. 179–186 (1997)
53. Kuncheva, L.: Combining Pattern Classifiers: Methods and Algorithms. Wiley, Chichester (2004)
54. Kuncheva, L.I.: Diversity in multiple classifier systems. Inform. Fus. **6**, 3–4 (2005)
55. Lorena, A., de Carvalho, A., Gama, J.: A review on the combination of binary classifiers in multiclass problems. Artif. Intell. Rev. **30**, 19–37 (2008)
56. Maclin, R., Opitz, D.: An empirical evaluation of bagging and boosting. In: Proceedings of the fourteenth national conference on artificial intelligence and ninth conference on Innovative applications of artificial intelligence, pp. 546–551 (1997)
57. Malossini, A., Blanzieri, E., Ng, R.T.: Detecting potential labeling errors in microarrays by data perturbation. Bioinformatics **22**(17), 2114–2121 (2006)
58. Mandler, E., Schuermann, J.: Combining the classification results of independent classifiers based on the Dempster/Shafer theory of evidence. In: Gelsema E.S., Kanal L.N. (eds.) Pattern Recognition and Artificial Intelligence, pp. 381–393. Amsterdam: North-Holland (1988)
59. Manwani, N., Sastry, P.S.: Noise tolerance under risk minimization. IEEE Trans. Cybern. **43**(3), 1146–1151 (2013)
60. Maulik, U., Chakraborty, D.: A robust multiple classifier system for pixel classification of remote sensing images. Fundamenta Informaticae **101**(4), 286–304 (2010)
61. Mayoraz, E., Moreira, M.: On the decomposition of polychotomies into dichotomies (1996)
62. Mazurov, V.D., Krivonogov, A.I., Kazantsev, V.S.: Solving of optimization and identification problems by the committee methods. Pattern Recognit. **20**, 371–378 (1987)
63. Mclachlan, G.J.: Discriminant Analysis and Statistical Pattern Recognition (Wiley Series in Probability and Statistics). Wiley-Interscience, New York (2004)
64. Melville, P., Shah, N., Mihalkova, L., Mooney, R.J.: Experiments on ensembles with missing and noisy data. In: Roli F., Kittler J., Windeatt T. (eds.) Multiple Classifier Systems, Lecture Notes in Computer Science, vol. 3077, pp. 293–302. Springer, Heidelberg (2004)
65. Miranda, A.L.B., Garcia, L.P.F., Carvalho, A.C.P.L.F., Lorena, A.C.: Use of classification algorithms in noise detection and elimination. In: Corchado E., Wu X., Oja E., Herrero l., Baruque B. (eds.) HAIS, Lecture Notes in Computer Science, vol. 5572, pp. 417–424. Springer, Heidelberg (2009)
66. Muhlenbach, F., Lallich, S., Zighed, D.A.: Identifying and handling mislabelled instances. J. Intell. Inf. Syst. **22**(1), 89–109 (2004)
67. Napierala, K., Stefanowski, J., Wilk, S.: Learning from imbalanced data in presence of noisy and borderline examples. Rough Sets and Current Trends in Computing. LNCS, vol. 6086, pp. 158–167. Springer, Berlin (2010)
68. Nath, R.K.: Fingerprint recognition using multiple classifier system. Fractals **15**(3), 273–278 (2007)
69. Nettleton, D., Orriols-Puig, A., Fornells, A.: A Study of the Effect of Different Types of Noise on the Precision of Supervised Learning Techniques. Artif. Intell. Rev. **33**, 275–306 (2010)
70. Pérez Carlos Javier, G.F.J.M.J.R.M.R.C.: Misclassified multinomial data: a Bayesian approach. RACSAM **101**(1), 71–80 (2007)
71. Pimenta, E., Gama, J.: A study on error correcting output codes. In: Portuguese Conference on Artificial Intelligence EPIA 2005, 218–223 (2005)
72. Polikar, R.: Ensemble based systems in decision making. IEEE Circ. Syst. Mag. **6**(3), 21–45 (2006)

73. Qian, B., Rasheed, K.: Foreign exchange market prediction with multiple classifiers. J. Forecast. **29**(3), 271–284 (2010)
74. Quinlan, J.R.: Induction of decision trees. Mach. Learn. **1**(1), 81–106 (1986)
75. Quinlan, J.R.: C4.5: Programs for Machine Learning. Morgan Kaufmann Publishers, San Francisco (1993)
76. Rifkin, R., Klautau, A.: In defense of one-vs-all classification. J. Mach. Learn. Res. **5**, 101–141 (2004)
77. Sáez, J.A., Luengo, J., Herrera, F.: Predicting noise filtering efficacy with data complexity measures for nearest neighbor classification. Pattern Recognit. **46**(1), 355–364 (2013)
78. Sánchez, J.S., Barandela, R., Marqués, A.I., Alejo, R., Badenas, J.: Analysis of new techniques to obtain quality training sets. Pattern Recognit. Lett. **24**(7), 1015–1022 (2003)
79. Segata, N., Blanzieri, E., Delany, S.J., Cunningham, P.: Noise reduction for instance-based learning with a local maximal margin approach. J. Intell. Inf. Syst. **35**(2), 301–331 (2010)
80. Shapley, L., Grofman, B.: Optimizing group judgmental accuracy in the presence of interdependencies. Pub. Choice **43**, 329–343 (1984)
81. Smith, M.R., Martinez, T.R.: Improving classification accuracy by identifying and removing instances that should be misclassified. In: IJCNN, pp. 2690–2697 (2011)
82. Sun, J., ying Zhao, F., Wang, C.J., Chen, S.: Identifying and correcting mislabeled training instances. In: FGCN (1), pp. 244–250. IEEE (2007)
83. Sun, Y., Wong, A.K.C., Kamel, M.S.: Classification of Imbalanced Data: a Review. Int. J. Pattern Recognit. Artif. Intell. **23**(4), 687–719 (2009)
84. Teng, C.M.: Correcting Noisy Data. In: Proceedings of the Sixteenth International Conference on Machine Learning, pp. 239–248. Morgan Kaufmann Publishers, San Francisco, USA (1999)
85. Teng, C.M.: Polishing blemishes: Issues in data correction. IEEE Intell. Syst. **19**(2), 34–39 (2004)
86. Thongkam, J., Xu, G., Zhang, Y., Huang, F.: Support vector machine for outlier detection in breast cancer survivability prediction. In: Ishikawa Y., He J., Xu G., Shi Y., Huang G., Pang C., Zhang Q., Wang G. (eds.) APWeb Workshops, Lecture Notes in Computer Science, vol. 4977, pp. 99–109. Springer (2008)
87. Titterington, D.M., Murray, G.D., Murray, L.S., Spiegelhalter, D.J., Skene, A.M., Habbema, J.D.F., Gelpke, G.J.: Comparison of discriminant techniques applied to a complex data set of head injured patients. J. R. Stat. Soc. Series A (General) **144**, 145–175 (1981)
88. Tomek, I.: Two Modifications of CNN. IEEE Tran. Syst. Man Cybern. **7**(2), 679–772 (1976)
89. Verbaeten, S., Assche, A.V.: Ensemble methods for noise elimination in classification problems. In: Fourth International Workshop on Multiple Classifier Systems, pp. 317–325. Springer, Heidelberg (2003)
90. Wang, R.Y., Storey, V.C., Firth, C.P.: A framework for analysis of data quality research. IEEE Trans. Knowl. Data Eng. **7**(4), 623–640 (1995)
91. Wemecke, K.D.: A coupling procedure for the discrimination of mixed data. Biometrics **48**, 497–506 (1992)
92. Wheway, V.: Using boosting to detect noisy data. In: Revised Papers from the PRICAI 2000 Workshop Reader, Four Workshops Held at PRICAI 2000 on Advances in Artificial Intelligence, pp. 123–132. Springer (2001)
93. Wilson, D.R., Martinez, T.R.: Instance pruning techniques. In: Proceedings of the Fourteenth International Conference on Machine Learning, ICML '97, pp. 403–411. Morgan Kaufmann Publishers Inc. (1997)
94. Woźniak, M., Graña, M., Corchado, E.: A survey of multiple classifier systems as hybrid systems. Inform. Fus. **16**, 3–17 (2013)
95. Wu, T.F., Lin, C.J., Weng, R.C.: Probability estimates for multi-class classification by pairwise coupling. J. Mach. Learn. Res. **5**, 975–1005 (2004)
96. Wu, X.: Knowledge Acquisition From Databases. Ablex Publishing Corp, Norwood (1996)
97. Xu, L., Krzyzak, A., Suen, C.Y.: Methods of combining multiple classifiers and their applications to handwriting recognition. IEEE Trans. Syst. Man Cybern. **22**(3), 418–435 (1992)

98. Zhang, C., Wu, C., Blanzieri, E., Zhou, Y., Wang, Y., Du, W., Liang, Y.: Methods for labeling error detection in microarrays based on the effect of data perturbation on the regression model. Bioinformatics **25**(20), 2708–2714 (2009)
99. Zhong, S., Khoshgoftaar, T.M., Seliya, N.: Analyzing software measurement data with clustering techniques. IEEE Intell. Syst. **19**(2), 20–27 (2004)
100. Zhu, X., Wu, X.: Class noise vs. attribute noise: a quantitative study. Artif. Intell. Rev. **22**, 177–210 (2004)
101. Zhu, X., Wu, X.: Class noise handling for effective cost-sensitive learning by cost-guided iterative classification filtering. IEEE Trans. Knowl. Data Eng. **18**(10), 1435–1440 (2006)
102. Zhu, X., Wu, X., Chen, Q.: Eliminating class noise in large datasets. In: Proceeding of the Twentieth International Conference on Machine Learning, pp. 920–927 (2003)
103. Zhu, X., Wu, X., Chen, Q.: Bridging local and global data cleansing: Identifying class noise in large, distributed data datasets. Data Min. Knowl. Discov. **12**(2–3), 275–308 (2006)
104. Zhu, X., Wu, X., Yang, Y.: Error detection and impact-sensitive instance ranking in noisy datasets. In: Proceedings of the Nineteenth National Conference on Artificial Intelligence, pp. 378–383. AAAI Press (2004)

Chapter 6
Data Reduction

Abstract The most common tasks for data reduction carried out in Data Mining consist of removing or grouping the data through the two main dimensions, examples and attributes; and simplifying the domain of the data. A global overview to this respect is given in Sect. 6.1. One of the well-known problems in Data Mining is the "curse of dimensionality", related with the usual high amount of attributes in data. Section 6.2 deals with this problem. Data sampling and data simplification are introduced in Sects. 6.3 and 6.4, respectively, providing the basic notions on these topics for further analysis and explanation in subsequent chapters of the book.

6.1 Overview

Currently, it is not difficult to imagine the disposal of a data warehouse for an analysis which contains millions of samples, thousands of attributes and complex domains. Data sets will likely be huge, thus the data analysis and mining would take a long time to give a respond, making such analysis infeasible and even impossible.

Data reduction techniques can be applied to achieve a reduced representation of the data set,it is much smaller in volume and tries to keep most of the integrity of the original data [11]. The goal is to provide the mining process with a mechanism to produce the same (or almost the same) outcome when it is applied over reduced data instead of the original data, at the same time as when mining becomes efficient. In this section, we first present an overview of data reduction procedures. A closer look at each individual technique will be provided throughout this chapter.

Basic data reduction techniques are usually categorized into three main families: *DR*, *sample numerosity reduction* and *cardinality reduction*.

DR ensures the reduction of the number of attributes or random variables in the data set. DR methods include *FS* and *feature extraction/construction* (Sect. 6.2 and Chap. 7 of this book), in which irrelevant dimensions are detected, removed or combined. The transformation or projection of the original data onto a smaller space can be done by *PCA* (Sect. 6.2.1), *factor analysis* (Sect. 6.2.2), *MDS* (Sect. 6.2.3) and *LLE* (Sect. 6.2.4), being the most relevant techniques proposed in this field.

S. García et al., *Data Preprocessing in Data Mining*,
Intelligent Systems Reference Library 72, DOI 10.1007/978-3-319-10247-4_6

Sample numerosity reduction methods replace the original data by an alternative smaller data representation. They can be either parametric or non-parametric methods. The former requires a model estimation that fits the original data, using parameters to represent the data instead of the actual data. They are closely-related DM techniques (regression and log-linear models are common parametric data reduction techniques) and we consider their explanation to be out of the scope of this book. However, non-parametric methods work directly with data itself and return other data representations with similar structures. They include *data sampling* (Sect. 6.3), different forms of data grouping, such as *data condensation*, *data squashing* and *data clustering* (Sects. 6.3.1, 6.3.2 and 6.3.3, respectively) and IS as a more intelligent form of sample reduction (Chap. 8 of this book).

Cardinality reduction comprises the transformations applied to obtain a reduced representation of the original data. As we have mention at the beginning of this book, there may be a high level of overlapping between data reduction techniques and data preparation techniques, this category being a representative example with respect to data transformations. As data reduction, we include the *binning* process (Sect. 6.4) and the more general discretization approaches (Chap. 9 of this book).

In the next sections, we will define the main aspects of each one of the aforementioned strategies.

6.2 The Curse of Dimensionality

A major problem in DM in large data sets with many potential predictor variables is the *the curse of dimensionality*. Dimensionality becomes a serious obstacle for the efficiency of most of the DM algorithms, because of their computational complexity. This statement was coined by Richard Bellman [4] to describe a problem that increases as more variables are added to a model.

High dimensionality of the input increases the size of the search space in an exponential manner and also increases the chance to obtain invalid models. It is well known that there is a linear relationship between the required number of training samples with the dimensionality for obtaining high quality models in DM [8]. But when considering non-parametric learners, such as those instance-based or decision trees, the situation is even more severe. It has been estimated that as the number of dimensions increase, the sample size needs to increase exponentially in order to have an effective estimate of multivariate densities [13].

It is evident that the curse of dimensionality affects data differently depending on the following DM task or algorithm. For example, techniques like decision trees could fail to provide meaningful and understandable results when the number of dimensions increase, although the speed in the learning stage is barely affected. On the contrary, instance-based learning has high dependence on dimensionality affecting its order of efficiency.

In order to alleviate this problem, a number of dimension reducers have been developed over the years. As linear methods, we can refer to factor analysis [18] and

PCA [7]. Nonlinear models are LLE [25], ISOMAP [26] and derivatives. They are concerned with the transformation of the original variables into a smaller number of projections. The underlying assumptions are that the variables are numeric and that the dimensions can be expressed as combinations of the actual variables, and vice versa. Further analysis on this type of techniques will be given in this chapter, especially for the two most popular techniques: PCA and LLE.

A set of methods are aimed at eliminating irrelevant and redundant features, reducing the number of variables in the model. They belong to the FS family of methods. They have the following immediate positive effects on the analysis and mining:

- Speed up the processing of the DM algorithm.
- Improve data quality.
- Increase the performance of the DM algorithm.
- Make the results easier to understand.

Formally, the problem of FS can be defined as follows [14]: Let A be the original set of features, with cardinality m. Let f represent the desired number of features in the selected subset B, $B \subset A$. Let the FS criterion function for the set B be represented by $J(B)$. Without any loss of generality, a lower value of J is considered to be a better feature subset, thus, J could represent the generalization error. The problem of FS is to find an optimal subset B that solves the following optimization problem:

$$\begin{gathered} \min J(Z) \\ s.t. \\ Z \subset A \\ |Z| = d \end{gathered}$$

A brute force search would require examining all $\frac{m!}{d! \cdot (m-d)!}$ possible combinations of the feature set A. A vast number of FS approaches, trends and applications have been proposed over the years, and therefore FS deserves a complete chapter of this book: Chap. 7.

Other forms of widely used DR also deserve to be described in this section. They are slightly more complicated than that previously seen, but also very widely used in conjunction with advanced DM approaches and real applications.

6.2.1 Principal Components Analysis

In this subsection, we introduction the Principal Components Analysis (PCA) as a DR method [17]. A detailed theoretical explanation is out of the scope of this book, hence we intend to give details on the basic idea, the method of operation and the objectives this technique pursues. PCA is one of the oldest and most used methods for reduction of multidimensional data.

The basic idea is to find a set of linear transformations of the original variables which could describe most of the variance using a relatively fewer number of variables. Hence, it searches for k n-dimensional orthogonal vectors that can best represent the data, where $k \leq n$. The new set of attributes are derived in a decreasing order of contribution, letting the first obtained variable, the one called *principal component* contain the largest proportion of the variance of the original data set. Unlike FS, PCA allows the combination of the essence of original attributes to form a new smaller subset of attributes.

The usual procedure is to keep only the first few principal components that may contain 95 % or more of the variance of the original data set. PCA is particularly useful when there are too many independent variables and they show high correlation.

The basic procedure is as follows:

- To normalize the input data, equalizing the ranges among attributes.
- To compute k orthonormal vectors to provide a basis for the normalized input data. These vectors point to a direction that is perpendicular to the others and are called *principal components*. The original data is in linear combination of the principal components. In order to calculate them, the eigenvalue-eigenvectors of the covariance matrix from the sample data are needed.
- To sort the principal components according to their strength, given by their associated eigenvalues. The principal components serve as a new set of axes for the data, adjusted according the variance of the original data. In Fig. 6.1, we show an illustrative example of the first two principal components for a given data set.

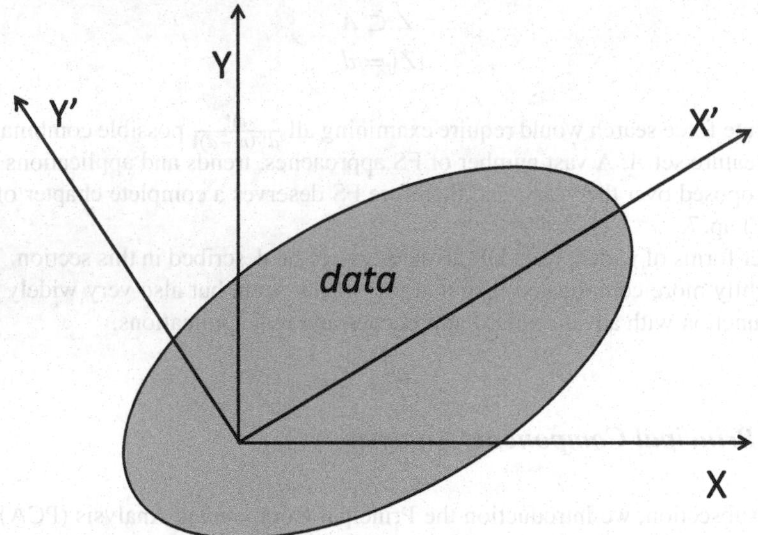

Fig. 6.1 PCA. X' and Y' are the first two principal components obtained

- To reduce the data by removing weaker components, with low variance. A reliable reconstruction of the data could be possible by using only the strongest principal components.

The final output of PCA is a new set of attributes representing the original data set. The user would use only the first few of these new variables because they contain most of the information represented in the original data. PCA can be applied to any type of data. It is also used as a data visualization tool by reducing any multidimensional data into two- or three-dimensional data.

6.2.2 Factor Analysis

Factor analysis is similar to PCA in the sense that it leads to the deduction of a new, smaller set of variables that practically describe the behaviour given in the original data. Nevertheless, factor analysis is different because it does not seek to find transformations for the given attributes. Instead, its goal is to discover hidden factors in the current variables [17]. Although factor analysis has an important role as a process of data exploration, we limit its description to a data reduction method.

In factor analysis, it is assumed that there are a set of unobservable *latent factors* $z_j, j = 1, \ldots, k$; which when acting together generate the original data. Here, the objective is to characterize the dependency among the variables by means of a smaller number of factors.

The basic idea behind factor analysis is to attempt to find a set of hidden factors so that the current attributes can be recovered by performing a set of linear transformations over these factors. Given the set of attributes $a_1, a_2, \ldots, a_m$, factor analysis attempts to find the set of factors $f_1, f_2, \ldots, f_k$, so that

$$
\begin{aligned}
a_1 - \mu_1 &= l_{11} f_1 + l_{12} f_2 + \cdots + l_{1k} f_k + \varepsilon_1 \\
a_2 - \mu_2 &= l_{21} f_1 + l_{22} f_2 + \cdots + l_{2k} f_k + \varepsilon_2 \\
&\vdots \\
a_m - \mu_m &= l_{m1} f_1 + l_{m2} f_2 + \cdots + l_{mk} f_k + \varepsilon_m
\end{aligned}
$$

where $\mu_1, \mu_2, \ldots, \mu_m$ are the means of the attributes $a_1, a_2, \ldots, a_m$, and the terms $\varepsilon_1, \varepsilon_2, \ldots, \varepsilon_m$ represent the unobservable part of the attributes, also called *specific factors*. The terms $l_{ij}, i = 1, \ldots, m, j = 1, \ldots, k$ are known as the loadings. The factors $f_1, f_2, \ldots, f_k$ are known as the *common factors*.

The previous equation can be written in matrix form as:

$$\mathbf{A} - \mu = \mathbf{LF} + \varepsilon$$

Thus, the factor analysis problem can be stated as given the attributes $\mathbf{A}$, along with the mean μ, we endeavor to find the set of factors $\mathbf{F}$ and the associated loadings $\mathbf{L}$, and therefore the above equation is accurate.

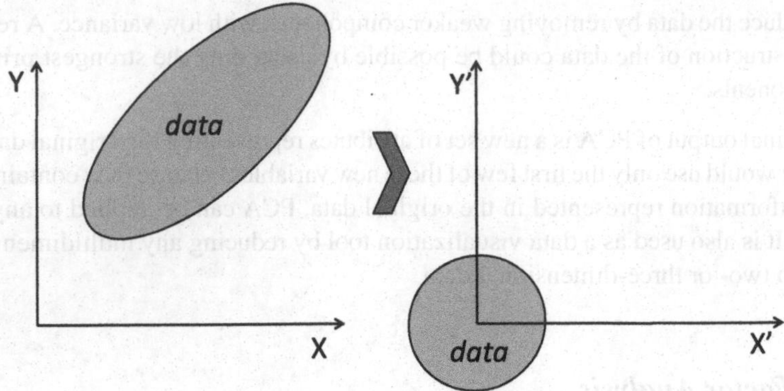

Fig. 6.2 Factors are independent unit normals that are scaled, rotated and translated to compose the inputs

To find **F** and **L**, three common restrictions on their statistical properties are adopted: (1) all the factors are independent, with zero mean and variance of unity, (2) all the error terms are also independent, with zero mean and constant variance, (3) the errors are independent of the factors.

There are two methods for solving the factor model equations for the matrix **K** and the factors **F**: (1) the maximum likelihood method and (2) the principal component method. The first assumes that original data is normally distributed and is computationally expensive. The latter is very fast, easy to interpret and guarantees to find a solution for all data sets.

1. Unlike PCA, factor analysis assumes and underlying structure that relates the factors to the observed data.
2. PCA tries to rotate the axis of the original variables, using a set of linear transformations. Factor analysis, instead, creates a new set of variables to explain the covariances and correlations between the observed variables.
3. In factor analysis, a two-factor model is completely different from a three-factor model, whereas in PCA, when we decide to use a third component, the two first principal components remain the same.
4. PC is fast and straightforward. However, in factor analyses, there are various alternatives to performing the calculations and some of them are complicated and time consuming.

Figure 6.2 exemplifies the process of factor analysis. The differences between PCA and factor analysis can be enumerated.

6.2.3 *Multidimensional Scaling*

Let us assume N points, and that we know the distances between the pairs of points, d_{ij}, for all $i, j = 1, \ldots, N$. Moreover, we do not know the precise coordinates of the points, their dimensionality or the way the distances between them were computed. *Multidimensional scaling* (MDS) is the method for situating these points in a low space such that a classical distance measure (like Euclidean) between them is as close as possible to each d_{ij}. There must be a projection from some unknown dimensional space to another space whose number of dimensions is known.

One of the most typical examples of MDS is to draw an approximation of the map that represents the travel distances between cities, knowing only the distance matrix. Obviously, the outcome is distorted due to the differences between the distances measured taking into account the geographical obstacles and the actual distance in a straight line between the cities. It common for the map to be stretched out to accommodate longer distances and that the map also is centered on the origin. However, the solution is not unique, we can get any rotating view of it.

MDS is within the DR techniques because we can compute the distances in a d-dimensional space of the actual data points and then to give as input this distance matrix to MDS, which then projects it in to a lower-dimensional space so as to preserve these distances.

Formally, let us say we have a sample $X = \{x^t\}_{t=1}^N$ as usual, where $x^t \in \mathbb{R}^d$. For the two points r and s, the squared Euclidean distance between them is

$$d_{rs}^2 = ||x^r - x^s||^2 = \sum_{j=1}^{d}(x_j^r - x_j^s)^2 = \sum_{j=1}^{d}(x_j^r)^2 - 2\sum_{j=1}^{d} x_j^r x_j^s + \sum_{j=1}^{d}(x_j^s)^2$$

$$= b_{rr} + b_{ss} - 2b_{rd}$$

where b_{rs} is defined as

$$b_{rs} = \sum_{j=1}^{d} x_j^r x_j^s$$

To constrain the solution, we center the data at the origin and assume

$$\sum_{t=1}^{N} x_j^t = 0, \quad \forall j = 1, \ldots, d$$

Then, summing up the previous equation on r, s, and defining

$$T = \sum_{t=1}^{n} b_{tt} = \sum_t \sum_j (x_j^t)^2$$

we get

$$\sum_r d_{rs}^2 = T + N b_{ss}$$

$$\sum_s d_{rs}^2 = Nb_{rr} + T$$

$$\sum_r \sum_s d_{rs}^2 = 2NT$$

When we define

$$d_{\cdot s}^2 = \frac{1}{N}\sum_r d_{rs}^2,\ d_{r\cdot}^2 = \frac{1}{N}\sum_s d_{rs}^2,\ d_{\cdot\cdot}^2 = \frac{1}{N^2}\sum_r \sum_s d_{rs}^2$$

and using the first equation, we get

$$b_{rs} = \frac{1}{2}(d_{r\cdot}^2 + d_{\cdot s}^2 - d_{\cdot\cdot}^2 - d_{rs}^2)$$

Having now calculated b_{rs} and knowing that $\mathbf{B} = \mathbf{X}\mathbf{X}^T$, we look for an approximation. We know from the spectral decomposition that $\mathbf{X} = \mathbf{C}\mathbf{D}^{1/2}$ can be used as an approximation for $\mathbf{X}$, where $\mathbf{C}$ is the matrix whose columns are the eigenvectors of $\mathbf{B}$ and $\mathbf{D}^{1/2}$ is a diagonal matrix with square roots of the eigenvalues on the diagonals. Looking at the eigenvalues of $\mathbf{B}$ we decide on a dimensionality k lower than that of d. Let us say $\mathbf{c}_j$ are the eigenvectors with λ_j as the corresponding eigenvalues. Note that $\mathbf{c}_j$ is N-dimensional. Then we get the new dimension as

$$z_j^t = \sqrt{\lambda_j c_j^t}, \quad j = 1, \ldots, k,\ t = 1, \ldots, N$$

That is, the new coordinates of instance t are given by the tth elements of the eigenvectors, $\mathbf{c}_j$, $j = 1, \ldots, k$, after normalization.

In [5], it has been shown that the eigenvalues of $\mathbf{X}\mathbf{X}^T$ $(N \times N)$ are the same as those of $\mathbf{X}^T\mathbf{X}$ $(d \times d)$ and the eigenvectors are related by a simple linear transformation. This shows that PCA does the same work with MDS and does it more easily.

In the general case, we want to find a mapping $\mathbf{z} = g(\mathbf{x}|\theta)$, where $\mathbf{z} \in \mathbb{R}^k$, $\mathbf{x} \in \mathbb{R}^d$, and $g(\mathbf{x}|\theta)$ is the mapping function from d to k dimensions defined up to a set of parameters θ. Classical MDS we discussed previously corresponds to a linear transformation

$$\mathbf{z} = g(\mathbf{x}|\mathbf{W}) = \mathbf{W}^T\mathbf{x}$$

but in a general case, nonlinear mapping can also be used: this is called *Sammon mapping*. the normalized error in mapping is called the *Sammon stress* and is defined as

$$E(\theta|X) = \sum_{r,s} \frac{(||\mathbf{z}^r - \mathbf{z}^s|| - ||\mathbf{x}^r - \mathbf{x}^s||)^2}{||\mathbf{x}^r - \mathbf{x}^s||^2}$$

$$= \sum_{r,s} \frac{(||g(\mathbf{x}^r|\theta) - g(\mathbf{x}^s|\theta)|| - ||\mathbf{x}^r - \mathbf{x}^s||^2)}{||\mathbf{x}^r - \mathbf{x}^s||^2}$$

In the case of classification, the class information can be included in the distance as

$$d'_{rs} = (1-\alpha)d_{rs} + \alpha c_{rs}$$

where c_{rs} is the "distance" between the classes $\mathbf{x}^r$ and $\mathbf{x}^s$ belong to. This interclass distance should be supplied subjectively and α could be optimized using CV.

6.2.4 Locally Linear Embedding

Locally Linear Embedding (LLE) recovers global nonlinear structure from locally linear fits [25]. Its main idea is that each local patch of the manifold can be approximated linearly and given enough data, each point can be written as a linear, weighted sum of its neighbors.

The LLE algorithm is based on simple geometric intuitions. Suppose the data consists of N real-valued vectors $\mathbf{X_i}$, each of dimensionality D, sampled from some smooth underlying manifold. It is expected that each data point and its neighbors to lie on or close to a locally linear patch of the manifold. The local geometry of these patches can be characterized by linear coefficients that reconstruct each data point from its neighbors. In the simplest formulation of LLE, the KNN are estimated per data point, as measured by Euclidean distance. Reconstruction errors are then measured by the cost function:

$$\varepsilon(W) = \sum_i \left| \mathbf{X_i} - \sum_j W_{ij}\mathbf{X_j} \right|^2$$

which adds up the squared distances between all the data points and their reconstructions. The weights W_{ij} summarize the contribution of the jth data point to the $i$sth reconstruction. To compute the weights W_{ij}, it is necessary to minimize the cost function subject to two constraints: first, that each data point $\mathbf{X_i}$ is reconstructed only from its neighbors, enforcing $W_{ij} = 0$ if $\mathbf{X_j}$ does not belong to this set; second, that the rows of the weight matrix sum to one: $\sum_j W_{ij} = 1$ s. The optimal weights W_{ij} subject to these constraints are found by solving a least squares problem.

The constrained weights that minimize these reconstruction errors are invariant to rotations, scaling, and translations of that data point and its neighbors. Suppose the data lie on or near a smooth nonlinear manifold of dimensionality $d \ll D$. To achieve a good approximation, then, there exists a linear mapping that maps the high dimensional coordinates of each neighborhood to global internal coordinates on the manifold. By design, the reconstruction weights W_{ij} reflect intrinsic geometric properties of the data that are invariant to exactly such transformations. We therefore expect their characterization of local geometry in the original data space to be equally valid for local patches on the manifold. In particular, the same weights W_{ij} that reconstruct the ith data point in D dimensions should also reconstruct its embedded manifold coordinates in d dimensions.

LLE constructs a neighborhood preserving mapping based on the above idea. In the final step of the algorithm, each high dimensional observation $\mathbf{X_i}$ is mapped to a low dimensional vector $\mathbf{Y_i}$ representing global internal coordinates on the manifold. This is done by choosing d-dimensional coordinates $\mathbf{Y_i}$ to minimize the embedding cost function:

$$\Phi(Y) = \sum_i \left| \mathbf{Y_i} - \sum_j W_{ij} \mathbf{Y_j} \right|^2$$

This cost function, like the previous one, is based on locally linear reconstruction errors, but here, the weights W_{ij} are fixed while optimizing the coordinates $\mathbf{Y_i}$. Now, the embedding cost can be minimized by solving a sparse $N \times N$ eigenvector problem, whose bottom d non-zero eigenvectors provide an ordered set of orthogonal coordinates centered on the origin.

It is noteworthy that while the reconstruction weights for each data point are computed from its local neighborhood, the embedding coordinates are computed by an $N \times N$ eigensolver, a global operation that couples all data points in connected components of the graph defined by the weight matrix. The different dimensions in the embedding space can be computed successively; this is done simply by computing the bottom eigenvectors from previous equation one at a time. But the computation is always coupled across data points. This is how the algorithm leverages overlapping local information to discover global structure. Implementation of the algorithm is fairly straightforward, as the algorithm has only one free parameter: the number of neighbors per data point, K.

6.3 Data Sampling

Sampling is used to ease the analysis and modeling of large data sets. In DM, data sampling serves four purposes:

- *To reduce the number of instances submitted to the DM algorithm.* In many cases, predictive learning can operate with 10–20 % of cases without a significant deterioration of the performance. After that, the addition of more cases should have expected outcomes. However, in descriptive analysis, it is better to have as many cases as possible.
- *To support the selection of only those cases in which the response is relatively homogeneous.* When you have data sets where different trends are clearly observable or the examples can be easily separated, you can partition the data for different types of modelling. For instance, imagine the learning of the approving decision of bank loans depending on some economic characteristics of a set of customers.

If data includes consumer loans and mortgages, it seems logical to partition both types of loans because the parameters and quantities involved in each one are completely different. Thus, it is a good idea to build separate models on each partition.

- *To assist regarding the balance of data and occurrence of rare events*. Predictive DM algorithms like ANNs or decision trees are very sensitive to imbalanced data sets. An imbalanced data set is one in which one category of the target variable is less represented compared to the other ones and, usually, this category has is more important from the point of view of the learning task. Balancing the data involves sampling the imbalanced categories more than average (over-sampling) or sampling the common less often (under-sampling) [3].
- *To divide a data set into three data sets to carry out the subsequent analysis of DM algorithms*. As we have described in Chap. 2, the original data set can be divided into the training set and testing set. A third kind of division can be performed within the training set, to aid the DM algorithm to avoid model over-fitting, which is a very common strategy in ANNs and decision trees. This partition is usually known as validation set, although, in various sources, it may be denoted as the testing set interchangeably [22]. Whatever the nomenclature used, some learners require an internal testing process and, in order to evaluate and compare a set of algorithms, there must be an external testing set independent of training and containing unseen cases.

Various forms of data sampling are known in data reduction. Suppose that a large data set, T, contains N examples. The most common ways that we could sample T for data reduction are [11, 24]:

- **Simple random sample without replacement (SRSWOR) of size** s: This is created by drawing s of the N tuples from T ($s < N$), where the probability of drawing any tuple in T is $1/N$, that is, all examples have equal chance to be sampled.
- **Simple random sample with replacement (SRSWR) of size** s: This is similar to SRSWOR, except that each time a tuple is drawn from T, it is recorded and replaced. In other words, after an example is drawn, it is placed back in T and it may be drawn again.
- **Balanced sample**: The sample is designed according to a target variable and is forced to have a certain composition according to a predefined criterion. For example, 90 % of customers who are older tah or who are 21 years old, and 10 % of customers who are younger than 21 years old. One of the most successful application of this type of sampling has been shown in imbalanced learning, as we have mentioned before.
- **Cluster sample**: If the tuples in T are grouped into G mutually disjointed groups or clusters, then an SRS of s clusters can be obtained, where $s < G$. For example, in spatial data sets, we may choose to define clusters geographically based on how closely different areas are located.
- **Stratified sample**: If T is divided into mutually disjointed parts called *strata*, a stratified sample of T is generated by obtaining an SRS at each stratum. This

assists in ensuring a representative sample. It is frequently used in classification tasks where the class imbalance is present. It is very closely related with balanced sample, but the predefined composition of the final results depends on the natural distribution of the target variable.

An important preference of sampling for data reduction is that the cost of obtaining a sample is proportionate to the size of the sample s, instead of being proportionate to N. So, the sampling complexity is sub-linear to the size of data and there is no need to conduct a complete pass of T to make decisions in order to or not to include a certain example into the sampled subset. Nevertheless, the inclusion of examples are made by unfounded decisions, allowing redundant, irrelevant, noisy or harmful examples to be included. A smart way to make decisions for sampling is known as IS, a topic that we will extend in Chap. 8.

Advanced schemes of data sampling deserve to be described in this section. As before, they are more difficult and allow better adjustments of data according to the necessities and applications.

6.3.1 Data Condensation

The selection of a small representative subset from a very large data set is known as data condensation. In some sources of DM, such as [22], this form of data reduction is differentiated from others. In this book, data condensation is integrated as one of the families of IS methods (see Chap. 8).

Data condensation emerges from the fact that naive sampling methods, such as random sampling or stratified sampling, are not suitable for real-world problems with noisy data since the performance of the algorithms may change unpredictably and significantly. The data sampling approach practically ignores all the information present in the samples which are not chosen in the reduced subset.

Most of the data condensation approaches are studied on classification-based tasks, and in particular, for the KNN algorithm. These methods attempt to obtain a minimal consistent set, i.e., a minimal set which correctly classifies all the original examples. The very first method of this kind was the condensed nearest neighbor rule (CNN) [12]. For a survey on data condensation methods for classification, we again invite the reader to check the Chap. 8 of this book.

Regarding the data condensation methods which are not affiliated with classification tasks, termed generic data condensation, condensation is performed by vector quantization, such as the well-known self-organizing map [19] and different forms of data clustering. Another group of generic data condensation methods are situated on the density-based techniques which consider the density function of the data for the aspiration of condensation instead of minimizing the quantization error. These approaches do not concern any learning process and, hence, are deterministic, (i.e., for a concrete input data set, the output condensed set is established). Clear examples of this kind of approaches are presented in [10, 21].

6.3.2 Data Squashing

A data squashing method seeks to compress, or "squash", the data in such a way that a statistical analysis carried out on the compressed data obtains the same outcome that the one obtained with the original data set; that is, the statistical information is preserved.

The first approach of data squashing was proposed in [6] and termed DS, as a solution of constructing a reduced data set. DS approach to squashing is model-free and relies on moment-matching. The squashed data set consists of a set of artificial data points chosen to replicate the moments of the original data within subsets of the actual data. DS studies various approaches to partitioning and ordering the moments and also provides a theoretical justification of their method by considering a Taylor series expansion of an arbitrary likelihood function. Since this relies upon the moments of the data, it should work well for any application in which the likelihood is well-approximated by the first few terms of a Taylor series. In practice, it is only proven with logistic regression.

In [20], the authors proposed the "likelihood-based data squashing" (LDS). LDS is similar to DS because it first partitions the data set and then chooses artificial data points corresponding to each subset of the partition. Nevertheless, the algorithms differ in how they build the partition and how they build the artificial data points. The DS algorithm partitions the data along certain marginal quartiles, and then matches moments. The LDS algorithm partitions the data using a likelihood-based clustering and then selects artificial data points so as to mimic the target sampling or posterior distribution. Both algorithms yield artificial data points with associated weights. The usage of squashed data requires algorithms that can use these weights conveniently. LDS is slightly more general than DS because it is also prepared for ANN-based learning.

A subsequent approach described in [23] presents a form of data squashing based on empirical likelihood. This method re-weights a random sample of data to match certain expected values to the population. The benefits of this method are the reduction of optimization cost in terms of computational complexity and the interest in enhancing the performance of boosted random trees.

6.3.3 Data Clustering

Clustering algorithms partition the data examples into groups, or *clusters*, so that data samples within a cluster are "similar" to one another and different to data examples that belong to other clusters. The similarity is usually defined by means of how near the examples are in space, according to a distance function. The quality of a cluster could be measured as a function of the length of its diameter, which is the maximum distance between any two samples belonging to the cluster. The average distance of each object within the cluster to the centroid is an alternative measure of cluster

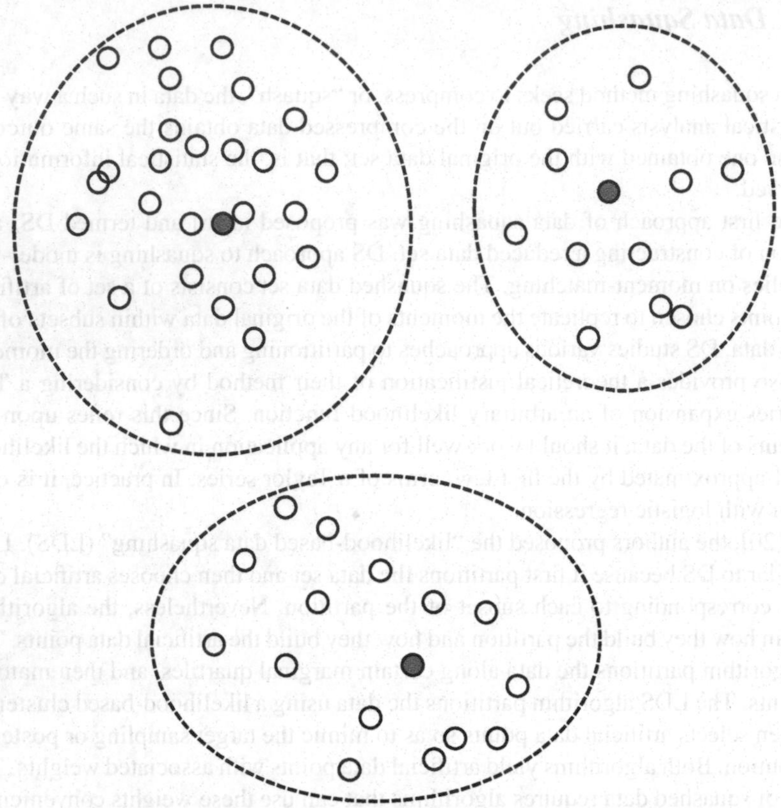

Fig. 6.3 Three clusters derived from a set of two-dimensional data

quality. An illustration of a three cluster derivation from a set of 2-D data points is depicted in Fig. 6.3.

In terms of data reduction, the cluster representations of the data are used instead of the actual data. In many applications, such as those in which data can be organized into distinct groups, this technique is higly effective.

There is a vast number of clustering techniques for defining clusters and for measuring their quality. In fact, clustering is surely the most popular and common form of unsupervised learning in DM, as we have mentioned in Chap. 1 of this book. For this reason, we have included it here due to the clear overlapping that clustering has with data reduction. Unfortunately, this book is not specifically devoted to learning and a deep study on clustering is beyond the scope of this book. However, the reader may consult the following references to an in-depth study: [1, 2, 9, 11, 15, 16, 27].

6.4 Binning and Reduction of Cardinality

Binning is the process of converting a continuous variable into a set of ranges. Then, each range can be treated as categories, with the choice of imposing order on them. This last choice is optional and depends on the further analysis to be made on the data. For example, we can bin the variable representing the annual income of a customer into ranges of 5,000 dollars (0–5,000; 5,001–10,000; 10,001–15,000, ..., etc.). Such a binning could allow the analysis in a business problem may reveal that customers in the first range have less possibility to get a loan than customers in the last range, grouping them within an interval that bounds a numerical variable. Therefore, it demonstrates that keeping the strict order of bins is not always necessary.

Cardinality reduction of nominal and ordinal variables is the process of combining two or more categories into one new category. It is well known that nominal variables with a high number of categories are very problematic to handle. If we perform a transformation of these large cardinality variables onto indicator variables, that is, binary variables that indicate whether or not a category is set for each example; we will produce a large number of new variables, almost all equal to zero. On the other hand, if we do not perform this conversion and use them just as they are in with the algorithm that can tolerate them, such as decision trees, we run into the problem of over-fitting the model. It is realistic to consider reducing the number of categories in such variables.

Both processes are two common transformations used to achieve two objectives:

- Reduce the complexity of independent and possible dependent variables.
- Improve the predictive power of the variable, by carefully binning or grouping the categories in such a way that we model the dependencies regarding the target variable in both estimation and classification problems.

Binning and cardinality reduction are very similar procedures, differing only in the type of variable that we want to process. In fact, both processes are distinctively grouped within the term **discretization**, which constitutes the most popular notation in the literature. It is also very common to distinguish between binning and discretization depending on the ease of the process performed. Binning is usually associated with a quick and easy discretization of a variable. In [11], the authors distinguish among three types of discretization: binning, histogram analysis-based and advanced discretization. The first corresponds to a splitting technique based on the specification of the number of bins. The second family is related with unsupervised discretization and finally, a brief inspection of the rest of the methods is drawn.

Regardless of the above, and under the *discretization* nomenclature, we will discuss all related issues and techniques in Chap. 9 of this book.

References

1. Aggarwal, C., Reddy, C.: Data clustering: recent advances and applications. Chapman and Hall/CRC Data Mining and Knowledge Discovery Series. Taylor & Francis Group, Boca Raton (2013)
2. Aggarwal, C.C., Reddy, C.K. (eds.): Data Clustering: Algorithms and Applications. CRC Press, New York (2014)
3. Batista, G.E.A.P.A., Prati, R.C., Monard, M.C.: A study of the behavior of several methods for balancing machine learning training data. SIGKDD Explor. Newsl. **6**(1), 20–29 (2004)
4. Bellman, R.E.: Adaptive control processes—a guided tour. Princeton University Press, Princeton (1961)
5. Chatfield, C., Collins, A.J.: Introduction to Multivariate Analysis. Chapman and Hall, London (1980)
6. DuMouchel, W., Volinsky, C., Johnson, T., Cortes, C., Pregibon, D.: Squashing flat files flatter. In: Proceedings of the Fifth ACM SIGKDD International Conference on Knowledge Discovery and Data Mining, KDD '99, pp. 6–15 (1999)
7. Dunteman, G.: Principal Components Analysis. SAGE Publications, Newbury Park (1989)
8. Fukunaga, K.: Introduction to Statistical Pattern Recognition, 2nd edn. Academic Press Professional, Inc., San Diego (1990)
9. Gan, G., Ma, C., Wu, J.: Data Clustering—Theory, Algorithms, and Applications. SIAM, Philadelphia (2007)
10. Girolami, M., He, C.: Probability density estimation from optimally condensed data samples. IEEE Trans. Pattern Anal. Mach. Intell. **25**(10), 1253–1264 (2003)
11. Han, J., Kamber, M.: Data Mining: Concepts and Techniques. Morgan Kaufmann Publishers Inc., San Francisco (2011)
12. Hart, P.E.: The condensed nearest neighbor rule. IEEE Trans. Inf. Theory **14**, 515–516 (1968)
13. Hwang, J., Lay, S., Lippman, A.: Nonparametric multivariate density estimation: a comparative study. IEEE Trans. Signal Process. **42**, 2795–2810 (1994)
14. Jain, A., Zongker, D.: Feature selection: evaluation, application, and small sample performance. IEEE Trans. Pattern Anal. Mach. Intell. **19**(2), 153–158 (1997)
15. Jain, A.K., Murty, M.N., Flynn, P.J.: Data clustering: A review. ACM Comput. Surv. **31**(3), 264–323 (1999)
16. Jain, A.K.: Data clustering: 50 years beyond k-means. Pattern Recogn. Lett. **31**(8), 651–666 (2010)
17. Johnson, R.A., Wichern, D.W.: Applied Multivariate Statistical Analysis. Prentice-Hall, Englewood Cliffs (2001)
18. Kim, J.O., Mueller, C.W.: Factor Analysis: Statistical Methods and Practical Issues (Quantitative Applications in the Social Sciences). Sage Publications, Inc, Beverly Hills (1978)
19. Kohonen, T.: The self organizing map. Proc. IEEE **78**(9), 1464–1480 (1990)
20. Madigan, D., Raghavan, N., DuMouchel, W., Nason, M., Posse, C., Ridgeway, G.: Likelihood-based data squashing: a modeling approach to instance construction. Data Min. Knowl. Disc. **6**(2), 173–190 (2002)
21. Mitra, P., Murthy, C.A., Pal, S.K.: Density-based multiscale data condensation. IEEE Trans. Pattern Anal. Mach. Intell. **24**(6), 734–747 (2002)
22. Nisbet, R., Elder, J., Miner, G.: Handbook of Statistical Analysis and Data Mining Applications. Academic Press, Boston (2009)
23. Owen, A.: Data squashing by empirical likelihood. Data Min. Knowl. Disc. **7**, 101–113 (2003)
24. Refaat, M.: Data Preparation for Data Mining Using SAS. Morgan Kaufmann Publishers Inc., San Francisco (2007)
25. Roweis, S., Saul, L.: Nonlinear dimensionality reduction by locally linear embedding. Science **290**(5500), 2323–2326 (2000)
26. Tenenbaum, J.B., Silva, V., Langford, J.C.: A global geometric framework for nonlinear dimensionality reduction. Science **290**(5500), 2319–2323 (2000)
27. Xu, R., Wunsch, D.: Survey of clustering algorithms. IEEE Trans. Neural Networks **16**(3), 645–678 (2005)

Chapter 7
Feature Selection

Abstract In this chapter, one of the most commonly used techniques for dimensionality and data reduction will be described. The feature selection problem will be discussed and the main aspects and methods will be analyzed. The chapter starts with the topics theoretical background (Sect. 7.1), dividing it into the major perspectives (Sect. 7.2) and the main aspects, including applications and the evaluation of feature selections methods (Sect. 7.3). From this point on, the successive sections make a tour from the classical approaches, to the most advanced proposals, in Sect. 7.4. Focusing on hybridizations, better optimization models and derivatives methods related with feature selection, Sect. 7.5 provides a summary on related and advanced topics, such as feature construction and feature extraction. An enumeration of some comparative experimental studies conducted in the specialized literature is included in Sect. 7.6.

7.1 Overview

In Chap. 6, we have seen that dimensionality constitutes a serious obstacle to the competence of most learning algorithms, especially due to the fact that they usually are computationally expensive. Feature selection (FS) is an effective form of dealing with DR.

We have to answer what is the result of FS and why we need FS. For the first question, the effect is to have a reduced subset of features from the original set; for the latter, the purposes can vary: (1) to improve performance (in terms of speed, predictive power, simplicity of the model); (2) to visualize the data for model selection; (3) to reduce dimensionality and remove noise. Combining all these issues, we can define FS as follows [29]:

Definition 7.1 *Feature Selection* is a process that chooses an optimal subset of features according to a certain criterion.

The criterion determines the details of evaluating feature subsets. The selection of the criterion must be done according to the purposes of FS. For example, an optimal subset could be a minimal subset that could give the best estimate of predictive accuracy.

S. García et al., *Data Preprocessing in Data Mining*,
Intelligent Systems Reference Library 72, DOI 10.1007/978-3-319-10247-4_7

Generally, the objective of FS is to identify the features in the data set which are important, and discard others as redundant or irrelevant. Since FS reduces the dimensionality of the data, DM algorithms, especially the predictive ones, can operate faster and obtain better outcomes by using FS. The main reason for this achieved improvement is mainly raised by an easier and more compact representation of the target concept [6].

Reasons for performing FS may include [48]:

- removing irrelevant data;
- increasing predictive accuracy of learned models;
- reducing the cost of the data;
- improving learning efficiency, such as reducing storage requirements and computational cost;
- reducing the complexity of the resulting model description, improving the understanding of the data and the model.

7.2 Perspectives

Although FS is used for all types and paradigms of learning, the most well known and commonly used field is classification. We will focus our efforts mainly on classification. The problem of FS can be explored in many perspectives. The four most important are (1) searching for the best subset of features, (2) criteria for evaluating different subsets, (3) principle for selecting, adding, removing or changing new features during the search and (4) applications.

First of all, FS is considered as a search problem for an optimal subset of features for general or specific purposes, depending on the learning task and kind of algorithm. Secondly, there must be a survey of evaluation criteria to determine proper applications. Third, the method used to evaluate the features is crucial to categorize methods according to the direction of the search process. The consideration of univariate or multivariate evaluation is also a key factor in FS. Lastly, we will specifically study the interaction between FS and classification.

7.2.1 The Search of a Subset of Features

FS can be considered as a search problem, where each state of the search space corresponds to a concrete subset of features selected. The selection can be represented as a binary array, with each element corresponding to the value 1, if the feature is currently selected by the algorithm and 0, if it does not occur. Hence, there should be a total of 2^M subsets where M is the number of features of a data set. A simple case of the search space for three features is depicted in Fig. 7.1. The optimal subset would be between the beginning and the end of this graph.

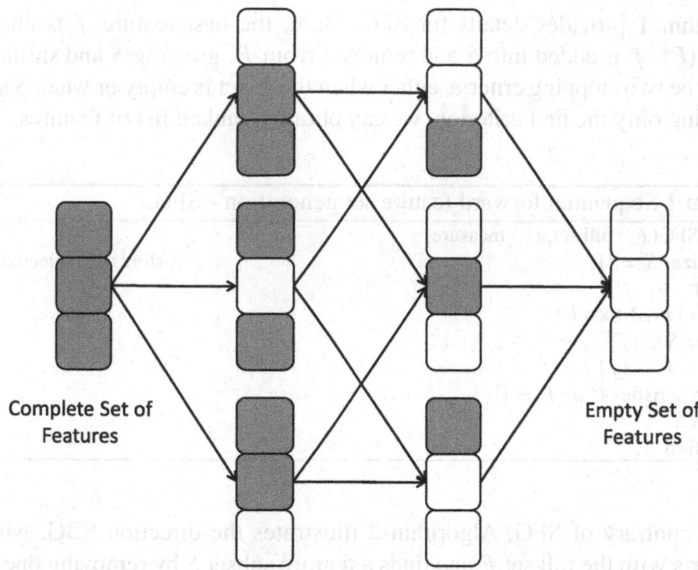

Fig. 7.1 Search space for FS

More realistic problems in DM do not only have three features, thus the search should not start with the full set of features. A search direction must be specified and different search strategies should be adopted to try to achieve optimal subsets in real problems.

7.2.1.1 Search Directions

With no prior knowledge about the problem, the search for a optimal subset can be achieved from the empty set, by inserting new features, or from the full set, by removing features, with the same probability. These two are the directions of search:

- *Sequential Forward Generation (SFG)*: It starts with an empty set of features S. As the search starts, features are added into S according to some criterion that distinguish the best feature from the others. S grows until it reaches a full set of original features. The stopping criteria can be a threshold for the number of relevant features m or simply the generation of all possible subsets in brute force mode.
- *Sequential Backward Generation (SBG)*: It starts with a full set of features and, iteratively, they are removed one at a time. Here, the criterion must point out the worst or least important feature. By the end, the subset is only composed of a unique feature, which is considered to be the most informative of the whole set. As in the previous case, different stopping criteria can be used.

Algorithm 1 provides details for SFG. Here, the best feature f is chosen by **FindNext(F)**. f is added into S and removed from F, growing S and shrinking F. There can be two stopping criteria, either when the F set is empty or when S satisfies U. Adopting only the first criterion, we can obtain a ranked list of features.

Algorithm 1 Sequential forward feature set generation - SFG.

```
function SFG(F - full set, U - measure)
    initialize: S = {}                          ▷ S stores the selected features
    repeat
        f = FINDNEXT(F)
        S = S ∪ {f}
        F = F − {f}
    until S satisfies U or F = {}
    return S
end function
```

To the contrary of SFG, Algorithm 2 illustrates the direction SBG, where the search starts with the full set F and finds a feature subset S by removing one feature at a time. The function **GetNext(F)** finds out the least relevant feature f which will be removed from F. In this case, also F shrinks and S grows, but S only stores the irrelevant features.

Algorithm 2 Sequential backward feature set generation - SBG.

```
function SBG(F - full set, U - measure)
    initialize: S = {}                          ▷ S holds the removed features
    repeat
        f = GETNEXT(F)
        F = F − {f}
        S = S ∪ {f}
    until S does not satisfy U or F = {}
    return F ∪ {f}
end function
```

There are other search directions that base their existence in the usual case in which the optimal subset should be in the middle range of the beginning and the end of the search space. So, it is very intuitive to start from both ends and perform a bidirectional search. The probability of finding the optimal subset is increased because a search in one direction usually finishes faster than in the other direction.

- *Bidirectional Generation (BG)*: Begins the search in both directions, performing SFG and SBG concurrently. They stop in two cases: (1) when one search finds the best subset comprised of m features before it reaches the exact middle, or (2) both searches achieve the middle of the search space. It takes advantage of both SFG and SBG.

When the number of relevant features m is smaller than $M/2$, SFG is faster, otherwise if m is greater than $M/2$, then SBG is faster. As the value of m is usually unknown, it would be impossible to know which scheme would be faster. Thus, the bidirectional generation makes sense, and it is shown in operation in Algorithm 3. In it, SFG and SBG are run in parallel and it stops if either find a satisfactory subset.

Algorithm 3 Bidirectional feature set generation - BG.

```
function BG(F_f, F_b - full set, U - measure)
    initialize: S_f = {}                          ▷ S_f holds the selected features
    initialize: S_b = {}                          ▷ S_b holds the removed features
    repeat
        f_f = FINDNEXT(F_f)
        f_b = GETNEXT(F_b)
        S_f = S_f ∪ {f_f}
        F_b = F_b − {f_b}
        F_f = F_f − {f_f}
        S_b = S_b ∪ {f_b}
    until (a) S_f satisfies U or F_f = {} or (b) S_b does not satisfy U or F_b = {}
    return S_f if (a) or F_b ∪ {f_b} if (b)
end function
```

Finally, there is another search direction used to not pursue any particular direction, instead, the direction is randomly chosen.

- *Random Generation (RG)*: It starts the search in a random direction. The choice of adding or removing a features is a random decision. RG tries to avoid the stagnation into a local optima by not following a fixed way for subset generation. Unlike SFG or SBG, the size of the subset of features cannot be stipulated.

A random generation scheme produces subsets at random. Based on a good random number generator attached with a function called **RandGen(F)** in such a way that every combination of features F has a chance to occur and only once. This scheme is summarized in Algorithm 4, where S is a subset of features.

From now on, we can combine these search directions with a suitable search strategy to design the best FS algorithm for a certain problem we may encounter.

7.2.1.2 Search Strategies

Brute force search doesn't make sense when M is large. The more resources we spend during the search process, the better the subset we may find. However, when the resources are finite, as usual, we have to reduce the optimality of the selected subsets. So, the purpose is to get a good trade-off between this optimality and the lesser quantity of resources required. Since it is not an easy task, three main categories summarize the search strategies:

Algorithm 4 Random feature set generation - RG.

```
function RG(F - full set, U - measure)
    initialize: S = S_best = {}                         ▷ S - subset set
    initialize: C_best = #(F)                           ▷ # - cardinality of a set
    repeat
        S = RANDGEN(F)
        C = #(S)
        if C ≤ C_best and S satisfies U then
            S_best = S
            C_best = C
        end if
    until some stopping criterion is satisfied
    return S_best                                       ▷ Best set found so far
end function
```

- *Exhaustive Search*: It corresponds to explore all possible subsets to find the optimal ones. As we said before, the space complexity is $O(2^M)$. If we establish a threshold m of minimum features to be selected and the direction of search, the search space is $\binom{M}{0} + \binom{M}{1} + \cdots + \binom{M}{m}$, independent of the forward or backward generation. Only exhaustive search can guarantee the optimality, however we can find an optimal subset without visiting all possible states, by using exact algorithms such as backtracking and branch and bound [38]. Nevertheless, they are also impractical in real data sets with a high M.
- *Heuristic Search*: It employs heuristics to carry out the search. Thus, it prevents brute force search, but it will surely find a non-optimal subset of features. It draws a path connecting the beginning and the end in Fig. 7.1, such in a way of a depth-first search. The maximum length of this path is M and the number of subsets generated is $O(M)$. The choice of the heuristic is crucial to find a closer optimal subset of features in a faster operation.
- *Nondeterministic Search*: This third category arises from a complementary combination of the previous two. It is also known as random search strategy and can generate best subsets constantly and keep improving the quality of selected features as time goes by. In each step, the next subset is obtained at random. There are two properties of this type of search strategiy: (1) it is unnecessary to wait until the search ends and (2) we do not know when the optimal set is obtained, although we know which one is better than the previous one and which one is the best at the moment.

7.2.2 Selection Criteria

After studying the essential approaches of search strategies and directions, the next issue to tackle is the measurement of the quality or goodness of a feature. It is necessary to distinguish between best or optimal subset of features and to be common in all FS techniques. Usually, the evaluation metrics work in two ways: (1) supporting the performance in terms of efficacy and (2) supporting the performance in terms of efficiency or yielding more understandable outcomes. For instance, in classification, as the main problem addressed in this chapter, the primary objective is to maximize predictive accuracy.

7.2.2.1 Information Measures

Information serves to measure the uncertainty of the receiver when she/he receives a message. If the receiver understands the message, her/his associated uncertainty is low, but if the receiver is not able to completely understand the message, all messages have almost equal probability of being received and the uncertainty increases. Under the context of predictive learning, the message is the output feature or class in classification. An information measure U is defined so that larger values for U represent higher levels of uncertainty.

Given an uncertainty function U and the prior class probabilities $P(c_i)$ where $i = 1, 2, \ldots, C$, being C the number of classes; the information gain from a feature A, $IG(A)$, is defined as the difference between the prior uncertainty $\sum_i U(P(c_i))$ and the expected posterior uncertainty using A, i.e.,

$$IG(A) = \sum_i U(P(c_i)) - \mathbf{E}\left[\sum_i U(P(c_i|A))\right]$$

where $\mathbf{E}$ represents the expectations. By Bayes' theorem, we have

$$P(c_i|\mathbf{x}) = \frac{P(c_i)P(\mathbf{x}|c_i)}{P(\mathbf{x})}$$

$$P(\mathbf{x}) = \sum P(c_i)P(\mathbf{x}|c_i)$$

A feature evaluation model inferred from the concept of information gain states that feature A_i is chosen instead of feature A_j if $IG(A_i) > IG(A_j)$; that is, if A_i reduces more uncertainty than A_j. Since $\sum_i U(P(c_i))$ is independent of features, we can rewrite the rule as A_i is preferred to A_j if $U'(A_i) < U'(A_j)$, where $U' = \mathbf{E}\left[\sum_i U(P(c_i|A))\right]$. This idea is used in C4.5 [43] for selecting a feature to generate new branches.

A commonly used uncertainty function is Shannon's entropy,

$$-\sum_i P(c_i)\log_2 P(c_i).$$

For example, considering a feature A, data D is split by A into p partitions $D_1, D_2, \ldots, D_p$, and C the number of classes. The information for D at the root amounts to

$$I(D) = -\sum_{i=1}^{C} P_D(c_i) \log_2 P_d(c_i),$$

the information for D_j due to partitioning D at A is

$$I(D_j^A) = -\sum_{i=1}^{C} P_{D_j^A}(c_i) \log_2 P_{D_j^A}(c_i),$$

and the information gain due to the feature A is defined as

$$IG(A) = I(D) - \sum_{j=1}^{p} \frac{|D_j|}{|D|} I(D_j^A),$$

where $|D|$ is the number of instances in D, and $P_D(c_i)$ are the prior probabilities for data D.

Information gain has a tendency to choose features with more distinct values. Instead, information gain ratio was suggested in [43] to balance the effect of many values. It is worthy mentioning that it is only applied to discrete features. For continuous ones, we have to find a split point with the highest gain or gain ratio among the sorted values in order to split the values into two segments. Then, information gain can be computed as usual.

7.2.2.2 Distance Measures

Also known as measures of separability, discrimination or divergence measures . The most typical is derived from distance between the class conditional density functions. For example, in a two-class problem, if $D(A)$ is the distance between $P(A|c_1)$ and $P(A|c_2)$, a feature evaluation rule based on distance $D(A)$ states that A_i is chosen instead A_j if $D(A_i) > D(A_j)$. The rationale behind this is that we try to find the best feature that is able to separate the two classes as far as possible.

Distance functions between the prior and posterior class probabilities are similar to the information gain approach, except that the functions are based on distances instead of uncertainty. Anyway, both have been proposed for feature evaluation.

Two popular distance measures are used in FS: directed divergence DD and variance V. We show their computation expressions as

$$DD(A_j) = \int \left[\sum P(c_i|A_j = a) \log \frac{P(c_i|A_j = a)}{P(c_i)} \right] P(A_j = a) dx.$$

Table 7.1 Distance measures for numeric variables (between X and Y)

	Mathematical form
Euclidean distance	$D_e = \{\sum_{i=1}^{m} (x_i - y_i)^2\}^{\frac{1}{2}}$
City-block distance	$D_{cb} = \sum_{i=1}^{m} \|x_i - y_i\|$
Cebyshev distance	$D_{ch} = \max_i \|x_i - y_i\|$
Minkowski distance of order m	$D_M = \{\sum_{i=1}^{m} (x_i - y_i)^m\}^{\frac{1}{m}}$
Quadratic distance Q, positive definite	$D_q = \sum_{i=1}^{m} \sum_{j=1}^{m} (x_i - y_i) Q_{ij} (x_j - y_j)$
Canberra distance	$D_{ca} = \sum_{i=1}^{m} \frac{\|x_i - y_i\|}{x_i + y_i}$
Angular separation	$D_{as} = \frac{\sum_{i=1}^{m} x_i \cdot y_i}{[\sum_{i=1}^{m} x_i^2 \sum_{i=1}^{m} y_i^2]^{\frac{1}{2}}}$

$$V(A_j) = \int \left[\sum P(c_i)(P(c_i|A_j = a) - P(c_i))^2 \right] P(A_j = a) dx.$$

Moreover, some of the most common distance measures for numeric variables used in FS are summarized in Table 7.1.

7.2.2.3 Dependence Measures

They are also known as measures of association or correlation. Its main goal is to quantify how strongly two variables are correlated or present some association with each other, in such way that knowing the value of one of them, we can derive the value for the other. In feature evaluation, the common procedure is to measure the correlation between any feature with the class. Denoting by $R(A)$ a dependence measure between feature A and class C, we choose feature A_i over feature A_j of $R(A_i) > R(A_j)$. In other words, the feature most correlated with the class is chosen. If A and C are statistically independent, they are not correlated and removing A should not affect the class separability regarding the rest of the features. In a contrary case, the feature should be selected because it could somewhat explain the trend of the class.

One of the most used dependence measures is the *Pearson correlation* coefficient, which measures the degree of linear correlation between two variables. For two variables X and Y with measurements $\{x_i\}$ and $\{y_i\}$, means $\bar{x}$ and $\bar{y}$, this is given by

$$\rho(X, Y) = \frac{\sum_i (x_i - \bar{x})(y_i - \bar{y})}{[\sum_i (x_i - \bar{x})^2 \sum_i (y_i - \bar{y})^2]^{\frac{1}{2}}}$$

If two variables are very correlated ($\rho \approx \pm 1$), one of them could be removed. However, linear correlations are not able to detect relationships that are not linear. Correlations with respect to the target variable can also be computed in order to estimate the relationship between any attribute to the class. This requires the coding of the target class as a binary vector.

Again, these types of measures are closely related to information and distance measures. Features with a strong association with the class are good features for predictive tasks. One of the most used dependence measures is the Bhattacharyya dependence measure B, defined as:

$$B(A_j) = \sum -\log\left[P(c_i)\int\sqrt{P(A_j = a|c_i)P(a_j = a)dx}\right]$$

7.2.2.4 Consistency Measures

The previous measures attempt to find the best features that can explain maximally one class from the others, but are not able to detect whether one of them is redundant. On the other hand, consistency measures attempt to find a minimum number of features that separate classes as the full set of features can. They aim to achieve $P(C|FullSet) = P(C|SubSet)$. Feature evaluation rules derived from consistency measures state that we should select the minimum subset of features that can maintain the consistency of data as observed by the full set of features. An inconsistency is defined as the case of two examples with the same inputs (same feature values) but with different output feature values (classes in classification). Using them, both irrelevant and redundant features can be removed.

7.2.2.5 Accuracy Measures

This form of evaluation relies on the classifier or learner. Among various possible subsets of features, the subset which yields the best predictive accuracy is chosen. This family is distinguished from the previous four due to the fact that is directly focused on improving the accuracy of the same learner used in the DM task. However, we have to take some considerations into account. Firstly, how to truly estimate the predictive accuracy avoiding the problem of over-fitting. Secondly, it is important to contemplate the required time taken by the DM model to complete learning from the data (usually, classifiers perform more complex tasks than the computation of any of the four measures seen above). Lastly, the subset of features could be biased towards an unique model of learning, producing subsets of features that are not generalized.

Table 7.2 summarizes the computation of the accuracy metric and some derivatives that have been used in FS. The notation used is: tp, true positives; fp, false positives; fn, false negatives; tn, true negatives; $tpr = tp/(tp + fn)$, sample true positive rate; $fpr = fp/(fp + tn)$, sample false positive rate; $precision = tp/(tp + fp)$; $recall = tpr$.

Table 7.2 Accuracy metric and derivatives for a two-class (positive class and negative class) problem

	Mathematical form
Accuracy	$\frac{tp+fp}{tp+tn+fp+fn}$
Error rate	$1-$ Accuracy
Chi-squared	$\frac{n(fp \times fn - tp \times tn)^2}{(tp+fp)(tp+fn)(fp+tn)(tn+fn)}$
Information gain	$e(tp+fn, fp+tn) - \frac{(tp+fp)e(tp,fp)+(tn+fn)e(fn,tn)}{tp+fp+tn+fn}$ where $e(x,y) = -\frac{x}{x+y}\log_2\frac{x}{x+y} - \frac{y}{x+y}\log_2\frac{y}{x+y}$
Odds ratio	$\frac{tpr}{1-tpr} \Big/ \frac{fpr}{1-fpr} = \frac{tp \times tn}{fp \times fn}$
Probability ratio	$\frac{tpr}{fpr}$

7.2.3 Filter, Wrapper and Embedded Feature Selection

It is surely the most known and employed categorization made in FS methods for years [33]. In the following, we will detail the three famous categories of feature selectors: filter, wrapper and embedded.

7.2.3.1 Filters

There is an extensive research effort in the development of indirect performance measures, mostly based on the four evaluation measures described before (information, distance, dependency and consistency), for selecting features. This model is called the *filter* model.

The filter approach operates independently of the DM method subsequently employed. The name "filter" proceeds from filtering the undesirable features out before learning. They use heuristics based on general characteristics of the data to evaluate the goodness of feature subsets.

Some authors differentiate a sub-category from filtering called rankers. It includes methods that apply some criteria on which to score each feature and provide a ranking. Using this ordering, the following learning process or user-defined threshold can decide the number of useful features.

The reasons that influence the use of filters are those related to noise removal, data simplification and increasing the performance of any DM technique. They are prepared for dealing with high dimensional data and provide general subsets of features that can be useful for any kind of learning process; rule induction, bayesian models or ANNs .

A filter model of FS consists of two stages (see Fig. 7.2): (1) FS using measures such as information, distance, dependence or consistency, with independence of the learning algorithm; (2) learning and testing, the algorithm learns from the training data with the best feature subset obtained and tested over the test data. Stage 2 is the

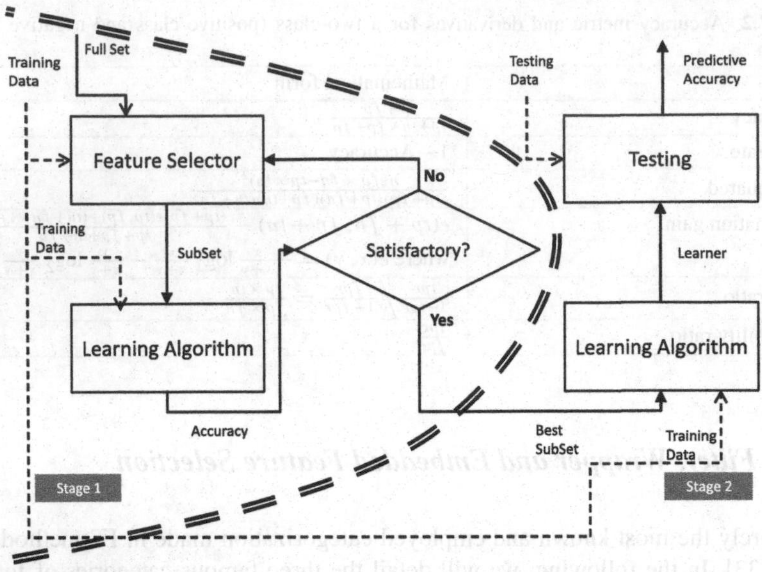

Fig. 7.2 A filter model for FS

usual learning and testing process in which we obtain the predictive accuracy on test data.

The filter model has several properties:

- measuring uncertainty, distances, dependence or consistency is usually cheaper than measuring the accuracy of a learning process. Thus, filter methods are usually faster.
- it does not rely on a particular learning bias, in such a way that the selected features can be used to learn different models from different DM techniques.
- it can handle larger sized data, due to the simplicity and low time complexity of the evaluation measures.

7.2.3.2 Wrappers

One can think that the simplest form of FS consists of engaging a classifier as an evaluator method for deciding the insertion or deletion of a certain feature in the subset, by using any metric for predictive performance. The aim is straightforward; to achieve the highest predictive accuracy possible by selecting the features that accomplish this for a fixed learning algorithm. This model is the so called the *wrapper* model.

In other words, the wrapper approach [23] uses a learning algorithm as a black box together with statistical validation (CV for example) to avoid over-fitting to select the best feature subset, agreeing on a predictive measure.

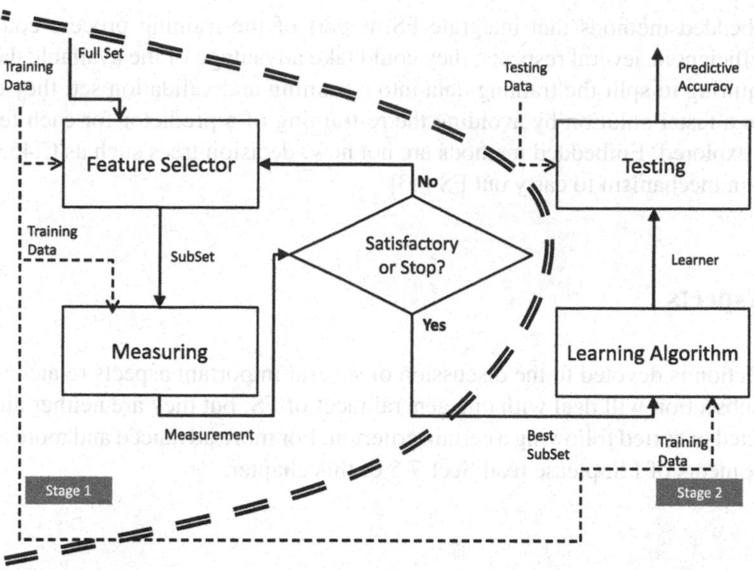

Fig. 7.3 A wrapper model for FS

Wrapper models can achieve the purpose of improving the particular learner's predictive performance. A wrapper model consists of two stages (see Fig. 7.3): (1) feature subset selection, which selects the best subset using the accuracy provided by the learning algorithm (on training data) as a criterion; (2) is the same as in the filter model. Since we keep only the best subset in stage 1, we need to learn again the DM model with the best subset. Stage 1 corresponds with the data reduction task.

It is well-known that the estimated accuracy using the training data may not reflect the same accuracy on test data, so the key issue is how to truly estimate or generalize the accuracy over non-seen data. In fact, the goal is not exactly to increase the predictive accuracy, rather to improve overall the DM algorithm in question. Solutions come from the usage of internal statistical validation to control the over-fitting, ensembles of learners [41] and hybridizations with heuristic learning like Bayesian classifiers or Decision Tree induction [23]. However, when data size is huge, the wrapper approach cannot be applied because the learning method is not able to manage all data.

Regarding filter models, they cannot allow a learning algorithm to fully exploit its bias, whereas wrapper methods do.

7.2.3.3 Embedded Feature Selection

The *embedded* approach [16] is similar to the wrapper approach in the sense that the features are specifically selected for a certain learning algorithm. Moreover, in this approach, the features are selected during the learning process.

Embedded methods that integrate FS as part of the training process could be more efficient in several respects: they could take advantage of the available data by not requiring to split the training data into a training and validation set; they could achieve a faster solution by avoiding the re-training of a predictor for each feature subset explored. Embedded methods are not new: decision trees such as C.45, have a built-in mechanism to carry out FS [43].

7.3 Aspects

This section is devoted to the discussion of several important aspects related to FS. Each subsection will deal with one general facet of FS, but they are neither directly connected or sorted following a certain criterion. For more advanced and more recent developments of FS, please read Sect. 7.5 of this chapter.

7.3.1 Output of Feature Selection

From the point of view of the output of FS methods, they can be grouped into two categories. The first one consists of ranking features according to some evaluation criteria; the other consists of choosing a minimum set of features that satisfy an evaluation criterion. Next, explain both of them in more detail.

7.3.1.1 Feature Ranking Techniques

In this category of methods, we expect as the output a ranked list of features which are ordered according to evaluation measures. The measures can be of any type: information, distance, dependence, consistency or accuracy. Thus, a feature selector belonging to this family does not inform about the minimum subset of features; instead, they return the relevance of the features.

The basic idea consists of evaluating each feature with a measure and attaching the result values to each feature. Then, the features are sorted according to the values. The run time complexity of this algorithm is $O(MN + M^2)$, where M is the number of features and N the number of instances. There are many variations of this algorithm that draw different FS methods. The common property is the outcome based on a ranked list of features. Algorithm 5 summarizes the operation of a univariate feature ranking technique.

For performing actual FS, the simplest way is to choose the first m features for the task at hand, whenever we know the most appropriate m value. But this is not always true, there is not a straightforward procedure to obtain m. Solutions could proceed from building DM models repeatedly until the generalization error is decreased. This

Algorithm 5 A univariate feature ranking algorithm.

```
function RANKING ALGORITHM(x - features, U - measure)
    initialize: list L = {}                          ▷ L stores ordered features
    for each feature x_i, i ∈ {1, ..., M} do
        v_i = COMPUTE(x_i, U)
        position x_i into L according to v_i
    end for
    return L in decreasing order of feature relevance.
end function
```

operation allows us to obtain a minimum subset of features and is usually adopted. However, it is not recommended if the goal is to find the minimum feature subset.

7.3.1.2 Minimum Subset Techniques

The number of relevant features is a parameter that is often not known by the practitioner. There must be a second category of techniques focused on obtaining the minimum possible subset without ordering the features. An algorithm belonging to this category returns a minimum feature subset and no difference is made for features in the subset. So, whatever is relevant within the subset, is otherwise irrelevant.

The minimum subset algorithm is detailed in Algorithm 6. The **subsetGenerate**() function returns a subset following a certain search method, in which a stopping criterion determines when **stop** is set to *true*; function **legitimacy**() returns *true* if subset S_k satisfies measure U. Function **subsetGenerate**() can take one of the generation schemes.

Algorithm 6 A minimum subset algorithm.

```
function MIN- SET ALGORITHM(x - features, U - measure)
    initialize: L = {}, stop = false                 ▷ S holds the minimum set
    repeat
        S_k = SUBSETGENERATE(x)                          ▷ stop can be set here
        if LEGITIMACY(S_k, U) is true and #(S_k) < #(S) then
            S = S_k                                      ▷ S is replaced by S_k
        end if
    until stop = true
    return S - the minimum subset of features
end function
```

7.3.2 *Evaluation*

Several aspects must be taken into account to evaluate a FS method. Among them, it is important to known how well does a feature selector work or the conditions under which it may work. It is normal to compare one method with another or a subset of a previously proposed model to enhance and justify its new benefits and also to comment on it and find out when it does not work.

To measure the concerns described above, one have to appeal to quantitative measures that overall define the performance of a method. Performance can be seen as a list of objectives and, for FS, the list is basically composed by three main goals:

- Inferability: For predictive tasks, assumed as the main purpose for which FS is developed, considered as an improvement of the prediction of unseen examples with respect to the direct usage of the raw training data. In other words, the model or structural representation obtained from the subset of features by the DM algorithms obtained better predictive capability than that built from the original data.
- Interpretability: Again considering predictive tasks, related to the model generated by the DM algorithm. Given the incomprehension of raw data by humans, DM is also used for generating more understandable structure representation that can explain the behavior of the data. It is obvious to pursue the simplest possible structural representation because the simpler a representation is, the easier is to interpret. This goal is at odds with accuracy.
- Data Reduction: Closely related to the previous goal, but in this case referring to the data itself, without involving any DM algorithms. It is better and simpler, from any point of view, to handle data with lower dimensions in terms of efficiency and interpretability. However, evidence shows that it is not true that the greater the reduction of the number of features, the better the understandability.

Our expectation is to increase the three goals mentioned above at the same time. However, it is a multi-objective optimization problem with conflicting sub-objectives, and it is necessary to find a good trade-off depending on the practice or on the application in question. We can derive three assessment measures from these three goals to be evaluated independently:

- Accuracy: It is the most commonly used measure to estimate the predictive power and generalizability of a DM algorithm. A high accuracy shows that a learned model works well on unseen data.
- Complexity: It indirectly measures the interpretability of a model. A model is structured according to a union of simpler elements, thus if the number of such elements is low, the complexity is also low. For instance, a decision tree is composed by branches, leaves and nodes as its basic elements. In a standard decision tree, the number of leaves is equal to the number of branches, although there may be branches of different lengths. The number of nodes in a branch can define the complexity of this branch. Even for each node, the mathematical expression used inside for splitting data can have one or more comparisons or operators. All together, the count of all of these elements may define the complexity of a representation.
- Number of features selected: A measure for assessing the size of the data. Small data sets mean fewer potential hypotheses to be learned, faster learning and simpler results.

Finally, we should also consider two important practical factors in FS:

- Speed of the FS method: It is concerned with the complexity of the FS itself. When dealing with large data sets, some FS techniques are impractical to be run, especially the exhaustive ones. Wrapper methods are usually slower than filters and sometimes it may be crucial to determine the best FS choice under time constraints.
- Generality of the selected features: It is concerned with the case of the estimation of a good subset of features the as general as possible to be used with any DM algorithm. It is a data closer and allows us to detect the most relevant or redundant features of any application. Filters are thought to be more appropriate for this rather than wrapper based feature selectors.

7.3.3 Drawbacks

FS methods, independent of their popularity, have several limitations:

- The resulted subsets of many models of FS (especially those obtained by wrapper-based approaches) are strongly dependent on the training set size. In other words, if the training data set is small, the subset of features returned will also be small, producing a subsequent loss of important features.
- It is not true that a large dimensionality input can always be reduced to a small subset of features because the objective feature (class in classification) is actually related with many input features and the removal of any of them will seriously effect the learning performance.
- A backward removal strategy is very slow when working with large-scale data sets. This is because in the firsts stages of the algorithm, it has to make decisions funded on huge quantities of data.
- In some cases, the FS outcome will still be left with a relatively large number of relevant features which even inhibit the use of complex learning methods.

7.3.4 Using Decision Trees for Feature Selection

The usage of decision trees for FS has one major advantage known as “anytime” [47]. Decision trees can be used to implement a trade-off between the performance of the selected features and the computation time which is required to find a subset. Decision tree inducers can be considered as anytime algorithms for FS, due to the fact that they gradually improve the performance and can be stopped at any time, providing sub-optimal feature subsets. In fact, decision trees have been used as an evaluation methodology for directing the FS search.

7.4 Description of the Most Representative Feature Selection Methods

Based on the characteristics seen before, we can consider some combinations of aspects as a basic component, grouping them to create or characterize a FS method. The procedure to follow is straightforward:

- Select a feature generation scheme.
- Select a search strategy.
- Select an evaluation measure.
- Establish a stopping criterion.
- Combine the above four components.

Although there actually are many possible combinations due to the numerous alternatives for evaluation measures or stopping criteria, we can generalize them to have three major components to categorize combinations. Thus, a 3-tuple can be built, considering three dimensions: Search Direction, with values *Forward*, *Backward* and *Random*; Search strategy, with values *Complete*, *Heuristic* and *Nondeterministic* and Evaluation Measure, with values *Probability*, *Consistency* and *Accuracy*. The bidirectional search direction can be considered as a random scheme and the three evaluation measures of information, dependency and distance are grouped into the category *Probability* since the three can be formulated as computation using prior probabilities from data.

In total, there could be 27 combinations ($3 \times 3 \times 3$), but some of them are impractical. For example, the combination of random with exhaustive search or the combination nondeterministic with forward or backward generation of features. In summary, we have a total of 18 possible combinations. Table 7.3 provides a general framework on this issue. We have to notice that this categorization is thought for classical FS, leaving to some of the advanced methods to be out of this categorization, as we will see in Sect. 7.5.

Table 7.3 All possible combinations for FS algorithms

Search direction	Evaluation measure	Search strategy		
		Exhaustive	Heuristic	Nondeterministic
Forward	Probability	C1	C7	–
	Consistency	C2	C8	–
	Accuracy	C3	C9	–
Backward	Probability	C4	C10	–
	Consistency	C5	C11	–
	Accuracy	C6	C12	–
Random	Probability	–	C13	C16
	Consistency	–	C14	C17
	Accuracy	–	C15	C18

The rest of this section presents some representative and so-considered classical FS algorithms in the specialized literature. The algorithms are described and their components are identified by using the components described above. We categorize them according to the search strategy in the next subsections. Also, we provide a discussion on one variant: feature weighting schemes.

7.4.1 Exhaustive Methods

Exhaustive methods cover the whole search space. We find six combinations (C1–C6) within this category and corresponding to a forward or a backward search (growing or reducing gradually the subset of features selected, respectively) with any kind of evaluation measure which must be satisfied in each step.

In particular, we will detail the combination C2, corresponding with the consistency measure for forward search, respectively. Here, the most famous method is *Focus* [2]. It considers all the combinations among A features starting from an empty subset: $\binom{M}{1}$ subsets first, $\binom{M}{2}$ subsets next, etc. When Focus finds a subset that satisfies the consistency measure, it stops. The details of this method is shown in Algorithm 7. Focus needs to generate $\sum_{i}^{m}\binom{M}{i}$ subsets in order to find a minimum subset of m features that satisfies the consistency criterion. If m is not small, the runtime is quite prohibitive. Heuristics variations of Focus replaces the pure consistency objective with the definition of good subsets.

Algorithm 7 Focus algorithm.

```
function FOCUS(F - all features in data D, U - inconsistency rate as evaluation measure)
  initialize: S = {}
  for i = 1 to M do
    for each subset S of size i do
      if CALU(S,D) = 0 then                    ▷ CalU(S,D) returns inconsistency
        return S - a minimum subset that satisfies U
      end if
    end for
  end for
end function
```

Other exhaustive FS methods deserve to be mentioned in this section. An overview of these methods can be found in [10]:

- Automatic Branch and Bound (ABB) [30], belonging to the C5 class.
- Best First Search (BFS) [60], belonging to the C1 class.
- Beam Search [12], belonging to the C3 class.
- Branch and Bound (BB) [38], belonging to the C4 class.

7.4.2 Heuristic Methods

FS based on heuristics involve most of the existing proposal in specialized literature. Generally, they do not have any expectations of finding an optimal subset with a rapid solution, which tries to be closer to an optimal one. The simplest method is to learn a model and select the features used by the DM algorithm (if the algorithm allows us to do it, like a decision tree or an ANN [49]). It can be viewed as a simple version of a FS belonging to the C12 class.

A more sophisticated version of a C12 feature selector consists of doing a backward selection using a wrapper classifier. A sequential forward selection algorithm (C9) works as follows: begin with an empty set S, in each iteration (with a maximum of M), choose one feature from the unchosen set of features that gives the best accuracy combining with the already chosen features in S. Other hybridizations of search direction, such as bidirectional or floating selection, are described for wrapper methods in [23].

Regarding the other evaluation measures, we can find some proposals in literature:

- SetCover [10] belongs to the C8 class.
- The set of algorithms presented in [37] belong to the C7 class.
- The algorithm proposed in [24] belongs to the C10 class, specifically by using information theory measures.

How about combinations C13, C14 and C15? In them, the features are randomly generated, but the search is heuristic. One implementation of these combinations is adopting a heuristic search algorithm and in each sub-search space, randomly generate the features and these subsets of features form the possible sub-search spaces.

One of the heuristic methods that deserves mention is the *Mutual Information based FS (MIFS)* [5], which is based on the single computation of the MI measure between two features at the same time, and replacing the impossible exhaustive search with a greedy algorithm. Given already selected features, the algorithm chooses the next feature as the one that maximizes information about the class to the average MI with the selected features. MIFS is described in the Algorithm 8, and belongs to the C10 category.

7.4.3 Nondeterministic Methods

Also known as stochastic methods, they add or remove features to and from a subset without a sequential order, allowing the search to follow feature subsets that are randomly generated.

Common techniques in this category are genetic algorithms and simulated annealing . In [18], a comparison of genetic algorithms with sequential methods is presented, it is very difficult to fairly compare them because of the parameter adjustment. The

Algorithm 8 MIFS algorithm.

```
function MIFS(F - all features in data, S - set of selected features, k - desired size of S, β -
    regulator parameter)
    initialize: S = {}
    for each feature f_i in F do
        Compute I(C, f_i)
    end for
    Find f_max that maximizes I(C, f)
    F = F − {f_max}
    S = S ⋃ f_max
    repeat
        for all couples of features (f_i ∈ F, s_j ∈ S) do
            Compute I(f_i, s_j)
        end for
        Find f_max that maximizes I(C, f) − β ∑_{s∈S} I(f_i, s_j)
        F = F − {f_max}
        S = S ⋃ f_max
    until |S| = k
    return S
end function
```

three combinations C16, C17 and C18 constitute the stochastic methods differing on the evaluation measure. In them, features are randomly generated and a fitness function, closely related to the evaluation measures, is defined.

As the most typical stochastic techniques, we will discuss two methods here: LVF (C17) and LVW (C18).

LVF is the acronym of *Las Vegas Filter* FS [31]. It consists of a random procedure that generates random subsets of features and an evaluation procedure that checks if each subset satisfies the chosen measure. For more details about LVF, see the Algorithm 7. The evaluation measure used in LVF is inconsistency rate. It receives as parameter the *allowed inconsistency rate*, that can be estimated by the inconsistency rate of the data considering all features. The other parameter is the maximum number of subsets to be generated in the process, which acts as a stopping criterion.

In Algorithm 9, *maxTries* is a number proportional to the number of original features (i.e., $l \times M$, being l a pre-defined constant). The rule of thumb is that the more features the data has, the more difficult the FS task is. Another way for setting *maxTries* is to relate to the size of the search space we want to explore. If the complete search space is 2^M and if we want to cover a $p\,\%$ of the entire space, then $l = 2^M \cdot p\,\%$.

LVF can be easily modified due to the simplicity of the algorithm. Changing the evaluation measure is the only thing we can do to keep it within this category. If we decide to use the accuracy measure, we will obtain the LVW (*Las Vegas Wrapper* FS) method. For estimating the classifier's accuracy, we usually draw on statistical validation, such as 10-FCV. LVF and LVW could be very different in terms of run time as well as subsets selected. Another difference of LVW regarding LVF is that the learning algorithm *LA* requires its input parameters to be set. Function **estimate()** in

Algorithm 9 LVF algorithm.

```
function LVF(D - a data set with M features, U - the inconsistency rate, maxTries - stopping
    criterion, γ - an allowed inconsistency rate)
    initialize: list L = {}                                ▷ L stores equally good sets
    C_best = M
    for maxTries iterations do
        S = RANDOMSET(seed)
        C = #(S)                                           ▷ # - the cardinality of S
        if C < C_best and CALU(S,D) < γ then
            S_best = S
            C_best = C
            L = {S}                                        ▷ L is reinitialized
        else if C = C_best and CALU(S,D) < γ then
            L = APPEND(S,L)
        end if
    end for
    return L                          ▷ all equivalently good subsets found by LVF
end function
```

LVW replaces the function **CalU()** in LVF (see Algorithm 10). Probabilistic measures (category C16) can also be used in LVF.

Algorithm 10 LVW algorithm.

```
function LVW(D - a data set with M features, LA - a learning algorithm, maxTries - stopping
    criterion, F - a full set of features)
    initialize: list L = {}                      ▷ L stores sets with equal accuracy
    A_best = ESTIMATE(D,F,LA)
    for maxTries iterations do
        S = RANDOMSET(seed)
        A = ESTIMATE(D,S,LA)                               ▷ # - the cardinality of S
        if A > A_best then
            S_best = S
            A_best = A
            L = {S}                                        ▷ L is reinitialized
        else if A = A_best then
            L = APPEND(S,L)
        end if
    end for
    return L                          ▷ all equivalently good subsets found by LVW
end function
```

7.4.4 Feature Weighting Methods

This is a variation of FS and it is closely related to some related work described in Chap. 8 regarding IS, lazy learning [1] and similarity measures . The *Relief* algorithm must be mentioned in a FS topic review. Relief was proposed in [22] and selects features that are statistically relevant. Although its goal is still selecting features, it does not explicitly generate feature subsets and test them like the methods reviewed above. Instead of generating feature subsets, Relief focuses on sampling instances without an explicit search for feature subsets. This follows the idea that relevant features are those whose values can distinguish among instances that are close to each other. Hence, two nearest neighbors (belonging to different classes in a two-class problem) are found for each given instance I, one is the so-called near-hit H and the other is near-miss J. We expect a feature to be relevant if its values are the same between I and H, and different between I and J. This checking can be carried out in terms of some distance between feature's values, which should be minimum for I and H and maximum for I and J. The distance of each feature for each randomly chosen instance is accumulated in a weight vector $\mathbf{w}$ of the same number of dimensions as the number of features. The relevant features are those having their weights exceeding a relevance threshold τ, which can be statistically estimated. The parameter m is the sample size and larger m produces a more reliable approximation. The algorithm is presented in Algorithm 11. It does not fit into any of the categories described in the previous section, although it evaluates a feature using distance measures.

Algorithm 11 Relief algorithm.

```
function RELIEF(x - features, m - number of instances sampled, τ - relevance threshold)
    initialize: w = 0
    for i = 1 to m do
        randomly select an instance I
        find nearest-hit H and nearest-miss J
        for j = 1 to M do
            w(j) = w(j) − dist(j, I, H)^2/m + dist(j, I, J)^2/m    ▷ dist is a distance function
        end for
    end for
    return w greater than τ
end function
```

The main advantage of Relief is that it can handle discrete and continuous data, by using distance measures which can work with categorical values. On the other hand, its main weakness is that it is limited to two-class data, although some extensions for multiple classes have been proposed, such as ReliefF [25].

7.5 Related and Advanced Topics

This section is devoted to highlighting some recent developments on FS and to shortly discuss related paradigms such as feature extraction (Sect. 7.5.2) and feature construction (Sect. 7.5.3). It is noteworthy to mention that the current state of specialized literature is quite chaotic and is composed by hundreds of proposals, ideas and applications related to FS. In fact, FS has surely been the most well known technique for data preprocessing and data reduction for years, being also the most hybridized with many DM tasks and paradigms. As it would be impossible to summarize all of the literature on the topic, we will focus our efforts on the most successful and popular approaches.

7.5.1 Leading and Recent Feature Selection Techniques

FS is, for most researchers, the basic data preprocessing technique, especially after the year 2000. Unfortunately, the related literature is huge, quite chaotic and difficult to understand or categorize the differences among the hundreds of algorithms published, due to the different conventions or notations adopted. These are the major reasons that disable the possibility of summarizing all the feature selectors proposed in this book. Instead of describing individual approaches, we prefer to focus attention on the main ideas that lead to updates and improvements with respect to the classical FSs methods reviewed in the previous sections. We intend to describe the most influential methods and ideas (which are usually published in highly cited papers) and the most recent and promising techniques published in high quality journals on DM, ML and Pattern Recognition fields.

Modifications of classical feature selectors cover a vast number of proposals in the literature. Among most of the representatives, we could emphasize some relevant approaches. For example, in [28], the authors proposed an extension of the MIFS algorithm under uniform information distribution (MIFS-U), and the combination of the greedy search of MIFS with Taguchi method. The same authors, in [27] presented a speed up MIFS based on Parzen windows, allowing the computation of MI without requiring a large amount of memory. Advances on MI are the minimal-redundancy-maximal-relevance (mRMR) criterion for incremental FS [42]. Behind the idea that, in traditional feature selectors, MI is estimated on the whole sampling space, the authors in [32], proposed the evaluation by dynamic MI, which is only estimated on unlabeled instances. The normalized mutual information FS (NMIFS) is proposed in [14] as an enhancement over classical MIFS, MIFS-U, and mRMR methods. Here, the average normalized MI is proposed as a measure of redundancy among features. A unifying framework for information theoretic FS can be found in [8]. Another method widely studied is the *Relief* and its derivatives. In [45], a theoretical and empirical analysis of this family of methods is conducted, concluding that they are robust and noise tolerant, besides the can alleviate their computational complexity by parallelism. Wrapper methods have been extensively studied by using classifiers such as SVMs [34], or frameworks to jointly perform FS and SVM parameter learning [39].

Other criteria related to separability measures and recently developed for performing FS include the kernel class separability [57], which has been applied to a variety of selection modes and different search strategies. In [15], the authors propose two subspace based separability measures to determine the individual discriminatory power of the features, namely the common subspace measure and Fisher subspace measure, which can easily be used for detecting the discrimination capabilities for FS. After demonstrating that the existence of sufficiently correlated features can always prevent selecting the optimal feature set, in [64], the redundancy-constrained FS (RCFS) method was proposed. Recent studies include FS via dependence maximization [51], using the Hilbert-Schmidt independence criterion. Furthermore, the similarity preserving FS was presented in [62].

The use of meta-heuristics is widely extended in FS. In [44], a genetic algorithm is employed to optimize a vector of feature weights with the KNN classifier allowing both FS and extraction tasks. A tabu search algorithm is introduced in [61], using 0/1 bit string for representing solutions and an evaluation measure based on error rates. More advanced hybridizations of genetic algorithms with local search operations have been also applied to FS [40]. Similar to the one previoulsy mentioned, the approach defined in [65] combines a wrapper-based genetic algorithm with a filter-based local search. An iterative version of *Relief*, called I-RELIEF, is proposed in [52] by exploring the framework of the EM algorithm.

One of the most successful paradigms used in FS is the Rough Sets theory. Since the appearance of the application of rough sets in pattern recognition [54], lots of FS methods have based their evaluation criteria in reducts and approximations according to this theory. Due to the fact that complete searches are not feasible for large sized data sets, the stochastic approaches based on meta-heuristics combined with rough sets evaluation criteria have been also analyzed. In particular, Particle Swarm optimization has been used for this task [58]. However, the main limitation of rough set-based attribute selection in the literature is the restrictive requirement that all data is discrete. For solving this problem, the authors in [20] proposed an approach based on fuzzy-rough sets, fuzzy rough FS (FRFS). In a later paper, in [9], a generalization of the FS based on rough sets is showed using fuzzy tolerance relations. Another way of evaluating numerical features is to generalize the model with neighborhood relations and introduce a neighborhood rough set model [17]. The neighborhood model is used to reduce numerical and categorical features by assigning different thresholds for different kinds of attributes.

The fusion of filters and wrappers in FS has also been studied in the literature. In [56], the evaluation criterion merges dependency, coefficients of correlations and error estimation by KNN. As we have mentioned before, the memetic FS algorithm proposed in [65] also combines wrapper and filter evaluation criteria. The method GAMIFS [14] can be viewed as a genetic algorithm to form an hybrid filter/wrapper feature selector. On the other hand, the fusion of predictive models in form of ensembles can generate a compact subset of non-redundant features [55] when data is wide, dirty, mixed with both numerical and categorical predictors, and may contain interactive effects that require complex models. The algorithm proposed here follows a process divided into four stages and considers a Random Forest ensemble: (1) identification of important variables, (2) computation of masking scores, (3) removal of masked variables and (4) generation of residuals for incremental adjustment.

With the proliferation of extremely high-dimensional data, two issues occur at the same time: FS becomes indispensable in any learning process and the efficiency and stability of FS algorithms could be neglected. One of the earlier studies regarding this issue can be found in [21]. The reduction of the FS task to a quadratic optimization problem is addressed in [46]. In that paper, the authors presented the Quadratic Programming FS (QPFS) that uses the Nystrӧn method for approximate matrix diagonalization, making it possible to deal with very large data sets. In their experiments, it outperformed mRMR and ReliefF using two evaluation criteria: Pearson's correlation coefficient and MI . In the presence of a huge number of irrelevant features and complex data distributions, a local learning based approach could be useful [53]. Using a prior stage for eliminating class-dependent density-based features for the feature ranking process can alleviate the effects of high-dimensional data sets [19]. Finally, and closely related to the emerging Big Data solutions for large-scale business data, there is a recent approach for massively parallel FS described in [63]. High-performance distributed computing architectures, such as Message Passing Interface (MPI) and MapReduce are being applied to scale any kind of algorithms to large data problems.

When class labels of the data are available, we can use supervised FS, otherwise the unsupervised FS is the appropriate. This family of methods usually involve the maximization of a clustering performance or the selection of features based on feature dependence, correlation and relevance. The basic principle is to remove those features carrying little or no additional information beyond that subsumed by the rest of features. For instance, the proposal presented in [35] uses feature dependency/similarity for redundancy reduction, without requiring any search process. The process follows a clustering partitioning based on features and it is governed by a similarity measure called maximal information compression index. Other algorithms for unsupervised FS are the forward orthogonal search (FOS) [59] whose goal is to maximize the overall dependency on the data to detect significant variables. Ensemble learning was also used in unsupervised FS [13]. In clustering, Feature Weighting has also been applied with promising results [36].

7.5.2 Feature Extraction

In feature extraction, we are interested in finding new features that are calculated as a function of the original features. In this context, DR is a mapping of a multidimensional space into a space of fewer dimensions.

The reader should now be reminded that in Chap. 6 we denoted these techniques as DR techniques. The rationale behind this is that the literature has adopted this term in greater extent than feature extraction, although both designations are correct. In fact, the FS is a sub-family of the DR techniques, which seems logical. In this book, we have preferred to separate FS from the general DR task due to its influence in the research community. Furthermore, the aim of this section is to establish a link between the corresponding sections of Chap. 6 with the FS task.

As we have discussed in Chap. 6, *PCA*, *factor analysis*, *MDS* and *LLE*, are the most relevant techniques proposed in this field.

7.5.3 Feature Construction

The feature construction emerged from the replication problem observed in the models produced by DM algorithms. See for example the case of subtrees replication in decision tree based learning. The main goal was to attach to the algorithms some mechanism to compound new features from the original ones endeavouring to improve accuracy and the decrease in model complexity.

The definition of feature construction as a data preprocessing task is the application of a set of constructive operators to a set of existing features, resulting in the generation of new features intended for use in the description of the target concept. Due to the fact that the new features are constructed from the existing ones, no new information is yielded. They have been extensively applied on separate-and-conquer predictive learning approaches.

Many constructive operators have been designed and implemented. The most common operator used in decision trees is the *product* (see an illustration on the effect of this operator in Fig. 7.4). Other operators are equivalent (the value is *true* if two features $x = y$, and *false* otherwise), inequalities, maximum, minimum, average, addition, subtraction, division, count (which estimates the number of features satisfying a ceratin condition), and many more.

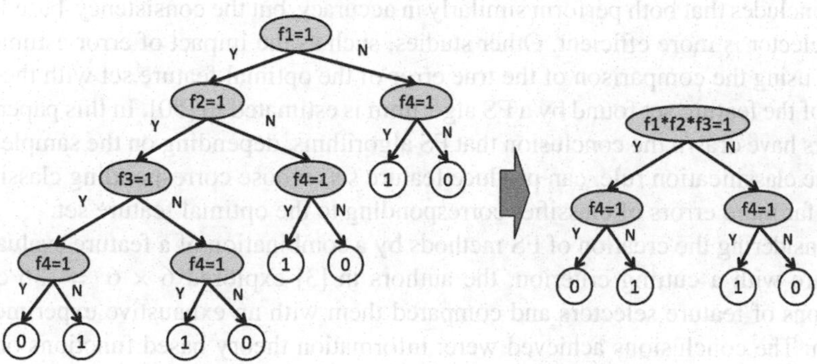

Fig. 7.4 The effect of using the product of features in decision tree modeling

7.6 Experimental Comparative Analyses in Feature Selection

As in the previous case, FS is considered as the data preprocessing technique in which more effort has been invested, resulting in a huge collection of papers and proposals that can be found in the literature. Thus, in this section we will refer to well-known comparative studies that have involved a large set of FS methods.

The first exhaustive comparison was done in [26], where the authors chose the 1-NN classifier and compare classical FS methods; a total of 18 methods, including 6 different versions of backward and forward selection, 2 bidirectional methods, 8 alternatives of branch and bounds methods and 2 genetic algorithm based approaches, a sequential and a parallel algorithm. The main conclusions point to the use of the bidirectional approaches for small and medium scale data sets (less than 50 features), the application of exhaustive methods such us branch and bound techniques being permissible up to medium scale data sets and the suitability of genetic algorithms for large-scale problems.

Regarding studies based on evaluation measures, we stress the ones devoted to the inconsistency criterion. In [11], this simple measure is compared with others under different search strategies as described in this chapter. The main characteristics extracted from this measure is that it is monotonic, fast, multivariate, able to remove redundant and/or irrelevant features, and capable of handling some noise. Using consistency in exhaustive, complete, heuristic, probabilistic and hybrid searches shows us the fact that it does not incorporate any search bias with regards to a particular classifier, enabling it to be used with a variety of different learning algorithms. In addition, in [4], the state of the art of consistency based FS methods is reviewed. An empirical evaluation is then conducted comparing them with wrapper approaches, and concludes that both perform similarly in accuracy, but the consistency-based feature selector is more efficient. Other studies, such as the impact of error estimation on FS using the comparison of the true error of the optimal feature set with the true error of the feature set found by a FS algorithm is estimated in [50]. In this paper, the authors have drawn the conclusion that FS algorithms, depending on the sample size and the classification rule, can produce feature sets whose corresponding classifiers cause far more errors of classifier corresponding to the optimal feature set.

Considering the creation of FS methods by a combination of a feature evaluation measure with a cutting criterion, the authors in [3] explored $6 \times 6 = 36$ combinations of feature selectors and compared them with an exhaustive experimental design. The conclusions achieved were: information theory based functions obtain better accuracy results; no cutting criterion can be generally recommended, although those independent from the measure are the best; and results vary among learners, recommending wrapper approaches for each kind of learner.

The use of synthetic data for studying the performance of FS methods has been addressed in [7]. The rationale behind this methodology is to analyze the methods in presence of a crescent number of irrelevant features, noise in the data, redundancy and interaction between attributes, as well as the ratio between number of instances and features. A total of nine feature selectors run over 11 artificial data sets are involved in the experimental study. According to them, ReliefF turned out to be the best option independent of the particulars of the data, adding that it is a filter with low computational cost. Like in the study mentioned before, wrapper approaches have proven to be an interesting choice in some domains, provided they can be applied with the same classifiers and taking into account that they require higher computational costs.

References

1. Aha, D.W. (ed.): Lazy Learning. Springer, Berlin (2010)
2. Almuallim, H., Dietterich, T.G.: Learning with many irrelevant features. In: Proceedings of the Ninth National Conference on Artificial Intelligence, pp. 547–552 (1991)
3. Arauzo-Azofra, A., Aznarte, J., Benítez, J.: Empirical study of feature selection methods based on individual feature evaluation for classification problems. Expert Syst. Appl. **38**(7), 8170–8177 (2011)
4. Arauzo-Azofra, A., Benítez, J., Castro, J.: Consistency measures for feature selection. J. Intell. Inf. Syst. **30**(3), 273–292 (2008)
5. Battiti, R.: Using mutual information for selection features in supervised neural net learning. IEEE Trans. Neural Netw. **5**(4), 537–550 (1994)
6. Blum, A.L., Langley, P.: Selection of relevant features and examples in machine learning. Artif. Intell. **97**(1–2), 245–271 (1997)
7. Bolón-Canedo, V., Sánchez-Maroño, N., Alonso-Betanzos, A.: A review of feature selection methods on synthetic data. Knowl. Inf. Syst. **34**(3), 483–519 (2013)
8. Brown, G., Pocock, A., Zhao, M.J., Luján, M.: Conditional likelihood maximisation: A unifying framework for information theoretic feature selection. J. Mach. Learn. Res. **13**, 27–66 (2012)
9. Cornelis, C., Jensen, R., Hurtado, G., Slezak, D.: Attribute selection with fuzzy decision reducts. Inf. Sci. **180**(2), 209–224 (2010)
10. Dash, M., Liu, H.: Feature selection for classification. Intell. Data Anal. **1**(3), 131–156 (1997)
11. Dash, M., Liu, H.: Consistency-based search in feature selection. Artif. Intell. **151**(1–2), 155–176 (2003)
12. Doak, J.: An Evaluation of Feature Selection Methods and Their Application to Computer Security. Tech. rep, UC Davis Department of Computer Science (1992)
13. Elghazel, H., Aussem, A.: Unsupervised feature selection with ensemble learning. Machine Learning, pp. 1–24. Springer, Berlin (2013)
14. Estévez, P., Tesmer, M., Perez, C., Zurada, J.: Normalized mutual information feature selection. IEEE Trans. Neural Netw. **20**(2), 189–201 (2009)
15. Gunal, S., Edizkan, R.: Subspace based feature selection for pattern recognition. Inf. Sci. **178**(19), 3716–3726 (2008)
16. Guyon, I., Elisseeff, A.: An introduction to variable and feature selection. J. Mach. Learn. Res. **3**, 1157–1182 (2003)
17. Hu, Q., Yu, D., Liu, J., Wu, C.: Neighborhood rough set based heterogeneous feature subset selection. Inf. Sci. **178**(18), 3577–3594 (2008)
18. Jain, A.: Feature selection: evaluation, application, and small sample performance. IEEE Trans. Pattern Anal. Mach. Intell. **19**(2), 153–158 (1997)
19. Javed, K., Babri, H., Saeed, M.: Feature selection based on class-dependent densities for high-dimensional binary data. IEEE Trans. Knowl. Data Eng. **24**(3), 465–477 (2012)
20. Jensen, R., Shen, Q.: Fuzzy-rough sets assisted attribute selection. IEEE Trans. Fuzzy Syst. **15**(1), 73–89 (2007)
21. Kalousis, A., Prados, J., Hilario, M.: Stability of feature selection algorithms: A study on high-dimensional spaces. Knowl. Inf. Syst. **12**(1), 95–116 (2007)

22. Kira, K., Rendell, L.A.: A practical approach to feature selection. In: Proceedings of the Ninth International Workshop on Machine Learning, ML92, pp. 249–256 (1992)
23. Kohavi, R., John, G.: Wrappers for feature subset selection. Artif. Intell. **97**(1–2), 273–324 (1997)
24. Koller, D., Sahami, M.: Toward optimal feature selection. In: Proceedings of the Thirteenth International Conference on Machine Learning, pp. 284–292 (1996)
25. Kononenko, I.: Estimating attributes: Analysis and extensions of relief. In: Proceedings of the European Conference on Machine Learning on Machine Learning, ECML-94, pp. 171–182 (1994)
26. Kudo, M., Sklansky, J.: Comparison of algorithms that select features for pattern classifiers. Pattern Recognit. **33**(1), 25–41 (2000)
27. Kwak, N., Choi, C.H.: Input feature selection by mutual information based on parzen window. IEEE Trans. Pattern Anal. Mach. Intell. **24**(12), 1667–1671 (2002)
28. Kwak, N., Choi, C.H.: Input feature selection for classification problems. IEEE Trans. Neural Netw. **13**(1), 143–159 (2002)
29. Liu, H., Motoda, H.: Feature Selection for Knowledge Discovery and Data Mining. Kluwer Academic, USA (1998)
30. Liu, H., Motoda, H., Dash, M.: A monotonic measure for optimal feature selection. In: Proceedings of European Conference of Machine Learning, Lecture Notes in Computer Science vol. 1398, pp. 101–106 (1998)
31. Liu, H., Setiono, R.: A probabilistic approach to feature selection - a filter solution. In: Proceedings of the International Conference on Machine Learning (ICML), pp. 319–327 (1996)
32. Liu, H., Sun, J., Liu, L., Zhang, H.: Feature selection with dynamic mutual information. Pattern Recognit. **42**(7), 1330–1339 (2009)
33. Liu, H., Yu, L.: Toward integrating feature selection algorithms for classification and clustering. IEEE Trans. Knowl. Data Eng. **17**(4), 491–502 (2005)
34. Maldonado, S., Weber, R.: A wrapper method for feature selection using support vector machines. Inf. Sci. **179**(13), 2208–2217 (2009)
35. Mitra, P., Murthy, C., Pal, S.: Unsupervised feature selection using feature similarity. IEEE Trans. Pattern Anal. Mach. Intell. **24**(3), 301–312 (2002)
36. Modha, D., Spangler, W.: Feature weighting on k-means clustering. Mach. Learn. **52**(3), 217–237 (2003)
37. Mucciardi, A.N., Gose, E.E.: A Comparison of Seven Techniques for Choosing Subsets of Pattern Recognition Properties, pp. 1023–1031. IEEE, India (1971)
38. Narendra, P.M., Fukunaga, K.: A branch and bound algorithm for feature subset selection. IEEE Trans. Comput. **26**(9), 917–922 (1977)
39. Nguyen, M., de la Torre, F.: Optimal feature selection for support vector machines. Pattern Recognit. **43**(3), 584–591 (2010)
40. Oh, I.S., Lee, J.S., Moon, B.R.: Hybrid genetic algorithms for feature selection. IEEE Trans. Pattern Anal. Mach. Intell. **26**(11), 1424–1437 (2004)
41. Opitz, D.W.: Feature selection for ensembles. In: Proceedings of the National Conference on Artificial Intelligence, pp. 379–384 (1999)
42. Peng, H., Long, F., Ding, C.: Feature selection based on mutual information: Criteria of max-dependency, max-relevance, and min-redundancy. IEEE Trans. Pattern Anal. Mach. Intell. **27**(8), 1226–1238 (2005)
43. Quinlan, J.R.: C4.5: Programs for Machine Learning. Morgan Kaufmann Publishers, USA (1993)
44. Raymer, M., Punch, W., Goodman, E., Kuhn, L., Jain, A.: Dimensionality reduction using genetic algorithms. IEEE Trans. Evolut. Comput. **4**(2), 164–171 (2000)
45. Robnik-Łikonja, M., Kononenko, I.: Theoretical and empirical analysis of relieff and rrelieff. Mach. Learn. **53**(1–2), 23–69 (2003)
46. Rodriguez-Lujan, I., Huerta, R., Elkan, C., Cruz, C.: Quadratic programming feature selection. J. Mach. Learn. Res. **11**, 1491–1516 (2010)

47. Rokach, L.: Data Mining with Decision Trees: Theory and Applications. Series in Machine Perception and Artificial Intelligence. World Scientific Publishing, USA (2007)
48. Saeys, Y., Inza, I., Larrañaga, P.: A review of feature selection techniques in bioinformatics. Bioinformatics **23**(19), 2507–2517 (2007)
49. Setiono, R., Liu, H.: Neural-network feature selector. IEEE Trans. Neural Netw. **8**(3), 654–662 (1997)
50. Sima, C., Attoor, S., Brag-Neto, U., Lowey, J., Suh, E., Dougherty, E.: Impact of error estimation on feature selection. Pattern Recognit. **38**(12), 2472–2482 (2005)
51. Song, L., Smola, A., Gretton, A., Bedo, J., Borgwardt, K.: Feature selection via dependence maximization. J. Mach. Learn. Res. **13**, 1393–1434 (2012)
52. Sun, Y.: Iterative relief for feature weighting: Algorithms, theories, and applications. IEEE Trans. Pattern Anal. Mach. Intell. **29**(6), 1035–1051 (2007)
53. Sun, Y., Todorovic, S., Goodison, S.: Local-learning-based feature selection for high-dimensional data analysis. IEEE Trans. Pattern Anal. Mach. Intell. **32**(9), 1610–1626 (2010)
54. Swiniarski, R., Skowron, A.: Rough set methods in feature selection and recognition. Pattern Recognit. Lett. **24**(6), 833–849 (2003)
55. Tuv, E., Borisov, A., Runger, G., Torkkola, K.: Feature selection with ensembles, artificial variables, and redundancy elimination. J. Mach. Learn. Res. **10**, 1341–1366 (2009)
56. Uncu, O., Trksen, I.: A novel feature selection approach: Combining feature wrappers and filters. Inf. Sci. **177**(2), 449–466 (2007)
57. Wang, L.: Feature selection with kernel class separability. IEEE Trans. Pattern Anal. Mach. Intell. **30**(9), 1534–1546 (2008)
58. Wang, X., Yang, J., Teng, X., Xia, W., Jensen, R.: Feature selection based on rough sets and particle swarm optimization. Pattern Recognit. Lett. **28**(4), 459–471 (2007)
59. Wei, H.L., Billings, S.: Feature subset selection and ranking for data dimensionality reduction. IEEE Trans. Pattern Anal. Mach. Intell. **29**(1), 162–166 (2007)
60. Xu, L., Yan, P., Chang, T.: Best first strategy for feature selection. In: Proceedings of the Ninth International Conference on Pattern Recognition, pp. 706–708 (1988)
61. Zhang, H., Sun, G.: Feature selection using tabu search method. Pattern Recognit. **35**(3), 701–711 (2002)
62. Zhao, Z., Wang, L., Liu, H., Ye, J.: On similarity preserving feature selection. IEEE Trans. Knowl. Data Eng. **25**(3), 619–632 (2013)
63. Zhao, Z., Zhang, R., Cox, J., Duling, D., Sarle, W.: Massively parallel feature selection: An approach based on variance preservation. Mach. Learn. **92**(1), 195–220 (2013)
64. Zhou, L., Wang, L., Shen, C.: Feature selection with redundancy-constrained class separability. IEEE Trans. Neural Netw. **21**(5), 853–858 (2010)
65. Zhu, Z., Ong, Y.S., Dash, M.: Wrapper-filter feature selection algorithm using a memetic framework. IEEE Trans. Syst. Man Cybern. Part B: Cybern. **37**(1), 70–76 (2007)

Chapter 8
Instance Selection

Abstract In this chapter, we consider instance selection as an important focusing task in the data reduction phase of knowledge discovery and data mining. First of all, we define a broader perspective on concepts and topics related with instance selection (Sect. 8.1). Due to the fact that instance selection has been distinguished over the years as two type of tasks, depending on the data mining method applied later, we clearly separate it into two processes: training set selection and prototype selection. Theses trends are explained in Sect. 8.2. Thereafter, and focusing on prototype selection, we present a unifying framework that covers existing properties obtaining as a result a complete taxonomy (Sect. 8.3). The description of the operation as the most well known and some recent instance and/or prototype selection methods are provided in Sect. 8.4. Advanced and recent approaches that incorporate novel solutions based of hybridizations with other types of data reduction techniques or similar solutions are collected in Sect. 8.5. Finally, we summarize example evaluation results for prototype selection in an exhaustive experimental comparative analysis in Sect. 8.6.

8.1 Introduction

Instance selection (IS) plays a pivotal role in the data reduction task due to the fact that it performs the complementary process regarding the FS. Although it is independent of FS, in most of the cases, both processes are jointly applied. Facing the enormous amounts of data may be achieved by scaling down the data as an alternative to improve the scaling-up of the DM algorithms. We have previously seen that FS already accomplishes this objective, through thc removal of irrelevant and unnecessary features. In an orthogonal way, the removal of instances can be considered the same or even more interesting from the point of view of scaling down the data in certain applications [108].

The major issue of scaling down the data is the selection or identification of relevant data from an immense pool of instances, and next to prepare it as input for a DM algorithm. Selection is synonymous of pressure in many scenarios, such as in

S. García et al., *Data Preprocessing in Data Mining*,
Intelligent Systems Reference Library 72, DOI 10.1007/978-3-319-10247-4_8

organization, business environments or nature evolution [109, 133]. It is conceived as a real necessity in the world surrounding us, thus also in DM. Many circumstances lead to perform a data selection, as we have enumerated previously. Remembering them, data is not pure and initially not prepared for DM; there is missing data; there is irrelevant redundant data; errors are more likely to occur during collecting or storing; data could be too overwhelming to manage.

IS is to choose a subset of data to achieve the original purpose of a DM application as if the whole data were used [42, 127]. However, from our point of view, data reduction by means of data subset selection is not always IS. We correspond IS with an intelligent operation of instance categorization, according to a degree of irrelevance or noise and depending on the DM task. In this way, for example, we do not consider data sampling as IS per se, because it has a more general purpose and the underlying purpose is to reduce the data randomly to enhance later learning tasks. Nevertheless, data sampling [49] also belongs to the data reduction family of methods and was mentioned in Chap. 6 of this book.

The optimal outcome of IS is a minimum data subset, model independent that can accomplish the same task with no performance loss. Thus, $P(DM_s) = P(DM_t)$, where P is the performance, DM is the DM algorithm, s is the subset of instance selected and t is the complete or training set of instances. According to Liu [109], IS has the following outstanding functions:

- *Enabling*: IS makes the impossible possible. When the data set is too huge, it may not be possible to run a DM algorithm or the DM task might not be able to be effectively performed. IS enables a DM algorithm to work with huge data.
- *Focusing*: The data are formed by a lot of information of almost everything in a domain, but a concrete DM task is focused on only one aspect of interest of the domain. IS focus the data on the relevant part.
- *Cleaning*: By selecting relevant instances, redundant as well as noisy instances are usually removed, improving the quality of the input data and, hence, expecting to also improve the DM performance.

In this chapter, we emphasize the importance of IS nowadays, since it is very common that databases exceed the size of data which DM algorithms can properly handle. As another topic for data reduction, it has recently been attracting more and more attention from researchers and practitioners. Experience has shown that when a DM algorithm is applied to the reduced data set, it still achieves sufficient and suitable results if the selection strategy has been well chosen taking into account the later situation. The situation will be conditioned by the learning task, DM algorithm and outcome expectations.

This book is especially oriented towards classification, thus we also focus the goal of an IS method on obtaining a subset $S \subset T$ such that S does not contain superfluous instances and $Acc(S) \cong Acc(T)$, where $Acc(X)$ is the classification accuracy obtained using X as a training set. Henceforth, S is used to denote the selected subset. As the training set is reduced, the runtime of the training process will be also reduced for the classifier, especially in those instance-based or lazy learning methods [68].

The global objective of this chapter can be summarized into three main purposes:

- To provide a unified description for a wide set of IS methods proposed in the literature.
- To present a complete taxonomy based on the main properties observed in IS. The taxonomy will allow us to learn the advantages and drawbacks from a theoretical point of view.
- To report an empirical study for analyzing the methods in terms of accuracy and reduction capabilities. Our goal is to identify the best methods in each family and to stress the relevant properties of each one. Moreover, some graphical representations of the main IS methods are depicted.

8.2 Training Set Selection Versus Prototype Selection

At first, several proposals for selecting the most relevant data from the training set were proposed thinking mostly in the KNN algorithm [32]. Later, when the term instance-based learning [1], also known as lazy learning [2], was minted for gathering all those methods that do not perform a training phase during learning, the *prototype selection* term arises from the literature (and many derivatives, such as prototype reduction, prototype abstraction, prototype generation, etc.). Nowadays, the family of IS methods also include the proposal which was thought to work with other learning methods, such as decision trees, ANNs or SVMs. However, there was no manner of appointing the concrete case in which an IS method is valid and can be applied to any type of DM algorithm (within the same learning paradigm, of course). For this reason, we distinguish between two types of processes: Prototype Selection (PS) [68] and Training Set Selection (TSS) [21, 130]. This section is devoted to detail and to explain both distinctions the as clearly as possible.

First, we provide a more detailed and formal definition of IS. It can be defined as follows: Let X_p be an instance where $X_p = (X_{p1}, X_{p2}, \ldots, X_{pm}, X_{pc})$, with X_p belonging to a class c given by X_{pc} and a m-dimensional space in which X_{pi} is the value of the i-th feature of the pth sample. Then, let us assume that there is a training set TR which consists of N instances X_p and a test set TS composed by t instances X_p. Let $S \subset TR$ be the subset of selected samples that resulted from the execution of an IS algorithm, then we classify a new pattern from TS by a DM algorithm acting over S. The whole data set is noted as D and it is composed of the union of TR and TS.

PS methods [68] are IS methods which expect to find training sets offering best classification accuracy and reduction rates by using instance based classifiers which consider a certain similarity or distance measure. Recently, PS methods have increased in popularity within the data reduction field. Various approaches to PS algorithms have been proposed in the literature (see [68, 88, 127, 167] for review). Figure 8.1 shows the basic steps of the PS process.

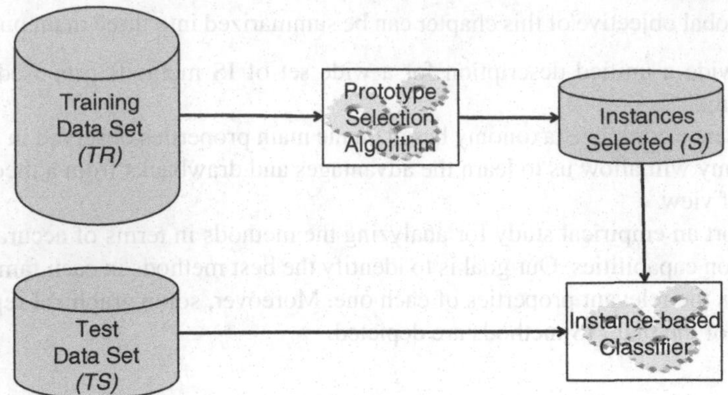

Fig. 8.1 PS process

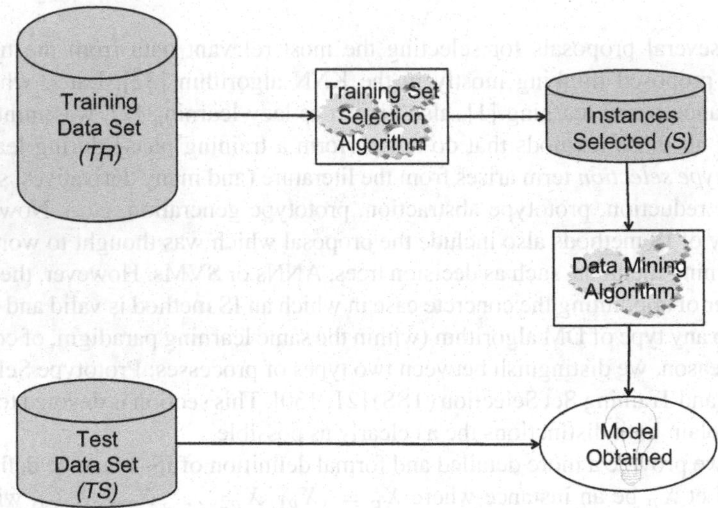

Fig. 8.2 TSS process

TSS methods are defined in a similar way. They are known as the application of IS methods over the training set used to build any predictive model. Thus, TSS can be employed as a way to improve the behavior of predictive models, precision and interpretability [135]. Figure 8.2 shows the basic steps of processing a decision tree (C4.5) on the TSS. Among others, ANNs [51, 94, 160], SVMs [31], decision trees [21, 85]; and even in other learning paradigms such as regression [8, 154], time series forecasting [79, 170], subgroup discovery [22, 23], imbalanced learning [11, 65, 66, 75, 110], multiple-instance learning [30, 58], and semi-supervised learning [80].

Due to the fact that most of the efforts in research are devoted to PS, we focus the rest of this chapter on this issue. It is very arduous to give an exact number of proposals belonging specifically to each of the two families mentioned above. When an IS method is proposed, the first step is to improve instance-based learners. When tackling TSS, the type of learning method is usually fixed to combine with the instance selector. Few methods are proposed thinking in both processes, although our experience allows us to suggest that any filter should work well for any DM model, mainly due to the low reductions rates achieved and the efforts in removing noise from the data with these kind of methods. An proper estimation of proposals reported in the specialized literature specifically considered for TSS may be around 10 % of the total number of techniques. Even though PS monopolizes almost all efforts in IS, TSS currently shows an upward trend.

8.3 Prototype Selection Taxonomy

This section presents the taxonomy of PS methods and the criteria used for building it. First, in Sec. 8.3.1, the main characteristics which will define the categories of the taxonomy will be outlined. In Sec. 8.3.2, we briefly enumerate all the PS methods proposed in the literature. The complete and abbreviated name will be given together with the reference. Finally, Sec. 8.3.3, presents the taxonomy.

8.3.1 Common Properties in Prototype Selection Methods

This section provides a framework for the discussion of the PS methods presented in the next subsection. The issues discussed include order of the search, type of selection and evaluation of the search. These mentioned issues are involved in the definition of the taxonomy, since they are exclusive to the operation of the PS algorithms. Other classifier-dependent issues such as distance functions or exemplar representation will be presented. Finally, some criteria will also be pointed out in order to compare PS methods.

8.3.1.1 Direction of Search

When searching for a subset S of prototypes to keep from training set TR, there are a variety of directions in which the search can proceed:

- **Incremental:** An incremental search begins with an empty subset S, and adds each instance in TR to S if it fulfills some criteria. In this case, the algorithm depends on the order of presentation and this factor could be very important. Under such a scheme, the order of presentation of instances in TR should be random because by

definition, an incremental algorithm should be able to handle new instances as they become available without all of them being present at the beginning. Nevertheless, some recent incremental approaches are order-independent because they add instances to S in a somewhat incremental fashion, but they examine all available instances to help select which instance to add next. This makes the algorithms not truly incremental as we have defined above, although we will also consider them as incremental approaches.
One advantage of an incremental scheme is that if instances are made available later, after training is complete, they can continue to be added to S according to the same criteria. This capability could be very helpful when dealing with data streams or online learning. Another advantage is that they can be faster and use less storage during the learning phase than non-incremental algorithms. The main disadvantage is that incremental algorithms must make decisions based on little information and are therefore prone to errors until more information is available.

- **Decremental:** The decremental search begins with $S = TR$, and then searches for instances to remove from S. Again, the order of presentation is important, but unlike the incremental process, all of the training examples are available for examination at any time.
One disadvantage with the decremental rule is that it presents a higher computational cost than incremental algorithms. Furthermore, the learning stage must be done in an off-line fashion because decremental approaches need all possible data. However, if the application of a decremental algorithm can result in greater storage reduction, then the extra computation during learning (which is done just once) can be well worth the computational savings during execution thereafter.
- **Batch:** Another way to apply a PS process is in batch mode. This involves deciding if each instance meets the removal criteria before removing any of them. Then all those that do meet the criteria are removed at once. As with decremental algorithms, batch processing suffers from increased time complexity over incremental algorithms.
- **Mixed:** A mixed search begins with a pre-selected subset S (randomly or selected by an incremental or decremental process) and iteratively can add or remove any instance which meets the specific criterion. This type of search allows rectifications to already done operations and its main advantage is to make it easy to obtain good accuracy-suited subsets of instances. It usually suffers from the same drawbacks reported in decremental algorithms, but this depends to a great extent on the specific proposal. Note that these kinds of algorithms are closely related to the order-independent incremental approaches but, in this case, instance removal from S is allowed.
- **Fixed:** A fixed search is a subfamily of mixed search in which the number of additions and removals remains the same. Thus, the number of final prototypes is determined at the beginning of the learning phase and is never changed.

8.3.1.2 Type of Selection

This factor is mainly conditioned by the type of search carried out by the PS algorithms, whether they seek to retain border points, central points or some other set of points.

- **Condensation:** This set includes the techniques which aim to retain the points which are closer to the decision boundaries, also called border points. The intuition behind retaining border points is that internal points do not affect the decision boundaries as much as border points, and thus can be removed with relatively little effect on classification. The idea is to preserve the accuracy over the training set, but the generalization accuracy over the test set can be negatively affected. Nevertheless, the reduction capability of condensation methods is normally high due to the fact that there are fewer border points than internal points in most of the data.
- **Edition:** These kinds of algorithms instead seek to remove border points. They remove points that are noisy or do not agree with their neighbors. This removes boundary points, leaving smoother decision boundaries behind. However, such algorithms do not remove internal points that do not necessarily contribute to the decision boundaries. The effect obtained is related to the improvement of generalization accuracy in test data, although the reduction rate obtained is low.
- **Hybrid:** Hybrid methods try to find the smallest subset S which maintains or even increases the generalization accuracy in test data. To achieve this, it allows the removal of internal and border points based on criteria followed by the two previous strategies. The KNN classifier is highly adaptable to these methods, obtaining great improvements even with a very small subset of instances selected.

8.3.1.3 Evaluation of Search

KNN is a simple technique and it can be used to direct the search of a PS algorithm. The objective pursued is to make a prediction on a non-definitive selection and to compare between selections. This characteristic influences the quality criterion and it can be divided into:

- **Filter:** When the kNN rule is used for partial data to determine the criteria of adding or removing and no leave-one-out validation scheme is used to obtain a good estimation of generalization accuracy. The fact of using subsets of the training data in each decision increments the efficiency of these methods, but the accuracy may not be enhanced.
- **Wrapper:** When the kNN rule is used for the complete training set with the leave-one-out validation scheme. The conjunction in the use of the two mentioned factors allows us to get a great estimation of generalization accuracy, which helps to obtain better accuracy over test data. However, each decision involves a complete computation of the kNN rule over the training set and the learning phase can be computationally expensive.

8.3.1.4 Criteria to Compare Prototype Selection Methods

When comparing PS methods, there are a number of criteria that can be used to evaluate the relative strengths and weaknesses of each algorithm. These include storage reduction, noise tolerance, generalization accuracy and time requirements.

- *Storage reduction:* One of the main goals of the PS methods is to reduce storage requirements. Furthermore, another goal closely related to this is to speed up classification. A reduction in the number of stored instances will typically yield a corresponding reduction in the time it takes to search through these examples and classify a new input vector.
- *Noise tolerance:* Two main problems may occur in the presence of noise. The first is that very few instances will be removed because many instances are needed to maintain the noisy decision boundaries. Secondly, the generalization accuracy can suffer, especially if noisy instances are retained instead of good instances.
- *Generalization accuracy:* A successful algorithm will often be able to significantly reduce the size of the training set without significantly reducing generalization accuracy.
- *Time requirements:* Usually, the learning process is done just once on a training set, so it seems not to be a very important evaluation method. However, if the learning phase takes too long it can become impractical for real applications.

8.3.2 Prototype Selection Methods

Almost 100 PS methods have been proposed in the literature. This section is devoted to enumerating and designating them according to a standard followed in this chapter. For more details on their descriptions and implementations, the reader can read the next section of this chapter. Implementations of some of the algorithms in Java can be found in KEEL software [3, 4], described in Chap. 10 of this book.

Table 8.1 presents an enumeration of PS methods reviewed in this chapter. The complete name, abbreviation and reference are provided for each one. In the case of there being more than one method in a row, they were proposed together and the best performing method (indicated by the respective authors) is depicted in bold.

8.3.3 Taxonomy of Prototype Selection Methods

The properties studied above can be used to categorize the PS methods proposed in the literature. The direction of the search, type of selection and evaluation of the search may differ among PS methods and constitute a set of properties which are exclusive to the way of operating of the PS methods. This section presents the taxonomy of PS methods based on these properties.

Table 8.1 IS methods reviewed

Complete name	Abbr. name	Reference
Condensed nearest neighbor	CNN	[83]
Reduced nearest neighbor	RNN	[76]
Edited nearest neighbor	ENN	[165]
No name specified	Ullmann	[156]
Selective nearest neighbor	SNN	[136]
Repeated edited Nearest neighbor	RENN	[149]
All-KNN	**AllKNN**	
Tomek condensed nearest neighbor	TCNN	[150]
Mutual neighborhood value	MNV	[78]
MultiEdit	MultiEdit	[47, 48]
Shrink	Shrink	[89]
Instance based 2	IB2	[1]
Instance based 3	**IB3**	
Monte carlo 1	MC1	[147]
Random mutation hill climbing	**RMHC**	
Minimal consistent set	MCS	[36]
Encoding length heuristic	ELH	[18]
Encoding length grow	ELGrow	
Explore	**Explore**	
Model class selection	MoCS	[16]
Variable similarity metric	VSM	[111]
Gabriel graph editing	GGE	[139]
Relative Neighborhood Graph Editing	**RNGE**	
Polyline functions	PF	[107]
Generational genetic algorithm	GGA	[100, 101]
Modified edited nearest neighbor	MENN	[84]
Decremental reduction optimization procedure 1	DROP1	[167]
Decremental reduction optimization procedure 2	DROP2	
Decremental reduction optimization procedure 3	**DROP3**	
Decremental reduction optimization procedure 4	DROP4	
Decremental reduction optimization procedure 5	DROP5	
Decremental encoding length	DEL	
Estimation of distribution algorithm	EDA	[146]
Tabu search	CerveronTS	[26]
Iterative case filtering	ICF	[15]
Modified condensed nearest neighbor	MCNN	[46]
Intelligent genetic algorithm	IGA	[86]
Prototype selection using relative certainty gain	PSRCG	[143, 144]
Improved KNN	IKNN	[168]

(continued)

Table 8.1 (continued)

Complete name	Abbr. name	Reference
Tabu search	ZhangTS	[171]
Iterative maximal nearest centroid neighbor	Iterative MaxNCN	[112]
Reconsistent	**Reconsistent**	
C-Pruner	CPruner	[173]
Steady-state genetic algorithm	SSGA	[19]
Population based incremental learning	PBIL	
CHC evolutionary algorithm	**CHC**	
Patterns by ordered projections	POP	[135]
Nearest centroid neighbor edition	NCNEdit	[140]
Edited normalized radial basis function	**ENRBF**	[88]
Edited normalized radial basis function 2	ENRBF2	
Edited nearest neighbor estimating class probabilistic	ENNProb	[158]
Edited nearest neighbor estimating	**ENNTh**	[158]
Class probabilistic and threshold		
Support vector based prototype selection	SVBPS	[104]
Backward sequential edition	BSE	[125]
Modified selective subset	MSS	[10]
Generalized condensed nearest neighbor	GCNN	[28]
Fast condensed nearest neighbor 1	**FCNN**	[7]
Fast condensed nearest neighbor 2	FCNN2	
Fast condensed nearest neighbor 3	FCNN3	
Fast condensed nearest neighbor 4	FCNN4	
Noise removing based on minimal consistent set	NRMCS	[161]
Genetic algorithm based on mean square error,	GA-MSE-CC-FSM	[77]
Clustered crossover and fast smart mutation		
Steady-state memetic algorithm	SSMA	[62]
Hit miss network C	HMNC	[116]
Hit miss network edition	HMNE	
Hit miss network edition iterative	**HMNEI**	
Template reduction for KNN	TRKNN	[53]
Prototype selection based on clustering	PSC	[126]
Class conditional instance selection	CCIS	[117]
Cooperative coevolutionary instance selection	CoCoIS	[72]

(continued)

Table 8.1 (continued)

Complete name	Abbr. name	Reference
Instance selection based on classification contribution	ISCC	[17]
Bayesian instance selection	EVA	[56]
Reward-Punishment editing	RP-Edit	[57]
Complete cross validation functional prototype selection	CCV	[87]
Sequential reduction algorithm	SeqRA	[132]
Local support vector machines noise reduction	LSVM	[145]
No name specified	Bien	[13]
Reverse nearest neighbor reduction	RNNR	[35]
Border-Edge pattern selection	BEPS	[105]
Class boundary preserving algorithm	CBP	[122]
Cluster-Based instance selectio	CBIS	[34]
RDCL profiling	RDCL	[38]
Multi-Selection genetic algorithm	MSGA	[73]
Ant colony prototype reduction	Ant-PR	[118]
Spectral instance reduction	SIR	[123]
Competence enhancement by Ranking-based instance selection	CRIS	[37]
Discriminative prototype selection	D-PS	[14]
Adaptive threshold-based instance selection algorithm	ATISA	[24]
InstanceRank based on borders for instance selection	IRB	[85]
Visualization-Induced self-organizing map for prototype reduction	VISOM	[106]
Support vector oriented instance selection	SVOIS	[154]
Dominant set clustering prototype selection	DSC	[157]
Fuzzy rough prototype selection	FRPS	[159]

Figure 8.3 illustrates the categorization following a hierarchy based on this order: type of selection, direction of search and evaluation of the search. It allows us to distinguish among families of methods and to estimate the size of each one.

One of the objectives in this chapter is to highlight the best methods depending on their properties, taking into account that we are conscious that the properties could determine the suitability of use of a specific scheme. To do this, in Sect. 8.6, we will conclude which methods perform best for each family considering several metrics of performance.

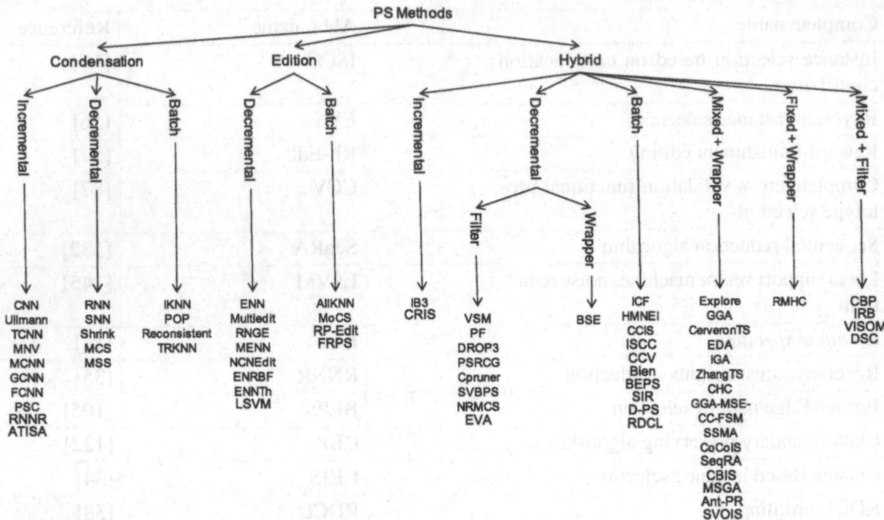

Fig. 8.3 PS taxonomy

8.4 Description of Methods

Algorithms for IS may be classified in three type groups: condensation algorithms, edition algorithms and hybrids.

8.4.1 Condensation Algorithms

This set includes the techniques which aim to retain the points which are closer to the decision boundaries, also called border points.

Considering their search direction they can be classified as:

8.4.1.1 Incremental

- **Condensed Nearest Neighbor (CNN)** [83]—This algorithm finds a subset S of the training set TR such that every member of TR is closer to a member of S of the same class than to a member of S of a different class. It begins by randomly selecting one instance belonging to each output class from TR and putting them in S. Then each instance in TR is classified using only the instances in S. If an instance is misclassified, it is added to S, thus ensuring that it will be classified correctly. This process is repeated until there are no instances in TR that are misclassified. This algorithm ensures that all instances in TR are classified correctly, though it does not guarantee a minimal set.

- **Tomek Condensed Nearest Neighbor (TCNN)** [150]—Tomek presents two modifications based on the CNN algorithm. The method 1 is similar to CNN, but, when an instance X_i is misclassified (because its nearest neighbor in S, s is from the opposite class), instead of adding it to S the method finds the nearest neighbor of s which is a member of the same class of X_i, and adds it to S.
  The method 2 is also a modification of CNN, where instead of use instances in TR to build S, only a subset of TR, F is employed. F is composed by the instances of TR which have the same class as its nearest neighbors.
- **Modified Condensed Nearest Neighbor (MCNN)** [46]—This algorithm is similar to CNN but, instead of adding a instance to the set S when it is misclassified, it flags all the instances misclassified and, when all the instances in TR have been tested, a representative example of each class is added to S, generating it as the centroid of the misclassified examples in each class. The process is conducted iteratively until no instance in TR is misclassified.
- **Generalized Condensed Nearest Neighbor (GCNN)** [28]—The GCNN algorithm tries to improve the CNN algorithm. Firstly, the initial prototypes are selected as the most voted from each class (considering a vote as to be the nearest instance to other of the same class). Then, the CNN rule is applied, but a new instance X is considered classified correctly only if its nearest neighbor X_i in S is from its same class, and the distance between X and X_i is lower than $dist$, where $dist$ is the distance between X and its nearest enemy in S.
- **Fast Condensed Nearest Neighbor family (FCNN)** [7]—The **FCNN1** algorithm starts by introducing in S the centroids of each class. Then, for each prototype p in S, its nearest enemy inside its Voronoi region is found, and add to S. This process is performed iteratively until no enemies are found on a single iteration.
  Fast Condensed Nearest Neighbor 2 (**FCNN2**): The FCNN2 algorithm is similar to FCNN1 but, instead of adding the nearest enemy on each Voronoi region, is added the centroid of the enemies found in the region.
  Fast Condensed Nearest Neighbor 3 (**FCNN3**): The FCNN3 algorithm is similar to FCNN1 but, instead of adding one prototype per region in each iteration, only one prototype is added (the one which belongs to the Voronoi region with most enemies). In FCNN3, S is initialized only with the centroid of the most populated class.
  Fast Condensed Nearest Neighbor 4 (**FCNN4**): The FCNN3 algorithm is similar to FCNN2 but, instead of adding one prototype per region in each iteration, only one centroid is added (the one which belongs to the Voronoi region with most enemies). In FCNN4, S is initialized only with the centroid of the most populated class.
- **Prototype Selection based on Clustering (PSC)** [126]—To build the S set, the PSC first employs the C-Means algorithm to extract clusters from the set TR of training prototypes. Then, for each cluster G, if it is homogeneous (all prototypes belongs to the same class), its centroid is added to S. If it is not homogenous, then their majority class G_m is computed, and every instance which do not belongs to G_m in the cluster is add to S, along with its nearest neighbor in class G_m.

8.4.1.2 Decremental

- **Reduced Nearest Neighbor (RNN)** [76]—RNN starts with $S = TR$ and removes each instance from S if such a removal does not cause any *other* instances in TR to be misclassified by the instances remaining in S. It will always generate a subset of the results of CNN algorithm.
- **Shrink (Shrink)** [89]—This algorithm starts with $S = TR$, and then removes any instances that would still be classified correctly by the remaining subset. This is similar to RNN, except that it only considers whether the *removed* instance would be classified correctly, whereas RNN considers whether the classification of *other* instances would be affected by the instance's removal.
- **Minimal Consistent Set (MCS)** [36]—The purpose of this algorithm is to find a Minimal consistent set of instances which will be able to classify all the training instances in TR. It performs the following steps:

1. Define an initial consistent set to be the given training data set, since the given set is by definition consistent with itself.
2. For a specific sample in the given training data set, determine the nearest sample distance among all the samples from all classes other than its own in the consistent set, i.e., identify and store the Nearest Unlike Neighbor (NUN) distance of the sample from the consistent set.
3. For this same sample, identify all the neighboring samples from its own class in the given data set which are closer than this NUN distance and cast an approval vote to each of these samples in the given set by incrementing the corresponding vote registers, while noting this voter's (sample) identity by updating the corresponding voter lists.
4. Repeat Step 2 and 3 for all samples in the given training set, which results in a list of the number of votes received by each sample in the given set along with the records of the identity of its voters.
5. Create a potential candidate consistent set consisting of all samples in the given set which are either (a) already present in the current consistent set or (b) whose inclusion will not create an inconsistency; i.e., the sample should not be nearer to any member of any other class other than that member's current NUN distance. In the first iteration, the entire consistent set (i.e., the given set) remains as the candidate consistent set as all samples satisfy condition (a)
6. Identify the most voted sample in this candidate consistent list and designate it as a member of a newly selected consistent set and identify all of its contributing voters.
7. Delete these voters from all the voter lists wherein they currently appear and correspondingly decrement the appropriate vote counts.
8. Repeat Step 6 and Step 7 until all the voters have been accounted for by the selected consistent set.
9. Now with this selected consistent set, the NUN distances of the input samples are likely to be greater than before as some of the original NUN samples may no longer be in the selected consistent set. Accordingly, repeat Step 2 using this

selected consistent set to determine the NUN distance thresholds for each sample in the given set.

10. Repeat Step 3 through 8 using all the samples in the given set to identify a new consistent set. This process of recursive application of step 2 through 8 is continued till the selected set is no longer getting smaller. It is easy to see that under this procedure this final subset remains consistent, i.e., is able to classify all samples in the original set correctly.

- **Modified Selective Algorithm (MSS)** [9]—Let R_i be the set of all X_i in TR such that X_j is of the same class of X_i and is closer to X_i than the nearest neighbor of X_i in TR of a different class than X_i. Then, MSS is defined as that subset of the TR containing, for every X_i in TR, that element of its R_i that is the nearest to a different class than that of X_i.

An efficient algorithmic representation of the *MSS* method is depicted as:

```
Q = TR
Sort the instances {X_j}_{j=1}^n according to increasing values
of enemy distance (D_j).
For each instance X_i do
    add ← FALSE
    For each instance X_j do
        If x_j ∈ Q ∧ d(X_i, X_j) < D_j then
            Q ← Q − {X_j}
            add ← TRUE
    If add then S ← S ∪ {X_i}
    If Q = ∅ then return S
```

8.4.1.3 Batch

- **Patterns by Ordered Projections (POP)** [135]—This algorithm consists of eliminating the examples that are not within the limits of the regions to which they belong. For it, each attribute is studied separately, sorting and increasing a value, called *weakness*, associated to each one of the instances, if it is not within a limit. The instances with a value of *weakness* equal to the number of attributes are eliminated.
- **Max Nearest Centroid Neighbor (Max-NCN)** [112] This algorithm is based on the Nearest Centroid Neighborhood (NCN) concept, defined by:

1. The first NCN of X_i is also its NN, Y_1.
2. The i-th NCN, Y_i, $i \geq 2$, is such that the centroid of this and previously selected NCN, Y_1, ..., Y_i is the closest to X_i.

MaxNCN algorithm can be written as follows:

```
For each instance X_i do
    neighbors_number[X_i] = 0
    neighbor = next_neighbor(X_i)
    While neighbor.class == X_i.class do
        neighbors_vector[X_i] = Id(neighbor)
        neighbors_number[X_i] + +
        neighbor = next_neighbor(X_i)
    End while
End for
While Max_neighbors() > 0 do
    EliminateNeighbors(id_Max_neighbors)
End while
```

- **Reconsistent** [112]—The Reconsistent algorithm is an enhanced version of the Iterative MaxNCN. When it has been applied to the set TR the subset resulting is processed by a condensing method (CNN), employing as reference set the original training set TR.
- **Template Reduction KNN (TRKNN)** [53]—The TRKNN method introduces the concept of nearest neighbors chains. Every chain is assigned to one instance, and it is built by finding the nearest neighbors of each element of the chain, which belong alternatively to the class of the starting instance, or to a different class. The chain is stopped when an element is selected twice to belong to the same chain. By building the distances between the patterns in the chain a non-increasing sequence is formed, thus the last elements of the chain will be near the decision boundaries. The TRKNN method will employ this property to drop all instances which are far away from the decision boundaries.

8.4.2 Edition Algorithms

These algorithms edit out noisy instances as well as close border class, leaving smoother decision boundaries. They also retain all internal points, because an internal instance may be labeled as the same class of its neighbors.

Considering their search direction they can be classified as:

8.4.2.1 Decremental

- **Edited Nearest Neighbor (ENN)** [165]—Wilson developed this algorithm which starts with $S = TR$ and then each instance in S is removed if it does not agree with the majority of its k nearest neighbors.

- **Repeated-ENN (RENN)** [165]—It applies the ENN algorithm repeatedly until all instances remaining have a majority of their neighbors with the same class.
- **Multiedit** [47, 48]—This method proceeds as follows.

```
Let S = TR.
Do
    Let R = S.
    Let S = ∅ and randomly split R into b blocks: R_1, ..., R_b,
    where b > 2
    For each b_i block
        Add to S the prototypes from R_{b_i} that are
        misclassified using the KNN rule with R_{(b_i+1) mod b}.
While S ≠ R.
```

- **Relative Neighborhood Graph Edition (RNGE)** [139]—A Proximity Graph (PG), $G = (V, E)$, is an undirected graph with a set of vertices $V = TR$, and a set of edges, E, such that $(t_i, t_j) \in E$ if ad only if t_i and t_j satisfy some neighborhood relation. In this case, we say that t_i and t_j are *graph neighbors*. The graph neighbors of a given point constitute its *graph neighborhood*. The graph neighborhood of a subset, $S \subseteq V$, consists of the union of all the graph neighbors of every node in S. The scheme of editing can be expressed in the following way.
  1. Construct the corresponding PG.
  2. Discard those instances that are misclassified by their graph neighbors (by the usual voting criterion).

  The PGs used are the **Gabriel Graph Edition (GGE)** and the **Relative Neighborhood Graph Edition (RNGE)**.

Gabriel Graph Editing (GGE)

–The *GGE* is defined as follows:

$$(t_i, t_j) \in E \Leftrightarrow d^2(t_i, t_j) \leq d^2(t_i, t_k) + d^2(t_j, t_k), \forall t_k \in T, k \neq i, j. \quad (8.1)$$

where $d(\cdot, \cdot)$ be the Euclidean distance.

Relative Nearest Graph Editing (RNGE)

–Analogously, the set of edges in the *RNGE* is defined as follows:

$$(t_i, t_j) \in E \Leftrightarrow d(t_i, t_j) \leq max(d(t_i, t_k), d(t_j, t_k)), \forall t_k \in T, k \neq i, j. \quad (8.2)$$

- **Modified Edited Nearest Neighbor (MENN)** [84]—This algorithm, similar to ENN, starts with $S = TR$ and then each instance X_i in S is removed if it does

not agree with all of its $k + l$ nearest neighbors, where l are all the instances in S which are at the same distance as the last neighbor of X_i.
In addition, MENN works with a prefixed number of pairs (k,k'). k is employed as the number of neighbors involved to perform the editing process, and k' is employed to validate the edited set S obtained. The best pair found is employed as the final reference set (if two or more sets are found as optimal, then both are employed in the classification of the test instances. A majority rule is used to decide the output of the classifier in this case).

- **Nearest Centroid Neighbor Edition (NCNEdit)** [140]—The NCN Editing algorithm applies the NCN classification rule to perform an edition process over the training set TR. The NCN classification rule can be defined as:
Having defined the NCN scheme, the editing process consists in set $S = TR$ an discard from S every prototype misclassified by the NCN rule.

8.4.2.2 Batch

- **AllKNN** [149]—All KNN is an extension of ENN. The algorithm, for $i = 0$ to k flags as bad any instance not classified correctly by its i nearest neighbors. When the loop is completed k times, it removes the instances flagged as bad.
- **Model Class Selection (MoCS)** [16]—Brodley's algorithm for reducing the size of the training set TR is to keep track of how many times each instance was one of the k nearest neighbors of another instance, and whether its class matched that of the instance being classified. If the number of times it was wrong is greater than the number of times it was correct then it is thrown out.

8.4.3 Hybrid Algorithms

Hybrids methods try to find the smallest subset S which lets keep or even increase the generalization accuracy in test data. For doing it, it allows the removal of internal and border points.

8.4.3.1 Incremental

- **Instance-Based Learning Algorithms Family (IB3)** [1]—A series of *instance-based* learning algorithms are presented. IB1 was simply the 1-NN algorithm, used as a baseline.

IB2

–It starts with S initially empty, and each instance in TR is added to S if it is not classified correctly by the instances already in S (with the first instance always added).

IB2 is similar to CNN, except that IB2 not seed S with one instance of each class and does not repeat the process after the first pass through the training set.

IB3

–The IB3 algorithm proceeds as follows:

```
For each instance X_i in TR
    Let a be the nearest acceptable instance in S to X_i.
    (if there are no acceptable instances in S,
    let a be a random instance in S)
    If class(a) ≠ class(X_i) then add X_i to S.
    For each instance s in S
        If s is at least as close to X_i as a is
            Update the classification record of s
            and remove s from S its classification
            record is significantly poor.
Remove all non-acceptable instance from S.
```

An instance is *acceptable* if the lower bound of its accuracy is statistically significantly higher (at a 90 % confidence level) than the upper bound on the frequency of its class. Similarly, an instance is dropped from *S* if the upper bound on its accuracy is statistically significantly lower (at a 70 % confidence level) than the lower bound on the frequency of its class. Other instances are kept in *S* during training, and then dropped at the end if they do not prove to be acceptable.

The expression for the upper and lower bounds of the confidence level interval is:

$$\frac{p + z^2/2g \pm z\sqrt{\frac{p(1-p)}{g} + \frac{z^2}{4g^2}}}{1 + z^2/g} \tag{8.3}$$

where for the *accuracy* of an instance in *S*, *g* is the number of classification attempts since introduction of the instance to *S* (i.e., the number of times it was at least as close to X_i as *a* was), *p* is the accuracy of such attempts (i.e., the number of times the instance's class matched X_i's class, divided by *g*), and *z* is the confidence (0.9 for acceptance, 0.7 for dropping). For the frequency of a class, *p* is the frequency (i.e. proportion of instances so far that are of this class), *g* is the number of previously processed instances, and *z* is the confidence.

8.4.3.2 Decremental

8.4.3.3 Filter

- **Variable Similarity Metric (VSM)** [111]—In order to reduce storage and remove noisy instances, an instance X_i is removed if all *k* of its neighbors are of the same class, even if they are of a different class than X_i. The instance is only removed if its neighbors are at least 60 % sure of their classification. The VSM method typically uses a fairly large k (i.e., $k = 10$).

- **Decremental Reduction Optimization Procedure Family (DROP)** [166] In order to present these reduction techniques, we need to define some concepts. Each instance X_i has k nearest neighbors where k is typically a small odd integer. X_i also has a nearest *enemy*, which is the nearest instance with a different output class. Those instances that have x_i as one of their k nearest neighbors are called *associates* of X_i.

DROP1

–Uses the following basic rule to decide if it is safe to remove an instance from the instance set S (where $S = TR$ originally):

Remove X_i if at least as many of its associates in S would be classified correctly without X_i.

The algorithm *DROP1* proceeds as follows.

```
DROP1 (Training set TR): Selection set S.
Let S = TR.
For each instance X_i in S:
  Find the k+1 nearest neighbors of X_i in S.
  Add X_i to each of its lists of associates.
For each instance X_i in S:
  Let with = # of associates of X_i classified
  correctly with X_i as a neighbor.
  Let without = # of associates of X_i classified
  correctly without X_i.
  If without ≥ with
     Remove X_i from S.
     For each associate a of X_i
        Remove X_i from a's list of neighbors.
        Find a new nearest neighbor for a.
        Add a to its new list of associates.
     For each neighbor b of X_i
        Remove X_i from b's lists of associates.
  Endif
Return S.
```

DROP2

–In this method, the removal criterion can be restated as:

Remove X_i if at least as many of its associates in TR would be classified correctly without X_i.

Using this modification, each instance X_i in the original training set TR continues to maintain a list of its $k+1$ nearest neighbors in S, even after X_i is removed from S. This means that instances in S have associates that are both in and out of S, while instances that have been removed from S have no associates.

DROP2 also changes the order of removal of instances. It initially sorts the instances in S by the distance to their nearest enemy. Instances are then checked for removal beginning at the instance furthest from its nearest enemy.

DROP3

–It is a combination of *DROP2* and *ENN* algorithms. *DROP3* uses a noise-filtering pass *before* sorting the instances in S (Wilson *ENN* editing). After this, it works identically to *DROP2*.

- **CPruner** [173]—First it is necessary to introduce some concepts underlying this algorithm.

Definition 8.1 For an instance X_i in TR, the k nearest neighbors of X_i make up its *k-reachability* set, denoted as *k-reachability*(X_i).

Definition 8.2 For an instance X_i in TR, those instances with similar class label to that of X_i, and have X_i as one of their nearest neighbors are called the *k-coverage* set of X_i, denoted as *k-coverage*(X_i).

Definition 8.3 For instance X_i in TR, if X_i can be classified correctly by *k-reachability*(X_i), then we say X_i is *implied* by *k-reachability*(X_i), and X_i is a *superfluous* instance in TR.

Definition 8.4 For instance X_i in TR, X_i is a *critical* instance, if the following conditions holds:

At least one instance X_j in *k-coverage*(X_i) is not implied by *k-reachability*(X_j), or

After X_i is deleted, at least one instance X_j in *k-coverage*(X_i) is not implied by *k-reachability*(X_j).

Definition 8.5 For instance X_i in TR, if X_i is not a superfluous instance and |*k-reachability*(X_i)| > |*k-coverage*(X_i)|, then X_i is a *noisy* instance.

Rule 8.1 *Instance pruning rule*

For an instance X_i in TR, if it can be pruned, it must satisfy one of the following two conditions:

It is a noisy instance;

It is a superfluous instance, but not a critical one.

Rule 8.2 *Rule for deciding the order of instances removal*

Let H-kNN(X_i) be the number of the instances of its class in kNN(X_i), and D-NE(X_i) be the distance of X_i to its nearest enemy.

For two prunable instances X_i and X_j in TR,
If H-kNN(X_i) > H-kNN(X_j), X_i should be removed before X_j;
If H-kNN(X_i) = H-kNN(X_j) and D-NE(X_i) > D-NE(X_j), X_j should be removed before x_i;
If H-kNN(X_i) = H-kNN(X_j) and D-NE(X_i) = D-NE(X_j), the order of removal is random decided.

Next, we present the *C-Pruner* algorithm.

```
S = TR
For all Xi ∈ S do
    Compute k-reachability(Xi) and k-coverage(Xi)
For all Xi ∈ S do
    If Xi is a noisy instance
        Remove Xi from S
        For all Xj ∈ k-coverage(Xi)
            Remove Xi from k-reachability(Xj)
            Update k-reachability(Xj)
        For all Xj ∈ k-reachability(Xi)
            Remove Xi from k-coverage(Xj)
Sort the order of instances in S according to rule 2
For all Xi ∈ S
    If Xi satisfies rule 1
        Remove Xi from S
        For all Xj ∈ k-coverage(Xi)
            Remove Xi from k-reachability(Xj)
            Update k-reachability(Xj)
Return S
```

- **Support Vector Based Prototype Selection (SVBPS)** [104]—The SVBPS method firstly learns a SVM employing a proper kernel function. The Support Vectors found in the procedure are post processed with the DROP2 algorithm, adopting its result as the final S set.
- **Class Conditional Instance Selection (CCIS)** [117]
This algorithm consists of the following two phases:
  - Class Conditional selection phase (CC). It removes from the training set outliers, isolated points and points close to the 1-NN decision boundary. This phase aims at enlarging the hypothesis margin and reducing the empirical error.

```
(X1, ..., Xn) = TR sorted in decreasing order of Score
S = (X1, ..., Xk0)
i = k0 + 1
go_on = 1
ub = n - |{a s.t. Score(a) ≤ θ }|
While i < ub and go_on do
```

```
Temp = S ⋃ {X_i}
If ε^S ≤ ε^A then
    go_on = θ
If ε^Temp < ε^S and go_on then
    S = Temp
    i= i +1
else
    go_on = θ
```

– Thin-out selection phase (THIN). It thins out points that are not important to the decision boundary of the resulting 1-NN rule. This phase aims at selecting a small number of instances without negatively affecting the 1NN empirical error.

```
S_f = { x∈S with in-degree G^S_bc > 0 }
S_prev = S
S_1 = S\S_f
go_on = 1
While go_on do
    S_t = { a∈S_1 with in-degree G^{S_t}_bc > θ
        and with in-degree G^{S_prev}_bc or in G^{S_prev}_wc > 0}
    go_on = ε^{S_f ⋃ S_t} < ε^{S_f}
    If go_on then
        S_f = S_f ⋃ S_t
        S_prev = S_1
        S_1 = S\S_f
```

8.4.3.4 Wrapper

- **Backward Sequential Edition (BSE)** [125]—The BSE algorithm starts with S=TR. Each instance X_i is tested to find how the performance of the KNN is increased when X_i is removed from S. The instance in which removal causes the best increase in the performance is finally deleted from S, and the process is repeated until no further increases in the performance of the classifier are found. To increase the efficiency of the method, the authors suggested the use of ENN or DROP procedures as a first stage of the BSE algorithm.

8.4.3.5 Batch

- **Iterative Case Filtering (ICF)** [15]—ICF defines *local set* $L(x)$ which contain all cases inside largest hypersphere centered in x_i such that the hypersphere contains only cases of the same class as instance x_i. Authors define two properties, *reachability* and *coverage*:

$$Coverage(X_i) = \{X'_i \in TR : X_i \in L(X'_i)\}, \tag{8.4}$$

$$Reachability(X_i) = \{X'_i \in TR : X'_i \in L(X_i)\}, \tag{8.5}$$

In the first phase ICF uses the ENN algorithm to remove the noise from the training set. In the second phase the ICF algorithm removes each instance X_i for which the $Reachability(X_i)$ is bigger than the $Coverage(X_i)$. This procedure is repeated for each instance in TR. After that ICF recalculates *reachability* and *coverage* properties and restarts the second phase (as long as any progress is observed).

- **Hit-Miss Network Algorithms (HMN)** [116]—Hit-Miss Networks are directed graphs where the points are the instances in TR and the edges are connections between an instance and its nearest neighbor from each class. The edge connecting a instance with a neighbor of its same class is called "Hit", and the rest of its edges are called "Miss".

  The **HMNC** method builds the network and removes all nodes not connected. The rest of the nodes are employed to build the final S set.

  Hit Miss Network Edition (**HMNE**): Starting from the output of HMNC, HMNE applies these four rules to prune the network:

  1. Every point with more "Miss"edges than "Hit" edges is flagged for removal.
  2. If the size of non flagged points of a class is too low, edges with at least one "Hit" from those classes are unflagged.
  3. If there are more than three classes, some points of each class with low number of "Miss" are unflagged.
  4. Points which are the "Hit" of a 25 % or more instances of a class are unflagged.

  Finally, the instances which remain flagged for removal are deleted from the network, in order to build the final S set.

  Hit Miss Network Edition Iterative (**HMNEI**): The HMNE method can be employed iteratively until the generalization accuracy of 1-NN on the original training set with the reduced set decreases.

8.4.3.6 Mixed+Wrapper

- **Encoding Length Familiy (Explore)** [18] Cameron-Jones used an *encoding length heuristic* to determine how good the subset S is in describing TR. His algorithms use cost function defined by:

$$COST(s, n, x) = F(s, n) + s \log_2(\Omega) + F(x, n - s) + x \log_2(\Omega - 1) \tag{8.6}$$

where n is the number of instances in TR, s is the number of instances in S, and x defines the number of badly classified instances (basing on S). Ω is the number of classes in the classification task. $F(s, n)$ is the cost of encoding which s instances if the n available are retained, and is defined as:

$$F(s, n) = \log^* \left(\sum_{j=0}^{s} \frac{n!}{j!(n-j)!} \right) \tag{8.7}$$

$\log^* n = arg\ min_k F(k) \geq n$, k is integer, and $F(0) = 1$, $F(i) = 2^{F(i-1)}$.
The **ELH** algorithm starts from the empty set and adds instances only if the cost function is minimized.
ELGrow additionally tries to remove instances if it helps to minimize the cost function. **Explore** extends ELGrow by 1000 mutations to try to improve the classifier. Each mutation tries adding an instance to S, removing one from S, or swapping one in S with one in TR - S, and keeps the change if it does not increase the cost of the classifier.

- **CHC (CHC)** [19]—During each generation the CHC develops the following steps.

1. It uses a parent population of size N to generate an intermediate population of N individuals, which are randomly paired and used to generate N potential offsprings.
2. Then, a survival competition is held where the best N chromosomes from the parent and offspring populations are selected to form the next generation.

   CHC also implements a form of heterogeneous recombination using HUX, a special recombination operator. HUX exchanges half of the bits that differ between parents, where the bit position to be exchanged is randomly determined. CHC also employs a method of incest prevention. Before applying HUX to two parents, the Hamming distance between them is measured. Only those parents who differ from each other by some number of bits (mating threshold) are mated. The initial threshold is set at $L/4$, where L is the length of the chromosomes. If no offspring are inserted into the new population then the threshold is reduced by one.

   No mutation is applied during the recombination phase. Instead, when the population converges or the search stops making progress, the population is reinitialized to introduce new diversity to the search. The chromosome representing the best solution found over the course of the search is used as a template to reseed the population. Reseeding of the population is accomplished by randomly changing 35 % of the bits in the template chromosome to form each of the other $N - 1$ new chromosomes in the population. The search is then resumed.

- **Steady-state memetic algorithm (SSMA)** [62]—The SSMA was proposed to cover a drawback of the conventional evolutionary PS methods that had appeared before: their lack of convergence when facing large problems. SSMA makes use of a local search or meme specifically developed for this PS problem. This interweaving of the global and local search phases allows the two to influence each other; i.e. SSGA chooses good starting points, and local search provides an accurate representation of that region of the domain. A brief pseudocode of the SSMA is shown as follows:

```
Initialize population
While (not termination-condition) do
    Use binary tournament to select two parents
    Apply crossover operator to create
    offspring (Off1,Off2)
    Apply mutation to Off1 and Off2
        Evaluate Off1 and Off2
        For each Offi
            Invoke Adaptive-PLS-mechanism to
            obtain PLSi for Offi
            If υ(0,1) < PLSi then
                Perform meme optimization for Offi
            End if
        End for
    Employ standard replacement for Off1 and Off2
End while
Return the best chromosome
```

- **COoperative COevolutionary Instance Selection (CoCoIS)** [72]—The cooperative algorithm presents the following steps:

```
Initialize population of combinations
Initialize subpopulations of selectors
While Stopping criterion not met do
    For N iterations do
        Evaluate population of combinations
        Select two individuals by roulette selection
            and perform two point crossover
        Offspring substitutes two worst individuals
        Perform mutation with probability Pmutation
    For M iterations do
        Foreach subpopulation i do
            Evaluate selectors of subpopulation i
            Copy Elitism% to new subpopulation i
            Fill (1-Elitism)% of subpopulation
            i by HUX crossover
            Apply random mutation with probability Prandom
            Apply RNN mutation with probability Prnn
        Evaluate population of combinations
```

Where N and M are the number of iterations.

8.4.3.7 Fixed+Wrapper

- **Random Mutation Hill Climbing (RMHC)** [147]—It randomly selects a subset S from TR which contains a fixed number of instances s ($s = \%|TR|$). In each iteration, the algorithm interchanges an instance from S with another from TR - S. The change is maintained if it offers better accuracy.

8.5 Related and Advanced Topics

Research in enhancing instance and PS through other data reduction and learning methods is common and in high demand nowadays. PS could represent a feasible and promising technique to obtain expected results, which justifies its relationship to other methods and problems. This section provides a wide review of other topics closely related to PS and describes other works and future trends which have been studied in the last few years. In each subsection, we provide a table that enumerates, in not an exhaustive way, the most relevant methods and papers in each of the topics. Although we do not extend them, it is included for informative purposes for the interested reader.

8.5.1 Prototype Generation

Prototype generation methods are not limited only to select examples from the training set. They could also modify the values of the samples, changing their position in the d-dimensional space considered. Most of them use merging or divide and conquer strategies to set new artificial samples [27], or are based on clustering approaches [12], LVQ [98] hybrids, advanced proposals [102, 113] and evolutionary algorithms based schemes [25, 151, 152]. A complete survey on this topic is [153].

Table 8.2 itemizes the main prototype generation methods proposed in the literature.

8.5.2 Distance Metrics, Feature Weighting and Combinations with Feature Selection

This area refers to the combination of IS and PS methods with other well-known schemes used for improving accuracy in classification problems. For example, the weighting scheme combines the PS with the FS [40, 147] or Feature Weighting [55, 129, 163], where a vector of weights associated with each attribute determines and influences the distance computations.

Table 8.2 Some of the most important prototype generation methods

Complete name	Abbr. name	Reference
Prototype nearest neighbor	PNN	[27]
Generalized editing using nearest neighbor	GENN	[99]
Learning vector quantization	LVQ	[98]
Chen algorithm	Chen	[29]
Modified Chang's algorithm	MCA	[12]
Integrated concept prototype learner	ICPL	[102]
Depuration algorithm	Depur	[140]
Hybrid LVQ3 algorithm	HYB	[90]
Reduction by space partitioning	RSP	[141]
Evolutionary nearest prototype classifier	ENPC	[54]
Adaptive condensing algorithm based on mixtures of Gaussians	MixtGauss	[113]
Self-generating prototypes	SGP	[52]
Adaptive Michigan PSO	AMPSO	[25]
Iterative prototype adjustment by differential evolution	IPADE	[151]
Differential evolution	DE	[152]

Several distance metrics have been used with kNN and PS, especially when working with categorical attributes [166]. There are some PS approaches which learn not only the subset of the selected prototype, but also the distance metric employed [59, 128]. Also, PS is suitable for use on other types of dissimilarity based classifiers [95, 131].

Table 8.3 enumerates the main advances in these topics proposed in the literature.

8.5.3 Hybridizations with Other Learning Methods and Ensembles

On the one hand, this family includes all the methods which simultaneously use instances and rules in order to compute the classification of a new object. If the values of the object are within the range of a rule, its consequent predicts the class; otherwise, if no rule matches the object, the most similar rule or instance stored in the data base is used to estimate the class. Similarity is viewed as the closest rule or instance based on a distance measure. In short, these methods can generalize an instance into a hyperrectangle or rule [50, 67, 114].

On the other hand, this area refers to ensemble learning, where an IS method is run several times and a classification decision is made according to the majority class obtained over several subsets and any performance measure given by a learner [5, 71].

Table 8.3 IS combined with FS and weighting

Description	Reference
Random mutation hill climbing for simultaneous instance and feature selection	[147]
Review of feature weighting methods for lazy learning algorithms	[163]
Distance functions for instance-based learning methods	[166]
Prototype reduction for sublinear space methods	[92]
Prototype reduction for sublinear space methods using ensembles	[93]
Learning feature weighting schemes for KNN	[129]
PS and feature weighting	[128]
PS for dissimilarity-based classifiers	[131]
Prototype reduction (selection and generation) for dissimilarity-based classifiers	[95]
Optimization of feature and instance selection with co-evolutionary algorithms	[59]
Prototype reduction and feature weighting	[55]
Instance and feature selection with cooperative co-evolutionary algorithms	[40]
Genetic algorithms for optimizing dissimilarity-based classifiers	[134]
Learning with weighted instances	[169]
Experimental review on prototype reduction for dissimilarity-based classifiers	[97]
Unification of feature and instance selection	[172]
Evolutionary IS with fuzzy rough FS	[43]
IS, instance weighting and feature weighting with co-evolutionary algorithms	[44]
Fuzzy rough IS for evolutionary FS	[45]
Feature and instance selection with genetic algorithms	[155]
Multi-objective genetic algorithm for optimizing instance weighting	[124]

Table 8.4 specifies the proposals in these issues proposed in the literature.

8.5.4 Scaling-Up Approaches

One of the disadvantages of the IS methods is that most of them report a prohibitive run time or even cannot be applied over large size data sets. Recent improvements in this field cover the stratification of data [20, 70, 81] and the development of distributed approaches for PS [6].

Table 8.5 draws the research works done in scaling-up for IS.

8.5.5 Data Complexity

This area studies the effect on the complexity of data when PS methods are applied previous to the classification [96] or how to make a useful diagnosis of the benefits of applying PS methods taking into account the complexity of the data [64, 119].

Table 8.4 Hybridizations with other learning approaches and ensembles

Description	Reference
First approach for nested generalized examples learning (hyperrectangle learning): EACH	[138]
Experimental review on nested generalized examples learning	[162]
Unification of rule induction with instance-based learning: RISE	[50]
Condensed nearest neighbour (CNN) ensembles	[5]
Inflating instances to obtain rules: INNER	[114]
Bagging for lazy learning	[174]
Evolutionary ensembles for classifiers selection	[142]
Ensembles for weighted IS	[71]
Boostrapping for KNN	[148]
Evolutionary optimization in hyperrectangles learning	[67]
Evolutionary optimization in hyperrectangles learning for imbalanced problems	[69]
Review of ensembles for data preprocessing in imbalanced problems	[60]
Boosting by warping of the distance metric for KNN	[121]
Evolutionary undersampling based on ensembles for imbalanced problems	[61]

Table 8.5 Scaling-up and distributed approaches

Description	Reference
Recursive subdivision of prototype reduction methods for tackling large data sets	[91]
Stratified division of training data sets to improve the scaling-up of PS methods	[20]
Usage of KD-trees for prototype reduction schemes	[120]
Distributed condensation for large data sets	[6]
Divide-and-conquer recursive division of training data for speed-up IS	[81]
Division of data based of ensembles with democratic voting for IS	[70]
Usage of stratification for scaling-up evolutionary algorithms for IS	[41]
Distributed implementation of the stratification process combined with k-means for IS	[33]
Scalable divide-and-conquer based on bookkeeping for instance and feature selection	[74]
Scaling-up IS based on the parallelization of small subsets of data	[82]

Table 8.6 collects the developments in data complexity related with IS found in the specialized literature.

8.6 Experimental Comparative Analysis in Prototype Selection

The aim of this section is to show all the factors and issues related to the experimental study. We specify the data sets, validation procedure, parameters of the algorithms, performance metrics and PS methods involved in the analysis. The experimental

Table 8.6 IS and data complexity

Description	Reference
Data characterization for effective edition and condensation schemes	[119]
Data characterization for effective PS	[64]
Usage of PS for enhance the computation of data complexity measures	[96]
Data characterization for effective under-sampling and over-sampling in imbalanced problems	[115]
Meta-learning framework for IS	[103]
Prediction of noise filtering efficacy with data complexity measures for KNN	[137]

conditions were discussed in Chap. 2 of this book. The data sets used are summarized in Table 8.7.

The data sets considered are partitioned using the 10-FCV procedure. The parameters of the PS algorithms are those recommended by their respective authors. We assume that the choice of the values of parameters is optimally chosen by their own authors. Nevertheless, in the PS methods that require the specification of the number of neighbors as a parameter, its value coincides with the k value of the KNN rule afterwards. But all edition methods consider a minimum of 3 nearest neighbors to operate (as recommended in [165]), although they were applied to a 1NN classifier. The Euclidean distance is chosen as the distance metric because it is well-known and the most used for KNN. All probabilistic methods (including incremental methods which depend on the order of instance presentation) are run three times and the final results obtained correspond to the average performance values of these runs.

Thus, the empirical study involves 42 PS methods from those listed in Table 8.1. We want to outline that the implementations are only based on the descriptions and specifications given by the respective authors in their papers. No advanced data structures and enhancements for improving the efficiency of PS methods have been carried out. All methods (including the slowest ones) are collected in KEEL software [3].

8.6.1 Analysis and Empirical Results on Small Size Data Sets

Table 8.8 presents the average results obtained by the PS methods over the 39 small size data sets. $Red.$ denotes reduction rate achieved, $tst\ Acc.$ and $tst\ Kap.$ denote the accuracy and kappa obtained in test data, respectively; $Acc. * Red.$ and $Kap. * Red.$ correspond to the product of accuracy/kappa and reduction rate, which is an estimator of how good a PS method is considering a tradeoff of reduction and success rate of classification. Finally, $Time$ denotes the average time elapsed in seconds to complete a run of a PS method.[1] In the case of 1NN, the time required is not displayed due to the fact that no PS stage is run before. For each type of result, the algorithms are ordered from the best to the worst. Algorithms highlighted in bold are those which obtain

[1] The machine used was an Intel Core i7 CPU 920 at 2.67GHz with 4GB of RAM.

Table 8.7 Enumeration of data sets used in the experimental study

Data set	Data set
Abalone	Appendicitis
Australian	Autos
Balance	Banana
Bands	Breast
Bupa	Car
Chess	Cleveland
Coil2000	Contraceptive
Crx	Dermatology
Ecoli	Flare-solar
German	Glass
Haberman	Hayes
Heart	Hepatitis
Housevotes	Iris
Led7digit	Lymphography
Magic	Mammographic
Marketing	Monk-2
Newthyroid	Nursery
Pageblocks	Penbased
Phoneme	Pima
Ring	Saheart
Satimage	Segment
Sonar	Spambase
Specfheart	Splice
Tae	Texture
Thyroid	Tic-tac-toe
Titanic	Twonorm
Vehicle	Vowel
Wine	Wisconsin
Yeast	Zoo

the best result in their corresponding family, according to the taxonomy illustrated in Fig. 8.3. They will make up the experimental study of medium size data sets, showed in the next subsection.

The Wilcoxon test [39, 63, 164] is adopted considering a level of significance of $\alpha = 0.1$. Table 8.9 shows a summary of all the possible comparisons employing the Wilcoxon test among all PS methods over small data sets. This table collects the statistical comparisons of the four main performance measures used in this chapter: $tst\ Acc.$, $tst\ Kap.$, $Acc. * Red.$ and $Kap. * Red.$. Table 8.9 shows, for each method in the row, the number of PS methods outperformed by using the Wilcoxon test under the column represented by the '+' symbol. The column with the '±' symbol indicates

Table 8.8 Average results obtained by the PS methods over small data sets

Red.		*tst Acc.*		*tst Kap.*		*Acc.* ∗ *Red.*		*Kap.* ∗ *Red.*		*Time*	
Explore	0.9789	**CHC**	0.7609	**SSMA**	0.5420	**CHC**	0.7399	**CHC**	0.5255	1NN	–
CHC	0.9725	SSMA	0.7605	CHC	0.5404	SSMA	0.7283	SSMA	0.5190	CNN	0.0027
NRMCS	0.9683	GGA	0.7566	GGA	0.5328	Explore	0.7267	GGA	0.5014	POP	0.0091
SSMA	0.9576	**RNG**	0.7552	**RMHC**	0.5293	GGA	0.7120	**RMHC**	0.4772	PSC	0.0232
GGA	0.9411	**RMHC**	0.7519	**HMNEI**	0.5277	**RMHC**	0.6779	Explore	0.4707	ENN	0.0344
RNN	0.9187	**MoCS**	0.7489	**RNG**	0.5268	**RNN**	0.6684	**RNN**	0.4309	IB3	0.0365
CCIS	0.9169	ENN	0.7488	**MoCS**	0.5204	NRMCS	0.6639	CoCoIS	0.4294	MSS	0.0449
IGA	0.9160	NCNEdit	0.7482	NCNEdit	0.5122	IGA	0.6434	IGA	0.4080	Multiedit	0.0469
CPruner	0.9129	AllKNN	0.7472	ENN	0.5121	CoCoIS	0.6281	**MCNN**	0.4051	ENNTh	0.0481
MCNN	0.9118	**HMNEI**	0.7436	AllKNN	0.5094	**MCNN**	0.6224	NRMCS	0.3836	FCNN	0.0497
RMHC	0.9015	ENNTh	0.7428	CoCoIS	0.4997	**CCIS**	0.6115	**DROP3**	0.3681	MoCS	0.0500
CoCoIS	0.8594	Explore	0.7424	ENNTh	0.4955	**CPruner**	0.6084	**CCIS**	0.3371	MCNN	0.0684
DROP3	0.8235	MENN	0.7364	1NN	0.4918	DROP3	0.5761	**IB3**	0.3248	MENN	0.0685
SNN	0.7519	1NN	0.7326	**POP**	0.4886	**IB3**	0.4997	CPruner	0.3008	AllKNN	0.0905
ICF	0.7160	CoCoIS	0.7309	MENN	0.4886	**ICF**	0.4848	**ICF**	0.2936	TRKNN	0.1040
IB3	0.7114	**POP**	0.7300	Explore	0.4809	PSC	0.4569	HMNEI	0.2929	CCIS	0.1090
PSC	0.7035	**RNN**	0.7276	Multiedit	0.4758	TCNN	0.4521	TCNN	0.2920	HMNEI	0.1234
SVBPS	0.6749	Multiedit	0.7270	**MSS**	0.4708	FCNN	0.4477	FCNN	0.2917	ENRBF	0.1438
Shrink	0.6675	MSS	0.7194	RNN	0.4691	SVBPS	0.4448	MNV	0.2746	PSRCG	0.1466
TCNN	0.6411	**FCNN**	0.7069	**FCNN**	0.4605	SNN	0.4324	CNN	0.2631	CPruner	0.1639
FCNN	0.6333	MCS	0.7060	**IB3**	0.4566	MNV	0.4266	SVBPS	0.2615	ICF	0.1708

(continued)

Table 8.8 (continued)

Red.		*tst Acc.*		*tst Kap.*		*Acc. ∗ Red.*		*Kap. ∗ Red.*		*Time*	
MNV	0.6071	CNN	0.7057	CNN	0.4560	HMNEI	0.4128	PSC	0.2594	Shrink	0.1811
CNN	0.5771	TCNN	0.7052	MCS	0.4559	CNN	0.4072	**MENN**	0.2443	VSM	0.1854
VSM	0.5669	IKNN	0.7027	TCNN	0.4555	**Reconsistent**	0.3840	**Reconsistent**	0.2406	IKNN	0.1920
Reconsistent	0.5581	MNV	0.7026	MNV	0.4523	**MENN**	0.3682	MCS	0.2348	NRMCS	0.2768
HMNEI	0.5551	**IB3**	0.7024	IKNN	0.4494	MCS	0.3637	ENNTh	0.2294	NCNEdit	0.3674
TRKNN	0.5195	IGA	0.7024	**DROP3**	0.4470	VSM	0.3600	TRKNN	0.2077	MCS	0.4126
MCS	0.5151	**DROP3**	0.6997	IGA	0.4455	TRKNN	0.3496	MSS	0.2073	DROP3	0.5601
PSRCG	0.5065	Reconsistent	0.6880	MCNN	0.4443	ENNTh	0.3439	PSRCG	0.2072	SNN	0.7535
MENN	0.5000	NRMCS	0.6856	Reconsistent	0.4310	PSRCG	0.3433	SNN	0.1983	SVBPS	1.0064
ENNTh	0.4629	ENRBF	0.6837	ICF	0.4101	Shrink	0.3411	VSM	0.1964	TCNN	1.9487
GCNN	0.4542	MCNN	0.6826	PSRCG	0.4092	MSS	0.3168	**AllKNN**	0.1799	Explore	2.1719
MSS	0.4404	PSRCG	0.6779	TRKNN	0.3999	GCNN	0.3022	GCNN	0.1774	MNV	2.5741
AllKNN	0.3532	ICF	0.6772	NRMCS	0.3962	**AllKNN**	0.2639	Multiedit	0.1657	Reconsistent	4.5228
Multiedit	0.3483	TRKNN	0.6729	GCNN	0.3905	Multiedit	0.2532	IKNN	0.1444	RNG	7.1695
IKNN	0.3214	CCIS	0.6669	SVBPS	0.3875	IKNN	0.2258	ENN	0.1293	RNN	16.1739
ENRBF	0.3042	CPruner	0.6664	PSC	0.3687	ENRBF	0.2080	RNG	0.1243	CHC	23.7252
ENN	0.2525	GCNN	0.6654	CCIS	0.3676	ENN	0.1891	Shrink	0.1152	SSMA	27.4869
RNG	0.2360	SVBPS	0.6591	VSM	0.3465	RNG	0.1782	NCNEdit	0.1146	RMHC	32.2845
NCNEdit	0.2237	PSC	0.6495	ENRBF	0.3309	NCNEdit	0.1674	ENRBF	0.1007	GCNN	61.4989
MoCS	0.1232	VSM	0.6350	CPruner	0.3295	MoCS	0.0923	MoCS	0.0641	GGA	84.9042
POP	0.0762	SNN	0.5751	SNN	0.2638	POP	0.0556	POP	0.0372	IGA	122.1011
1NN	–	Shrink	0.5110	Shrink	0.1726	1NN	–	1NN	–	CoCoIS	267.3500

Table 8.9 Wilcoxon test results over small data sets

	tst Acc.		tst Kap.		Acc. * Red.		Kappa. * Red.	
	+	±	+	±	+	±	+	±
AllKNN	25	40	22	40	7	16	10	25
CCIS	4	21	2	25	30	36	26	35
CHC	31	**41**	29	**41**	**41**	**41**	38	**41**
CNN	8	19	7	27	12	20	15	27
CoCoIS	22	37	20	38	29	36	27	38
CPruner	8	25	0	21	28	34	1	26
DROP3	7	28	5	31	29	32	29	35
ENN	24	39	22	38	3	9	6	20
ENNTh	22	**41**	19	**41**	10	27	13	27
ENRBF	20	37	0	29	3	13	0	7
Explore	22	38	11	35	38	40	33	40
FCNN	5	20	4	26	14	25	13	28
GCNN	4	27	5	31	2	14	1	14
GGA	27	40	25	40	37	39	38	40
HMNEI	25	39	22	**41**	9	27	15	30
IB3	5	23	5	29	23	29	22	31
ICF	4	21	2	24	22	28	18	30
IGA	7	25	5	28	30	34	32	35
IKNN	11	29	11	32	1	11	2	12
MCNN	2	15	5	24	29	34	29	34
MCS	7	23	9	29	16	28	13	30
MENN	27	**41**	21	**41**	11	27	12	29
MNV	6	20	6	26	14	26	16	29
ModelCS	26	39	26	**41**	1	2	1	8
MSS	17	29	17	32	2	12	4	20
Multiedit	23	35	7	34	7	15	7	18
NCNEdit	26	40	27	**41**	2	9	3	18
NRMCS	6	28	2	28	36	38	25	38
POP	16	33	19	38	0	0	0	4
PSC	0	10	3	11	17	26	9	27
PSRCG	5	15	4	19	5	16	2	20
Reconsistent	4	16	4	22	12	18	8	24
RMHC	27	38	26	40	33	37	34	39
RNG	**34**	**41**	29	**41**	3	9	5	18
RNN	15	30	7	30	33	36	33	37
Shrink	0	5	0	2	7	22	0	13
SNN	0	4	1	5	15	26	2	27
SSMA	28	**41**	**30**	**41**	39	40	**39**	**41**
SVBPS	2	17	3	24	18	27	12	27
TCNN	8	24	5	27	15	24	16	27
TRKNN	2	17	3	24	11	22	5	21
VSM	1	8	2	13	6	17	2	16

the number of wins and ties obtained by the method in the row. The maximum value for each column is highlighted in bold.

Observing Tables. 8.8 and 8.9, we can point out the best performing PS methods:

- In condensation incremental approaches, all methods are very similar in behavior, except PSC, which obtains the worst results. FCNN could be highlighted in accuracy/kappa performance and MCNN with respect to reduction rate with a low decrease in efficacy.
- Two methods can be emphasized in from the condensation decremental family: RNN and MSS. RNN obtains good reduction rates and accuracy/kappa performances, whereas MSS also offers good performance. RNN has the drawback of being quite slow.
- In general, the best condensation methods in terms of efficacy are the decremental ones, but their main drawback is that they require more computation time. POP and MSS methods are the best performing in terms of accuracy/kappa, although the reduction rates are low, especially those achieved by POP. However, no condensation method is more accurate than 1NN.
- With respect to edition decremental approaches, few differences can be observed. ENN, RNGE and NCNEdit obtain the best results in accuracy/kappa and MENN and ENNTh offers a good tradeoff considering the reduction rate. Multiedit and ENRBF are not on par with their competitors and they are below 1NN in terms of accuracy.
- AllKNN and MoCS, in edition batch approaches, achieve similar results to the methods belonging to the decremental family. AllKNN achieves better reduction rates.
- Within the hybrid decremental family, three methods deserve mention: DROP3, CPruner and NRMCS. The latter is the best, but curiously, its time complexity rapidly increases in the presence of larger data sets and it cannot tackle medium size data sets. DROP3 is more accurate than CPruner, which achieves higher reduction rates.
- Considering the hybrid mixed+wrapper methods, SSMA and CHC techniques achieve the best results.
- Remarkable methods belonging to the hybrid family are DROP3, CPruner, HMNEI, CCIS, SSMA, CHC and RMHC. Wrapper based approaches are slower.
- The best global methods in terms of accuracy or kappa are MoCS, RNGE and HMNEI.
- The best global methods considering the tradeoff reduction-accuracy/kappa are RMHC, RNN, CHC, Explore and SSMA.

8.6.2 Analysis and Empirical Results on Medium Size Data Sets

This section presents the study and analysis of medium size data sets and the best PS methods per family, which are those highlighted in bold in Table 8.8. The goal pursued is to study the effect of scaling up the data in PS methods. Table 8.10 shows the average results obtained in the distinct performance measures considered (it follows the same format as Table 8.8) and Table 8.11 summarizes the Wilcoxon test results over medium data sets.

We can analyze several details from the results collected in Tables. 8.10 and 8.11:

- Five techniques outperform 1NN in terms of accuracy/kappa over medium data sets: RMHC, SSMA, HMNEI, MoCS and RNGE. Two of them are edition schemes (MoCS and RNGE) and the rest are hybrid schemes. Again, no condensation method is more accurate than 1NN.
- Some methods present clear differences when dealing with larger data sets. This is the case with AllKNN, MENN and CHC. The first two, tend to try new reduction passes in the edition process, which is against the interests of accuracy and kappa, and in medium size problems this fact is more noticeable. Furthermore, CHC loses the balance between reduction and accuracy when data size increases, due to the fact that the reduction objective becomes easier.
- There are some techniques whose run could be prohibitive when the data scales up. This is the case for RNN, RMHC, CHC and SSMA.
- The best methods in terms of accuracy or kappa are RNGE and HMNEI.
- The best methods considering the tradeoff reduction-accuracy/kappa are RMHC, RNN and SSMA.

8.6.3 Global View of the Obtained Results

Assuming the results obtained, several PS methods could be emphasized according to the accuracy/kappa obtained (RMHC, SSMA, HMNEI, RNGE), the reduction rate achieved (SSMA, RNN, CCIS) and computational cost required (POP, FCNN). However, we want to remark that the choice of a certain method depends on various factors and the results are offered here with the intention of being useful in making this decision. For example, an edition scheme will usually outperform the standard kNN classifier in the presence of noise, but few instances will be removed. This fact could determine whether the method is suitable or not to be applied over larger data sets, taking into account the expected size of the resulting subset. We have seen that the PS methods which allow high reduction rates while preserving accuracy are usually the slowest ones (hybrid mixed approaches such as SSMA) and they may require an advanced mechanism to be applied over large size data sets or they may even be useless under these circumstances. Fast methods that achieve high reduction rates are the condensation approaches, but we have seen that they are not able to improve kNN in terms of accuracy. In short, each method has advantages and disadvantages and the results offered in this section allow an informed decision to be made within each category.

In short, and focusing on the objectives usually considered in the use of PS algorithms, we can suggest the following, to choose the proper PS algorithm:

Table 8.10 Average results obtained by the best PS methods per family over medium data sets

Red.		*tst Acc.*		*tst Kap.*		*Acc. ∗ Red.*		*Kap. ∗ Red.*		*Time*	
MCNN	0.9914	RMHC	0.8306	RMHC	0.6493	SSMA	0.8141	SSMA	0.6328	1NN	–
CHC	0.9914	SSMA	0.8292	SSMA	0.6446	CHC	0.8018	CHC	0.6006	POP	0.1706
SSMA	0.9817	RNG	0.8227	HMNEI	0.6397	RNN	0.7580	RMHC	0.5844	CNN	1.1014
CCIS	0.9501	HMNEI	0.8176	ModelCS	0.6336	RMHC	0.7476	RNN	0.5617	FCNN	3.2733
RNN	0.9454	ModelCS	0.8163	RNG	0.6283	GGA	0.7331	GGA	0.5513	MCNN	4.4177
GGA	0.9076	CHC	0.8088	1NN	0.6181	CCIS	0.6774	IB3	0.4615	IB3	6.6172
RMHC	0.9001	GGA	0.8078	POP	0.6143	CPruner	0.6756	FCNN	0.4588	MSS	7.9165
DROP3	0.8926	1NN	0.8060	MSS	0.6126	MCNN	0.6748	DROP3	0.4578	CCIS	12.4040
CPruner	0.8889	AllKNN	0.8052	GGA	0.6074	DROP3	0.6635	CPruner	0.4555	ModelCS	15.4658
ICF	0.8037	POP	0.8037	CHC	0.6058	IB3	0.6144	CCIS	0.5579	AllKNN	24.6167
IB3	0.7670	RNN	0.8017	FCNN	0.6034	FCNN	0.6052	CNN	0.4410	HMNEI	28.9782
FCNN	0.7604	IB3	0.8010	IB3	0.6018	CNN	0.5830	MCNN	0.4295	CPruner	35.3761
CNN	0.7372	MSS	0.8008	CNN	0.5982	ICF	0.5446	Reconsistent	0.3654	MENN	37.1231
Reconsistent	0.6800	FCNN	0.7960	AllKNN	0.5951	Reconsistent	0.5101	MSS	0.3513	ICF	93.0212
MSS	0.5735	CNN	0.7909	RNN	0.5941	MSS	0.4592	HMNEI	0.3422	DROP3	160.0486
HMNEI	0.5350	MENN	0.7840	MENN	0.5768	HMNEI	0.4374	ICF	0.3337	Reconsistent	1,621.7693
MENN	0.3144	CPruner	0.7600	Reconsistent	0.5373	MENN	0.2465	MENN	0.1814	RNG	1,866.7751
AllKNN	0.2098	Reconsistent	0.7501	DROP3	0.5129	AllKNN	0.1689	AllKNN	0.1248	SSMA	6,306.6313
RNG	0.1161	DROP3	0.7433	CPruner	0.5124	RNG	0.0955	RNG	0.0729	CHC	6,803.7974
POP	0.0820	CCIS	0.7130	CCIS	0.4714	POP	0.0659	POP	0.0504	RMHC	12,028.3811
ModelCS	0.0646	MCNN	0.6806	MCNN	0.4332	ModelCS	0.0527	ModelCS	0.0409	GGA	21,262.6911
1NN	–	ICF	0.6776	ICF	0.4152	1NN	–	1NN	–	RNN	24,480.0439

Table 8.11 Wilcoxon test results over medium data sets

	tst Acc.		tst Kap.		Acc. * Red.		Kappa. * Red.	
	+	±	+	±	+	±	+	±
AllKNN	9	19	10	19	3	4	3	6
CCIS	1	7	0	4	11	18	4	14
CHC	5	**20**	5	19	**19**	**20**	15	**20**
CNN	4	10	5	13	6	11	7	15
CPruner	3	14	0	5	8	14	2	15
DROP3	2	11	2	10	9	14	8	15
FCNN	4	12	5	14	6	16	9	17
GGA	5	13	4	14	12	18	12	17
HMNEI	10	19	12	**20**	5	10	5	14
IB3	2	11	4	9	9	17	9	16
ICF	0	4	0	7	6	11	3	11
MCNN	0	2	0	4	10	17	7	18
MENN	11	19	8	18	4	8	4	9
ModelCS	10	**20**	12	**20**	1	1	0	3
MSS	6	18	9	18	3	10	4	11
POP	7	**20**	10	**20**	0	0	0	2
Reconsistent	1	10	3	10	3	9	4	11
RMHC	11	19	9	19	13	18	14	19
RNG	**15**	**20**	**15**	**20**	2	3	0	4
RNN	4	14	4	12	14	18	15	19
SSMA	8	**20**	9	19	**19**	**20**	**19**	**20**

- For the tradeoff reduction-accuracy rate: The algorithms which obtain the best behavior are RMHC and SSMA. However, these methods achieve a significant improvement in the accuracy rate due to a high computation cost. The methods that harm the accuracy at the expense of a great reduction of time complexity are DROP3 and CCIS.
- If the interest is the accuracy rate: In this case, the best results are to be achieved with the RNGE as editor and HMNEI as hybrid method.
- When the key factor is the condensation: FCNN is the highlighted one, being one of the fastest.

8.6.4 Visualization of Data Subsets: A Case Study Based on the Banana Data Set

This section is devoted to illustrating the subsets selected resulting from some PS algorithms considered in our study. To do this, we focus on the banana data set, which

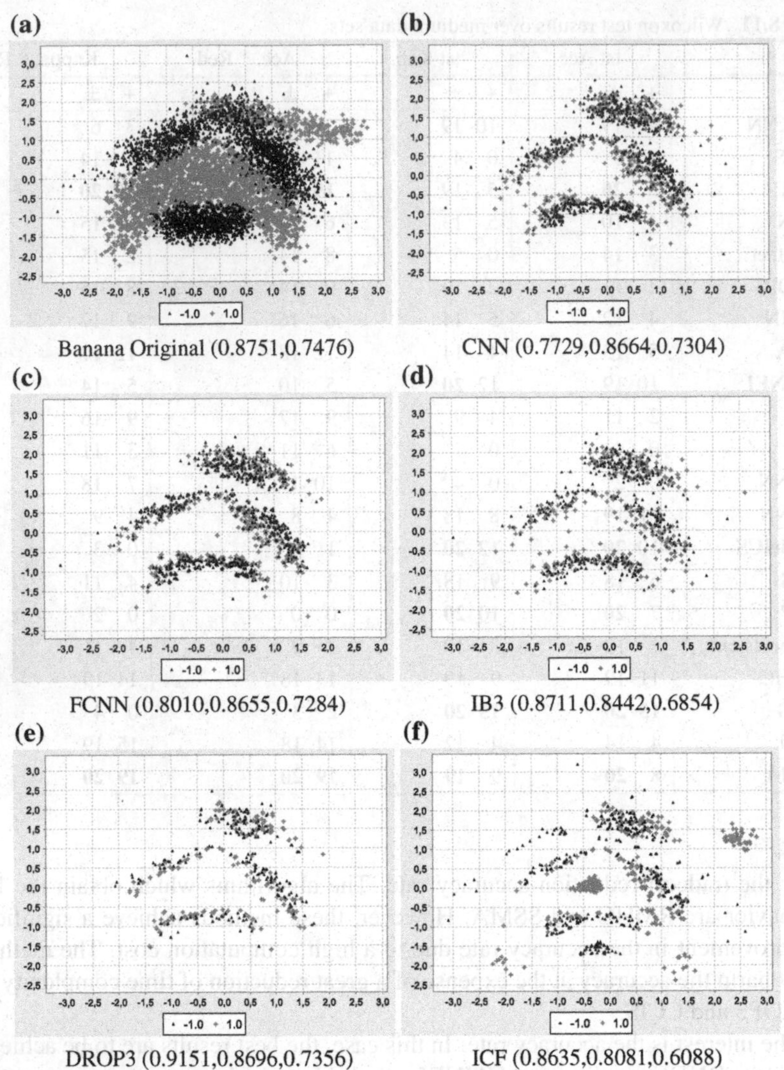

Fig. 8.4 Data subsets in banana data set (1). **a** Banana original (0.8751, 0.7476). **b** CNN (0.7729, 0.8664, 0.7304). **c** FCNN (0.8010, 0.8655, 0.7284). **d** IB3 (0.8711, 0.8442, 0.6854). **e** DROP3 (0.9151, 0.8696, 0.7356). **f** ICF (0.8635, 0.8081, 0.6088)

contains 5,300 examples in the complete set (Figs. 8.4 and 8.5). It is an artificial data set of 2 classes composed of three well-defined clusters of instances of the class -1 and two clusters of the class 1. Although the borders are clear among the clusters there is a high overlap between both classes. The complete data set is illustrated in Fig. 8.4a.

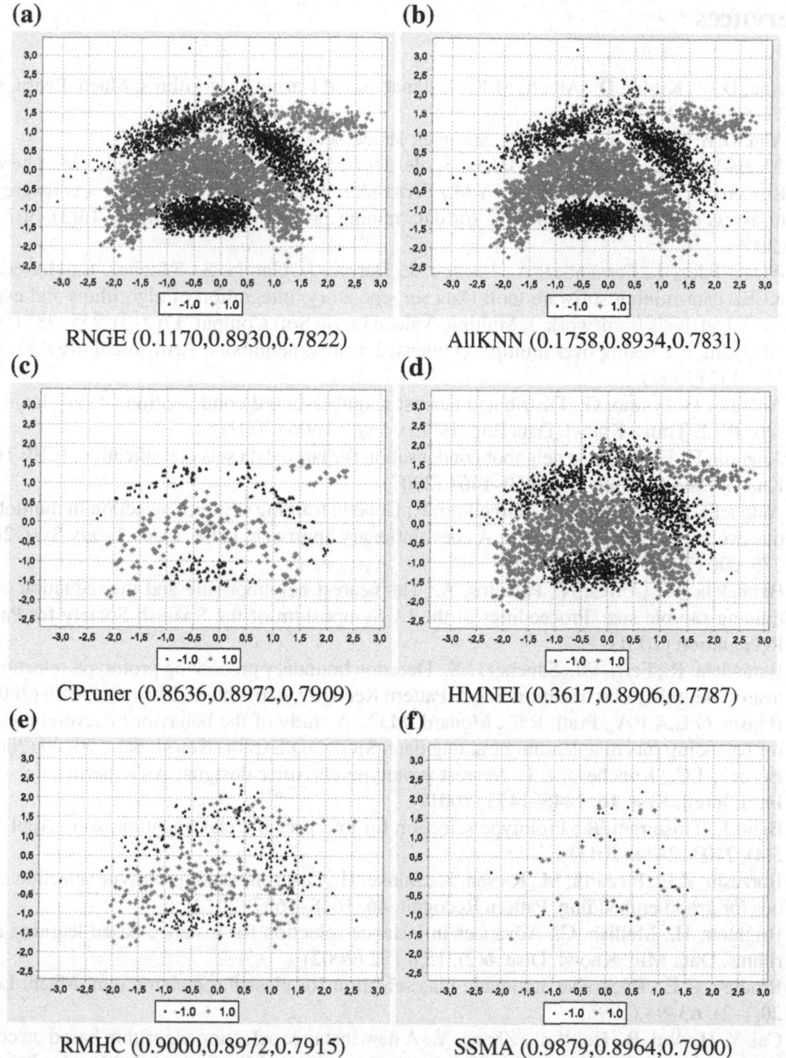

Fig. 8.5 Data subsets in banana data set (2). **a** RNGE (0.1170, 0.8930, 0.7822). **b** AllKNN (0.1758, 0.8934, 0.7831). **c** CPruner (0.8636, 0.8972, 0.7909). **d** HMNEI (0.3617, 0.8906, 0.7787). **e** RMHC (0.9000, 0.8972, 0.7915). **f** SSMA (0.9879, 0.8964, 0.7900)

The pictures of the subset selected by some PS methods could help to visualize and understand their way of working and the results obtained in the experimental study. The reduction rate, the accuracy and kappa values in test data registered in the experimental study are specified in this order for each one. In original data sets, the two values indicated correspond to accuracy and kappa with 1NN (Fig. 8.4a).

References

1. Aha, D.W., Kibler, D., Albert, M.K.: Instance-based learning algorithms. Mach. Learn. **6**(1), 37–66 (1991)
2. Aha, D.W. (ed.): Lazy Learning. Springer, Heidelberg (2010)
3. Alcalá-Fdez, J., Sánchez, L., García, S., del Jesus, M.J., Ventura, S., Garrell, J.M., Otero, J., Romero, C., Bacardit, J., Rivas, V.M., Fernández, J.C., Herrera, F.: KEEL: a software tool to assess evolutionary algorithms for data mining problems. Soft Comput. **13**(3), 307–318 (2009)
4. Alcalá-Fdez, J., Fernández, A., Luengo, J., Derrac, J., García, S., Sánchez, L., Herrera, F.: KEEL data-mining software tool: Data set repository, integration of algorithms and experimental analysis framework. J. Multiple-Valued Logic Soft Comput. **17**(2–3), 255–287 (2011)
5. Alpaydin, E.: Voting over multiple condensed nearest neighbors. Artif. Intell. Rev. **11**(1–5), 115–132 (1997)
6. Angiulli, F., Folino, G.: Distributed nearest neighbor-based condensation of very large data sets. IEEE Trans. Knowl. Data Eng. **19**(12), 1593–1606 (2007)
7. Angiulli, F.: Fast nearest neighbor condensation for large data sets classification. IEEE Trans. Knowl. Data Eng. **19**(11), 1450–1464 (2007)
8. Antonelli, M., Ducange, P., Marcelloni, F.: Genetic training instance selection in multiobjective evolutionary fuzzy systems: A coevolutionary approach. IEEE Trans. Fuzzy Syst. **20**(2), 276–290 (2012)
9. Barandela, R., Cortés, N., Palacios, A.: The nearest neighbor rule and the reduction of the training sample size. Proceedings of the IX Symposium of the Spanish Society for Pattern Recognition (2001)
10. Barandela, R., Ferri, F.J., Sánchez, J.S.: Decision boundary preserving prototype selection for nearest neighbor classification. Int. J. Pattern Recognit Artif Intell. **19**(6), 787–806 (2005)
11. Batista, G.E.A.P.A., Prati, R.C., Monard, M.C.: A study of the behavior of several methods for balancing machine learning training data. SIGKDD Explor. Newsl. **6**(1), 20–29 (2004)
12. Bezdek, J.C., Kuncheva, L.I.: Nearest prototype classifier designs: An experimental study. Int. J. Intell. Syst. **16**, 1445–1473 (2001)
13. Bien, J., Tibshirani, R.: Prototype selection for interpretable classification. Ann. Appl. Stat. **5**(4), 2403–2424 (2011)
14. Borzeshi, Z.E., Piccardi, M., Riesen, K., Bunke, H.: Discriminative prototype selection methods for graph embedding. Pattern Recognit. **46**, 1648–1657 (2013)
15. Brighton, H., Mellish, C.: Advances in instance selection for instance-based learning algorithms. Data Min. Knowl. Disc. **6**(2), 153–172 (2002)
16. Brodley, C.E.: Recursive automatic bias selection for classifier construction. Mach. Learn. **20**(1–2), 63–94 (1995)
17. Cai, Y.-H., Wu, B., He, Y.-L., Zhang, Y.: A new instance selection algorithm based on contribution for nearest neighbour classification. In: International Conference on Machine Learning and Cybernetics (ICMLC), pp. 155–160 (2010)
18. Cameron-Jones, R.M.: Instance selection by encoding length heuristic with random mutation hill climbing. In: Proceedings of the Eighth Australian Joint Conference on Artificial Intelligence, pp. 99–106 (1995)
19. Cano, J.R., Herrera, F., Lozano, M.: Using evolutionary algorithms as instance selection for data reduction in KDD: an experimental study. IEEE Trans. Evol. Comput. **7**(6), 561–575 (2003)
20. Cano, J.R., Herrera, F., Lozano, M.: Stratification for scaling up evolutionary prototype selection. Pattern Recogn. Lett. **26**(7), 953–963 (2005)
21. Cano, J.R., Herrera, F., Lozano, M.: Evolutionary stratified training set selection for extracting classification rules with trade off precision-interpretability. Data Knowl. Eng. **60**(1), 90–108 (2007)

22. Cano, J.R., García, S., Herrera, F.: Subgroup discover in large size data sets preprocessed using stratified instance selection for increasing the presence of minority classes. Pattern Recogn. Lett. **29**(16), 2156–2164 (2008)
23. Cano, J.R., Herrera, F., Lozano, M., García, S.: Making CN2-SD subgroup discovery algorithm scalable to large size data sets using instance selection. Expert Syst. Appl. **35**(4), 1949–1965 (2008)
24. Cavalcanti, G.D.C., Ren, T.I., Pereira, C.L.: ATISA: Adaptive threshold-based instance selection algorithm. Expert Syst. Appl. **40**(17), 6894–6900 (2013)
25. Cervantes, A., Galván, I.M., Isasi, P.: AMPSO: a new particle swarm method for nearest neighborhood classification. IEEE Trans. Syst. Man Cybern. B Cybern. **39**(5), 1082–1091 (2009)
26. Cerverón, V., Ferri, F.J.: Another move toward the minimum consistent subset: a tabu search approach to the condensed nearest neighbor rule. IEEE Trans. Syst. Man Cybern. B Cybern. **31**(3), 408–413 (2001)
27. Chang, C.L.: Finding prototypes for nearest neighbor classifiers. IEEE Trans. Comput. **23**(11), 1179–1184 (1974)
28. Chang, F., Lin, C.C., Lu, C.J.: Adaptive prototype learning algorithms: Theoretical and experimental studies. J. Mach. Learn. Res. **7**, 2125–2148 (2006)
29. Chen, C.H., Jóźwik, A.: A sample set condensation algorithm for the class sensitive artificial neural network. Pattern Recogn. Lett. **17**(8), 819–823 (1996)
30. Chen, Y., Bi, J., Wang, J.Z.: MILES: Multiple-instance learning via embedded instance selection. IEEE Trans. Pattern Anal. Mach. Intell. **28**(12), 1931–1947 (2006)
31. Chen, J., Zhang, C., Xue, X., Liu, C.L.: Fast instance selection for speeding up support vector machines. Knowl.-Based Syst. **45**, 1–7 (2013)
32. Cover, T.M., Hart, P.E.: Nearest neighbor pattern classification. IEEE Trans. Inf. Theory **13**(1), 21–27 (1967)
33. Czarnowski, I.: Prototype selection algorithms for distributed learning. Pattern Recognit. **43**(6), 2292–2300 (2010)
34. Czarnowski, I.: Cluster-based instance selection for machine classification. Knowl. Inf. Syst. **30**(1), 113–133 (2012)
35. Dai, B.R., Hsu, S.M.: An instance selection algorithm based on reverse nearest neighbor. In: PAKDD (1), Lecture Notes in Computer Science, vol. 6634, pp. 1–12 (2011)
36. Dasarathy, B.V.: Minimal consistent set (MCS) identification for optimal nearest neighbor decision system design. IEEE Trans. Syst. Man Cybern. B Cybern. **24**(3), 511–517 (1994)
37. de Santana Pereira, C., Cavalcanti, G.D.C.: Competence enhancement for nearest neighbor classification rule by ranking-based instance selection. In: International Conference on Tools with Artificial Intelligence, pp. 763–769 (2012)
38. Delany, S.J., Segata, N., Namee, B.M.: Profiling instances in noise reduction. Knowl.-Based Syst. **31**, 28–40 (2012)
39. Demšar, J.: Statistical comparisons of classifiers over multiple data sets. J. Mach. Learn. Res. **7**, 1–30 (2006)
40. Derrac, J., García, S., Herrera, F.: IFS-CoCo: Instance and feature selection based on cooperative coevolution with nearest neighbor rule. Pattern Recognit. **43**(6), 2082–2105 (2010)
41. Derrac, J., García, S., Herrera, F.: Stratified prototype selection based on a steady-state memetic algorithm: a study of scalability. Memetic Comput. **2**(3), 183–199 (2010)
42. Derrac, J., García, S., Herrera, F.: A survey on evolutionary instance selection and generation. Int. J. Appl. Metaheuristic Comput. **1**(1), 60–92 (2010)
43. Derrac, J., Cornelis, C., García, S., Herrera, F.: Enhancing evolutionary instance selection algorithms by means of fuzzy rough set based feature selection. Inf. Sci. **186**(1), 73–92 (2012)
44. Derrac, J., Triguero, I., García, S., Herrera, F.: Integrating instance selection, instance weighting, and feature weighting for nearest neighbor classifiers by coevolutionary algorithms. IEEE Trans. Syst. Man Cybern. B Cybern. **42**(5), 1383–1397 (2012)

45. Derrac, J., Verbiest, N., García, S., Cornelis, C., Herrera, F.: On the use of evolutionary feature selection for improving fuzzy rough set based prototype selection. Soft Comput. **17**(2), 223–238 (2013)
46. Devi, V.S., Murty, M.N.: An incremental prototype set building technique. Pattern Recognit. **35**(2), 505–513 (2002)
47. Devijver, P.A., Kittler, J.: A Statistical Approach Pattern Recognition. Prentice Hall, New Jersey (1982)
48. Devijver, P.A.: On the editing rate of the multiedit algorithm. Pattern Recogn. Lett. **4**, 9–12 (1986)
49. Domingo, C., Gavaldà, R., Watanabe, O.: Adaptive sampling methods for scaling up knowledge discovery algorithms. Data Min. Knowl. Disc. **6**, 131–152 (2002)
50. Domingos, P.: Unifying instance-based and rule-based induction. Mach. Learn. **24**(2), 141–168 (1996)
51. El-Hindi, K., Al-Akhras, M.: Smoothing decision boundaries to avoid overfitting in neural network training. Neural Netw. World **21**(4), 311–325 (2011)
52. Fayed, H.A., Hashem, S.R., Atiya, A.F.: Self-generating prototypes for pattern classification. Pattern Recognit. **40**(5), 1498–1509 (2007)
53. Fayed, H.A., Atiya, A.F.: A novel template reduction approach for the k-nearest neighbor method. IEEE Trans. Neural Networks **20**(5), 890–896 (2009)
54. Fernández, F., Isasi, P.: Evolutionary design of nearest prototype classifiers. J. Heuristics **10**(4), 431–454 (2004)
55. Fernández, F., Isasi, P.: Local feature weighting in nearest prototype classification. IEEE Trans. Neural Networks **19**(1), 40–53 (2008)
56. Ferrandiz, S., Boullé, M.: Bayesian instance selection for the nearest neighbor rule. Mach. Learn. **81**(3), 229–256 (2010)
57. Franco, A., Maltoni, D., Nanni, L.: Data pre-processing through reward-punishment editing. Pattern Anal. Appl. **13**(4), 367–381 (2010)
58. Fu, Z., Robles-Kelly, A., Zhou, J.: MILIS: multiple instance learning with instance selection. IEEE Trans. Pattern Anal. Mach. Intell. **33**(5), 958–977 (2011)
59. Gagné, C., Parizeau, M.: Coevolution of nearest neighbor classifiers. IEEE Trans. Pattern Anal. Mach. Intell. **21**(5), 921–946 (2007)
60. Galar, M., Fernández, A., Barrenechea, E., Bustince, H., Herrera, F.: A review on ensembles for the class imbalance problem: Bagging-, boosting-, and hybrid-based approaches. IEEE Trans. Syst. Man Cybern. C **42**(4), 463–484 (2012)
61. Galar, M., Fernández, A., Barrenechea, E., Herrera, F.: Eusboost: enhancing ensembles for highly imbalanced data-sets by evolutionary undersampling. Pattern Recognit. **46**(12), 3460–3471 (2013)
62. García, S., Cano, J.R., Herrera, F.: A memetic algorithm for evolutionary prototype selection: A scaling up approach. Pattern Recognit. **41**(8), 2693–2709 (2008)
63. García, S., Herrera, F.: An extension on "statistical comparisons of classifiers over multiple data sets" for all pairwise comparisons. J. Mach. Learn. Res. **9**, 2677–2694 (2008)
64. García, S., Cano, J.R., Bernadó-Mansilla, E., Herrera, F.: Diagnose of effective evolutionary prototype selection using an overlapping measure. Int. J. Pattern Recognit. Artif. Intell. **23**(8), 1527–1548 (2009)
65. García, S., Fernández, A., Herrera, F.: Enhancing the effectiveness and interpretability of decision tree and rule induction classifiers with evolutionary training set selection over imbalanced problems. Appl. Soft Comput. **9**(4), 1304–1314 (2009)
66. García, S., Herrera, F.: Evolutionary under-sampling for classification with imbalanced data sets: Proposals and taxonomy. Evol. Comput. **17**(3), 275–306 (2009)
67. García, S., Derrac, J., Luengo, J., Carmona, C.J., Herrera, F.: Evolutionary selection of hyperrectangles in nested generalized exemplar learning. Appl. Soft Comput. **11**(3), 3032–3045 (2011)
68. García, S., Derrac, J., Cano, J.R., Herrera, F.: Prototype selection for nearest neighbor classification: taxonomy and empirical study. IEEE Trans. Pattern Anal. Mach. Intell. **34**(3), 417–435 (2012)

69. García, S., Derrac, J., Triguero, I., Carmona, C.J., Herrera, F.: Evolutionary-based selection of generalized instances for imbalanced classification. Knowl.-Based Syst. **25**(1), 3–12 (2012)
70. García-Osorio, C., de Haro-García, A., García-Pedrajas, N.: Democratic instance selection: a linear complexity instance selection algorithm based on classifier ensemble concepts. Artif. Intell. **174**(5–6), 410–441 (2010)
71. García-Pedrajas, N.: Constructing ensembles of classifiers by means of weighted instance selection. IEEE Trans. Neural Networks **20**(2), 258–277 (2009)
72. García-Pedrajas, N., Romero del Castillo, J.A., Ortiz-Boyer, D.: A cooperative coevolutionary algorithm for instance selection for instance-based learning. Mach. Learn. **78**(3), 381–420 (2010)
73. García-Pedrajas, N., Pérez-Rodríguez, J.: Multi-selection of instances: a straightforward way to improve evolutionary instance selection. Appl. Soft Comput. **12**(11), 3590–3602 (2012)
74. García-Pedrajas, N., de Haro-García, A., Pérez-Rodríguez, J.: A scalable approach to simultaneous evolutionary instance and feature selection. Inf. Sci. **228**, 150–174 (2013)
75. García-Pedrajas, N., Pérez-Rodríguez, J.: OligoIS: scalable instance selection for class-imbalanced data sets. IEEE Trans. Cybern. **43**(1), 332–346 (2013)
76. Gates, G.W.: The reduced nearest neighbor rule. IEEE Trans. Inf. Theory **22**, 431–433 (1972)
77. Gil-Pita, R., Yao, X.: Evolving edited k-nearest neighbor classifiers. Int. J. Neural Syst. **18**(6), 459–467 (2008)
78. Gowda, K.C., Krishna, G.: The condensed nearest neighbor rule using the concept of mutual nearest neighborhood. IEEE Trans. Inf. Theory **29**, 488–490 (1979)
79. Guillén, A., Herrera, L.J., Rubio, G., Pomares, H., Lendasse, A., Rojas, I.: New method for instance or prototype selection using mutual information in time series prediction. Neurocomputing **73**(10–12), 2030–2038 (2010)
80. Guo, Y., Zhang, H., Liu, X.: Instance selection in semi-supervised learning. Canadian conference on AI, Lecture Notes in Computer Science, vol. 6657, pp. 158–169 (2011)
81. Haro-García, A., García-Pedrajas, N.: A divide-and-conquer recursive approach for scaling up instance selection algorithms. Data Min. Knowl. Disc. **18**(3), 392–418 (2009)
82. de Haro-García, A., García-Pedrajas, N., del Castillo, J.A.R.: Large scale instance selection by means of federal instance selection. Data Knowl. Eng. **75**, 58–77 (2012)
83. Hart, P.E.: The condensed nearest neighbor rule. IEEE Trans. Inf. Theory **14**, 515–516 (1968)
84. Hattori, K., Takahashi, M.: A new edited k-nearest neighbor rule in the pattern classification problem. Pattern Recognit. **33**(3), 521–528 (2000)
85. Hernandez-Leal, P., Carrasco-Ochoa, J.A., Trinidad, J.F.M., Olvera-López, J.A.: Instancerank based on borders for instance selection. Pattern Recognit. **46**(1), 365–375 (2013)
86. Ho, S.Y., Liu, C.C., Liu, S.: Design of an optimal nearest neighbor classifier using an intelligent genetic algorithm. Pattern Recogn. Lett. **23**(13), 1495–1503 (2002)
87. Ivanov, M.: Prototype sample selection based on minimization of the complete cross validation functional. Pattern Recognit. Image anal. **20**(4), 427–437 (2010)
88. Jankowski, N., Grochowski, M.: Comparison of instances selection algorithms I. algorithms survey. In: ICAISC, Lecture Notes in Computer Science, vol. 3070, pp. 598–603 (2004)
89. Kibler, D., Aha, D.W.: Learning representative exemplars of concepts: an initial case study. In: Proceedings of the Fourth International Workshop on Machine Learning, pp. 24–30 (1987)
90. Kim, S.W., Oomenn, B.J.: Enhancing prototype reduction schemes with LVQ3-type algorithms. Pattern Recognit. **36**, 1083–1093 (2003)
91. Kim, S.W., Oommen, B.J.: Enhancing prototype reduction schemes with recursion: a method applicable for large data sets. IEEE Trans. Syst. Man Cybern. B **34**(3), 1384–1397 (2004)
92. Kim, S.W., Oommen, B.J.: On using prototype reduction schemes to optimize kernel-based nonlinear subspace methods. Pattern Recognit. **37**(2), 227–239 (2004)
93. Kim, S.W., Oommen, B.J.: On using prototype reduction schemes and classifier fusion strategies to optimize kernel-based nonlinear subspace methods. IEEE Trans. Pattern Anal. Mach. Intell. **27**(3), 455–460 (2005)
94. Kim, K.J.: Artificial neural networks with evolutionary instance selection for financial forecasting. Expert Syst. Appl. **30**(3), 519–526 (2006)

95. Kim, S.W., Oommen, B.J.: On using prototype reduction schemes to optimize dissimilarity-based classification. Pattern Recognit. **40**(11), 2946–2957 (2007)
96. Kim, S.W., Oommen, B.J.: On using prototype reduction schemes to enhance the computation of volume-based inter-class overlap measures. Pattern Recognit. **42**(11), 2695–2704 (2009)
97. Kim, S.W.: An empirical evaluation on dimensionality reduction schemes for dissimilarity-based classifications. Pattern Recogn. Lett. **32**(6), 816–823 (2011)
98. Kohonen, T.: The self organizing map. Proc. IEEE **78**(9), 1464–1480 (1990)
99. Koplowitz, J., Brown, T.: On the relation of performance to editing in nearest neighbor rules. Pattern Recognit. **13**, 251–255 (1981)
100. Kuncheva, L.I.: Editing for the k-nearest neighbors rule by a genetic algorithm. Pattern Recogn. Lett. **16**(8), 809–814 (1995)
101. Kuncheva, L.I., Jain, L.C.: Nearest neighbor classifier: simultaneous editing and feature selection. Pattern Recogn. Lett. **20**(11–13), 1149–1156 (1999)
102. Lam, W., Keung, C.K., Liu, D.: Discovering useful concept prototypes for classification based on filtering and abstraction. IEEE Trans. Pattern Anal. Mach. Intell. **14**(8), 1075–1090 (2002)
103. Leyva, E., González, A., Pérez, R.: Knowledge-based instance selection: a compromise between efficiency and versatility. Knowl.-Based Syst. **47**, 65–76 (2013)
104. Li, Y., Hu, Z., Cai, Y., Zhang, W.: Support vector based prototype selection method for nearest neighbor rules. In: First International Conference on Advances in Natural Computation (ICNC), Lecture Notes in Computer Science, vol. 3610, pp. 528–535 (2005)
105. Li, Y., Maguire, L.P.: Selecting critical patterns based on local geometrical and statistical information. IEEE Trans. Pattern Anal. Mach. Intell. **33**(6), 1189–1201 (2011)
106. Li, I.J., Chen, J.C., Wu, J.L.: A fast prototype reduction method based on template reduction and visualization-induced self-organizing map for nearest neighbor algorithm. Appl. Intell. **39**(3), 564–582 (2013)
107. Lipowezky, U.: Selection of the optimal prototype subset for 1-nn classification. Pattern Recogn. Lett. **19**(10), 907–918 (1998)
108. Liu, H., Motoda, H.: Instance Selection and Construction for Data Mining. Kluwer Academic Publishers, Norwell (2001)
109. Liu, H., Motoda, H.: On issues of instance selection. Data Min. Knowl. Disc. **6**(2), 115–130 (2002)
110. López, V., Fernández, A., García, S., Palade, V., Herrera, F.: An insight into classification with imbalanced data: empirical results and current trends on using data intrinsic characteristics. Inf. Sci. **250**, 113–141 (2013)
111. Lowe, D.G.: Similarity metric learning for a variable-kernel classifier. Neural Comput. **7**(1), 72–85 (1995)
112. Lozano, M.T., Sánchez, J.S., Pla, F.: Using the geometrical distribution of prototypes for training set condensing. CAEPIA, Lecture Notes in Computer Science, vol. 3040, pp. 618–627 (2003)
113. Lozano, M., Sotoca, J.M., Sánchez, J.S., Pla, F., Pekalska, E., Duin, R.P.W.: Experimental study on prototype optimisation algorithms for prototype-based classification in vector spaces. Pattern Recognit. **39**(10), 1827–1838 (2006)
114. Luaces, O., Bahamonde, A.: Inflating examples to obtain rules. Int. J. Intell. syst. **18**, 1113–1143 (2003)
115. Luengo, J., Fernández, A., García, S., Herrera, F.: Addressing data complexity for imbalanced data sets: analysis of smote-based oversampling and evolutionary undersampling. Soft Comput. **15**(10), 1909–1936 (2011)
116. Marchiori, E.: Hit miss networks with applications to instance selection. J. Mach. Learn. Res. **9**, 997–1017 (2008)
117. Marchiori, E.: Class conditional nearest neighbor for large margin instance selection. IEEE Trans. Pattern Anal. Mach. Intell. **32**, 364–370 (2010)
118. Miloud-Aouidate, A., Baba-Ali, A.R.: Ant colony prototype reduction algorithm for knn classification. In: International Conference on Computational Science and Engineering, pp. 289–294 (2012)

119. Mollineda, R.A., Sánchez, J.S., Sotoca, J.M.: Data characterization for effective prototype selection. In: Proc. of the 2nd Iberian Conf. on Pattern Recognition and Image Analysis (ICPRIA), Lecture Notes in Computer Science, vol. 3523, pp. 27–34 (2005)
120. Narayan, B.L., Murthy, C.A., Pal, S.K.: Maxdiff kd-trees for data condensation. Pattern Recognit. Lett. **27**(3), 187–200 (2006)
121. Neo, T.K.C., Ventura, D.: A direct boosting algorithm for the k-nearest neighbor classifier via local warping of the distance metric. Pattern Recognit. Lett. **33**(1), 92–102 (2012)
122. Nikolaidis, K., Goulermas, J.Y., Wu, Q.H.: A class boundary preserving algorithm for data condensation. Pattern Recognit. **44**(3), 704–715 (2011)
123. Nikolaidis, K., Rodriguez-Martinez, E., Goulermas, J.Y., Wu, Q.H.: Spectral graph optimization for instance reduction. IEEE Trans. Neural Networks Learn. Syst. **23**(7), 1169–1175 (2012)
124. Nikolaidis, K., Mu, T., Goulermas, J.: Prototype reduction based on direct weighted pruning. Pattern Recognit. Lett. **36**, 22–28 (2014)
125. Olvera-López, J.A., Martínez-Trinidad, J.F., Carrasco-Ochoa, J.A.: Edition schemes based on BSE. In: 10th Iberoamerican Congress on Pattern Recognition (CIARP), Lecture Notes in Computer Science, vol. 3773, pp. 360–367 (2005)
126. Olvera-López, J.A., Carrasco-Ochoa, J.A., Martínez-Trinidad, J.F.: A new fast prototype selection method based on clustering. Pattern Anal. Appl. **13**(2), 131–141 (2010)
127. Olvera-López, J.A., Carrasco-Ochoa, J.A., Martínez-Trinidad, J.F., Kittler, J.: A review of instance selection methods. Artif. Intell. Rev. **34**(2), 133–143 (2010)
128. Paredes, R., Vidal, E.: Learning prototypes and distances: a prototype reduction technique based on nearest neighbor error minimization. Pattern Recog. **39**(2), 180–188 (2006)
129. Paredes, R., Vidal, E.: Learning weighted metrics to minimize nearest-neighbor classification error. IEEE Trans. Pattern Anal. Mach. Intell. **28**(7), 1100–1110 (2006)
130. García-Pedrajas, N.: Evolutionary computation for training set selection. Wiley Interdisc. Rev.: Data Min. Knowl. Disc. **1**(6), 512–523 (2011)
131. Pekalska, E., Duin, R.P.W., Paclík, P.: Prototype selection for dissimilarity-based classifiers. Pattern Recognit. **39**(2), 189–208 (2006)
132. Raniszewski, M.: Sequential reduction algorithm for nearest neighbor rule. In: ICCVG (2), Lecture Notes in Computer Science, vol. 6375, pp. 219–226. Springer, Heidelberg (2010)
133. Reinartz, T.: A unifying view on instance selection. Data Min. Knowl. Disc. **6**(2), 191–210 (2002)
134. Calana, Y.P., Reyes, E.G., Alzate, M.O., Duin, R.P.W.: Prototype selection for dissimilarity representation by a genetic algorithm. In: International Conference on Pattern Recogition (ICPR), pp. 177–180 (2010)
135. Riquelme, J.C., Aguilar-Ruiz, J.S., Toro, M.: Finding representative patterns with ordered projections. Pattern Recognit. **36**(4), 1009–1018 (2003)
136. Ritter, G.L., Woodruff, H.B., Lowry, S.R., Isenhour, T.L.: An algorithm for a selective nearest neighbor decision rule. IEEE Trans. Inf. Theory **25**, 665–669 (1975)
137. Sáez, J.A., Luengo, J., Herrera, F.: Predicting noise filtering efficacy with data complexity measures for nearest neighbor classification. Pattern Recognit. **46**(1), 355–364 (2013)
138. Salzberg, S.: A nearest hyperrectangle learning method. Mach. Learn. **6**, 251–276 (1991)
139. Sánchez, J.S., Pla, F., Ferri, F.J.: Prototype selection for the nearest neighbor rule through proximity graphs. Pattern Recognit. Lett. **18**, 507–513 (1997)
140. Sánchez, J.S., Barandela, R., Marqués, A.I., Alejo, R., Badenas, J.: Analysis of new techniques to obtain quality training sets. Pattern Recognit. Lett. **24**(7), 1015–1022 (2003)
141. Sánchez, J.S.: High training set size reduction by space partitioning and prototype abstraction. Pattern Recognit. **37**(7), 1561–1564 (2004)
142. Dos Santos, E.M., Sabourin, R., Maupin, P.: Overfitting cautious selection of classifier ensembles with genetic algorithms. Inf. Fusion **10**(2), 150–162 (2009)
143. Sebban, M., Nock, R.: Instance pruning as an information preserving problem. In: ICML '00: Proceedings of the Seventeenth International Conference on Machine Learning, pp. 855–862 (2000)

144. Sebban, M., Nock, R., Brodley, E., Danyluk, A.: Stopping criterion for boosting-based data reduction techniques: from binary to multiclass problems. J. Mach. Learn. Res. **3**, 863–885 (2002)
145. Segata, N., Blanzieri, E., Delany, S.J., Cunningham, P.: Noise reduction for instance-based learning with a local maximal margin approach. J. Intell. Inf. Sys. **35**(2), 301–331 (2010)
146. Sierra, B., Lazkano, E., Inza, I., Merino, M., Larrañaga, P., Quiroga, J.: Prototype selection and feature subset selection by estimation of distribution algorithms. a case study in the survival of cirrhotic patients treated with TIPS. In: AIME '01: Proceedings of the 8th Conference on AI in Medicine in Europe, Lecture Notes in Computer Science, vol. 2101, pp. 20–29 (2001)
147. Skalak, D.B.: Prototype and feature selection by sampling and random mutation hill climbing algorithms. In: Proceedings of the Eleventh International Conference on Machine Learning, pp. 293–301 (1994)
148. Steele, B.M.: Exact bootstrap k-nearest neighbor learners. Mach. Learn. **74**(3), 235–255 (2009)
149. Tomek, I.: An experiment with the edited nearest-neighbor rule. IEEE Trans. Syst. Man Cybern. **6**(6), 448–452 (1976)
150. Tomek, I.: Two modifications of CNN. IEEE Trans. Syst. Man Cybern. **6**(6), 769–772 (1976)
151. Triguero, I., García, S., Herrera, F.: IPADE: iterative prototype adjustment for nearest neighbor classification. IEEE Trans. Neural Networks **21**(12), 1984–1990 (2010)
152. Triguero, I., García, S., Herrera, F.: Differential evolution for optimizing the positioning of prototypes in nearest neighbor classification. Pattern Recognit. **44**(4), 901–916 (2011)
153. Triguero, I., Derrac, J., García, S., Herrera, F.: A taxonomy and experimental study on prototype generation for nearest neighbor classification. IEEE Trans. Syst. Man Cybern. C **42**(1), 86–100 (2012)
154. Tsai, C.F., Chang, C.W.: SVOIS: support vector oriented instance selection for text classification. Inf. Syst. **38**(8), 1070–1083 (2013)
155. Tsai, C.F., Eberle, W., Chu, C.Y.: Genetic algorithms in feature and instance selection. Knowl.-Based Syst. **39**, 240–247 (2013)
156. Ullmann, J.R.: Automatic selection of reference data for use in a nearest-neighbor method of pattern classification. IEEE Trans. Inf. Theory **24**, 541–543 (1974)
157. Vascon, S., Cristani, M., Pelillo, M., Murino, V.: Using dominant sets for k-nn prototype selection. In: International Conference on Image Analysis and Processing (ICIAP (2)), pp. 131–140 (2013)
158. Vázquez, F., Sánchez, J.S., Pla, F.: A stochastic approach to Wilson's editing algorithm. In: 2nd Iberian Conference on Pattern Recognition and Image Analysis (IbPRIA), Lecture Notes in Computer Science, vol. 3523, pp. 35–42 (2005)
159. Verbiest, N., Cornelis, C., Herrera, F.: FRPS: a fuzzy rough prototype selection method. Pattern Recognit. **46**(10), 2770–2782 (2013)
160. Wang, X., Miao, Q., Zhai, M.Y., Zhai, J.: Instance selection based on sample entropy for efficient data classification with elm. In: International Conference on Systems, Man and Cybernetics, pp. 970–974 (2012)
161. Wang, X.Z., Wu, B., He, Y.L., Pei, X.H.: NRMCS : Noise removing based on the MCS. In: Proceedings of the Seventh International Conference on Machine Learning and Cybernetics, pp. 89–93 (2008)
162. Wettschereck, D., Dietterich, T.G.: An experimental comparison of the nearest-neighbor and nearest-hyperrectangle algorithms. Mach. Learn. **19**(1), 5–27 (1995)
163. Wettschereck, D., Aha, D.W., Mohri, T.: A review and empirical evaluation of feature weighting methods for a class of lazy learning algorithms. Artif. Intell. Rev. **11**(1–5), 273–314 (1997)
164. Wilcoxon, F.: Individual comparisons by ranking methods. Biometrics **1**, 80–83 (1945)
165. Wilson, D.L.: Asymptotic properties of nearest neighbor rules using edited data. IEEE Trans. Syst. Man Cybern. B Cybern. **2**(3), 408–421 (1972)
166. Wilson, D.R., Martinez, T.R.: Improved heterogeneous distance functions. J. Artif. Intell. Res. **6**, 1–34 (1997)

167. Wilson, D.R., Martinez, T.R.: Reduction techniques for instance-based learning algorithms. Mach. Learn. **38**(3), 257–286 (2000)
168. Wu, Y., Ianakiev, K.G., Govindaraju, V.: Improved k-nearest neighbor classification. Pattern Recognit. **35**(10), 2311–2318 (2002)
169. Yang, T., Cao, L., Zhang, C.: A novel prototype reduction method for the k-nearest neighbor algorithm with k >= 1. In: PAKDD (2), Lecture Notes in Computer Science, vol. 6119, pp. 89–100 (2010)
170. Zhai, T., He, Z.: Instance selection for time series classification based on immune binary particle swarm optimization. Knowl.-Based Syst. **49**, 106–115 (2013)
171. Zhang, H., Sun, G.: Optimal reference subset selection for nearest neighbor classification by tabu search. Pattern Recognit. **35**(7), 1481–1490 (2002)
172. Zhang, L., Chen, C., Bu, J., He, X.: A unified feature and instance selection framework using optimum experimental design. IEEE Trans. Image Process. **21**(5), 2379–2388 (2012)
173. Zhao, K.P., Zhou, S.G., Guan, J.H., Zhou, A.Y.: C-pruner: An improved instance pruning algorithm. In: Proceeding of the 2th International Conference on Machine Learning and Cybernetics, pp. 94–99 (2003)
174. Zhu, X., Yang, Y.: A lazy bagging approach to classification. Pattern Recognit. **41**(10), 2980–2992 (2008)

Chapter 9
Discretization

Abstract Discretization is an essential preprocessing technique used in many knowledge discovery and data mining tasks. Its main goal is to transform a set of continuous attributes into discrete ones, by associating categorical values to intervals and thus transforming quantitative data into qualitative data. An overview of discretization together with a complete outlook and taxonomy are supplied in Sects. 9.1 and 9.2. We conduct an experimental study in supervised classification involving the most representative discretizers, different types of classifiers, and a large number of data sets (Sect. 9.4).

9.1 Introduction

As it was mentioned in the introduction of this book, data usually comes in different formats, such as discrete, numerical, continuous, categorical, etc. Numerical data, provided by discrete or continuous values, assumes that the data is ordinal, there is an order among the values. However, in categorical data, no order can be assumed amongst them. The domain and type of data is crucial to the learning task to be performed next. For example, in a decision tree induction process a feature must be chosen from a subset based on some metric gain associated with its values. This process usually requires inherent finite values and also prefers to perform a branch of values that are not ordered. Obviously, the tree structure is a finite structure and there is a need to split the feature to produce the associated nodes in further divisions. If data is continuous, there is a need to discretize the features either before the decision tree induction or throughout the process of tree modelling.

Discretization, as one of the basic data reduction techniques, has received increasing research attention in recent years [75] and has become one of the preprocessing techniques most broadly used in DM. The discretization process transforms quantitative data into qualitative data, that is, numerical attributes into discrete or nominal attributes with a finite number of intervals, obtaining a non-overlapping partition of a continuous domain. An association between each interval with a numerical discrete value is then established. In practice, discretization can be viewed as a data reduction method since it maps data from a huge spectrum of numeric values to a greatly

S. García et al., *Data Preprocessing in Data Mining*,
Intelligent Systems Reference Library 72, DOI 10.1007/978-3-319-10247-4_9

reduced subset of discrete values. Once the discretization is performed, the data can be treated as nominal data during any induction or deduction DM process. Many existing DM algorithms are designed only to learn in categorical data, using nominal attributes, while real-world applications usually involve continuous features. Numerical features have to be discretized before using such algorithms.

In supervised learning, and specifically classification, the topic of this survey, we can define the discretization as follows. Assuming a data set consisting of N examples and C target classes, a discretization algorithm would discretize the continuous attribute A in this data set into m discrete intervals $D = \{[d_0, d_1], (d_1, d_2], \ldots, (d_{m-1}, d_m]\}$, where d_0 is the minimal value, d_m is the maximal value and $d_i < d_{i+i}$, for $i = 0, 1, \ldots, m-1$. Such a discrete result D is called a discretization scheme on attribute A and $P = \{d_1, d_2, \ldots, d_{m-1}\}$ is the set of cut points of attribute A.

The necessity of using discretization on data can be caused by several factors. Many DM algorithms are primarily oriented to handle nominal attributes [36, 75, 123], or may even only deal with discrete attributes. For instance, three of the ten methods considered as the top ten in DM [120] require an embedded or an external discretization of data: C4.5 [92], Apriori [1] and Naïve Bayes [44, 122]. Even with algorithms that are able to deal with continuous data, learning is less efficient and effective [29, 94]. Other advantages derived from discretization are the reduction and the simplification of data, making the learning faster and yielding more accurate, with compact and shorter results; and any noise possibly present in the data is reduced. For both researchers and practitioners, discrete attributes are easier to understand, use, and explain [75]. Nevertheless, any discretization process generally leads to a loss of information, making the minimization of such information loss the main goal of a discretizer.

Obtaining the optimal discretization is NP-complete [25]. A vast number of discretization techniques can be found in the literature. It is obvious that when dealing with a concrete problem or data set, the choice of a discretizer will condition the success of the posterior learning task in accuracy, simplicity of the model, etc. Different heuristic approaches have been proposed for discretization, for example, approaches based on information entropy [36, 41], statistical χ^2 test [68, 76], likelihood [16, 119], rough sets [86, 124], etc. Other criteria have been used in order to provide a classification of discretizers, such as univariate/multivariate, supervised/unsupervised, top-down/bottom-up, global/local, static/dynamic and more. All these criteria are the basis of the taxonomies already proposed and they will be deeply elaborated upon in this chapter. The identification of the best discretizer for each situation is a very difficult task to carry out, but performing exhaustive experiments considering a representative set of learners and discretizers could help to make the best choice.

Some reviews of discretization techniques can be found in the literature [9, 36, 75, 123]. However, the characteristics of the methods are not studied completely, many discretizers, even classic ones, are not mentioned, and the notation used for categorization is not unified. In spite of the wealth of literature, and apart from the absence of a complete categorization of discretizers using a unified notation, it can be observed that, there are few attempts to empirically compare them. In this way, the algorithms proposed are usually compared with a subset of the complete family of discretizers and, in most of the studies, no rigorous empirical analysis has been carried out. In [51], it was noticed that the most compared techniques are EqualWidth, EqualFrequency, MDLP [41], ID3 [92], ChiMerge [68], 1R [59], D2 [19] and Chi2 [76].

These reasons motivate the global purpose of this chapter. We can summarize it into three main objectives:

- To provide an updated and complete taxonomy based on the main properties observed in the discretization methods. The taxonomy will allow us to characterize their advantages and drawbacks in order to choose a discretizer from a theoretical point of view.
- To make an empirical study analyzing the most representative and newest discretizers in terms of the number of intervals obtained and inconsistency level of the data.
- Finally, to relate the best discretizers to a set of representative DM models using two metrics to measure the predictive classification success.

9.2 Perspectives and Background

Discretization is a wide field and there have been many advances and ideas over the years. This section is devoted to provide a proper background on the topic, together with a set of related areas and future perspectives on discretization.

9.2.1 Discretization Process

Before starting, we must first introduce some terms used by different sources for the sake of unification.

9.2.1.1 Feature

Also called *attribute* or *variable* refers to an aspect of the data and it is usually associated to the columns in a data table. M stands for the number of features in the data.

9.2.1.2 Instance

Also called *tuple*, *example*, *record* or *data point* refers to a collection of feature values for all features. A set of instances constitute a data set and they are usually associated to row in a data table. According to the introduction, we will set N as the number of instances in the data.

9.2.1.3 Cut Point

Refers to a real value that divides the range into two adjacent intervals, being the first interval less than or equal to the cut point and the second interval greater than the cut point.

9.2.1.4 Arity

In discretization context, it refers to the number of partitions or intervals. For example, if the arity of a feature after being discretized is m, then there will be $m - 1$ cut points.

We can associate a typical discretization as an univariate discretization. Although this property will be reviewed in Sect. 9.3.1, it is necessary to introduce it here for the basic understanding of the basic discretization process. Univariate discretization operates with one continuous feature at a time while multivariate discretization considers multiple features simultaneously.

A typical discretization process generally consists of four steps (seen in Fig. 9.1): (1) *sorting* the continuous values of the feature to be discretized, (2) *evaluating* a cut

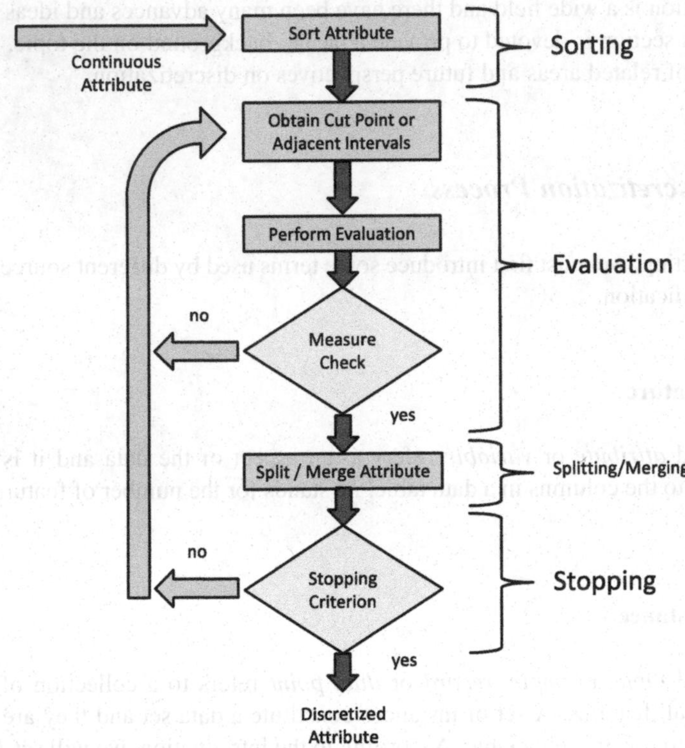

Fig. 9.1 Discretization process

point for splitting or adjacent intervals for merging, (3) *splitting or merging* intervals of continuous values according to some defined criterion, and (4) *stopping* at some point. Next, we explain these four steps in detail.

9.2.1.5 Sorting

The continuous values for a feature are sorted in either descending or ascending order. It is crucial to use an efficient sorting algorithm with a time complexity of $O(NlogN)$, for instance the well-known *Quick Sort* algorithm. Sorting must be done only once and for all the start of discretization. It is a mandatory treatment and can be applied when the complete instance space is used for discretization. However, if the discretization is within the process of other algorithms (such as decision trees induction), it is a local treatment and only a region of the whole instance space is considered for discretization.

9.2.1.6 Selection of a Cut Point

After sorting, the best cut point or the best pair of adjacent intervals should be found in the attribute range in order to split or merge in a following required step. An evaluation measure or function is used to determine the correlation, gain, improvement in performance and any other benefit according to the class label. There are numerous evaluation measures and they will be discussed in Sect. 9.3.1, the entropy and the statistical dependency being the most well known.

9.2.1.7 Splitting/Merging

Depending on operation method of the discretizers, intervals either can be split or merged. For splitting all the possible cut points from the whole universe within an attribute must be evaluated. The universe is formed from all the different real values presented in an attribute. Then, the best one is found and a split of the continuous range into two partitions is performed. Discretization continues with each part until a stopping criterion is satisfied. Similarly for merging, instead of finding the best cut point, the discretizer aims to find the best adjacent intervals to merge in each iteration. Discretization continues with the reduced number of intervals until the stopping criterion is satisfied.

9.2.1.8 Stopping Criteria

It specifies when to stop the discretization process. It should assume a trade-off between lower arity getting a better understanding or simplicity with high accuracy or consistency. For example, a threshold for m can be an upper bound for the arity of the resulting discretization. A stopping criterion can be very simple such as fixing the number of final intervals at the beginning of the process or more complex like estimating a function.

9.2.2 Related and Advanced Work

Research in improving and analyzing discretization is common and in high demand currently. Discretization is a promising technique to obtain the hoped results, depending on the DM task, which justifies its relationship to other methods and problems. This section provides a brief summary of topics closely related to discretization from a theoretical and practical point of view and describes other works and future trends which have been studied in the last few years.

- *Discretization Specific Analysis:* Susmaga proposed an analysis method for discretizers based on binarization of continuous attributes and rough sets measures [104]. He emphasized that his analysis method is useful for detecting redundancy in discretization and the set of cut points which can be removed without decreasing the performance. Also, it can be applied to improve existing discretization approaches.
- *Optimal Multisplitting:* Elomaa and Rousu characterized some fundamental properties for using some classic evaluation functions in supervised univariate discretization. They analyzed entropy, information gain, gain ratio, training set error, Gini index and normalized distance measure, concluding that they are suitable for use in the optimal multisplitting of an attribute [38]. They also developed an optimal algorithm for performing this multisplitting process and devised two techniques [39, 40] to speed it up.
- *Discretization of Continuous Labels:* Two possible approaches have been used in the conversion of a continuous supervised learning (regression problem) into a nominal supervised learning (classification problem). The first one is simply to use regression tree algorithms, such as CART [17]. The second consists of applying discretization to the output attribute, either statically [46] or in a dynamic fashion [61].
- *Fuzzy Discretization:* Extensive research has been carried out around the definition of linguistic terms that divide the domain attribute into fuzzy regions [62]. Fuzzy discretization is characterized by membership value, group or interval number and affinity corresponding to an attribute value, unlike crisp discretization which only considers the interval number [95].
- *Cost-Sensitive Discretization:* The objective of cost-based discretization is to take into account the cost of making errors instead of just minimizing the total sum of errors [63]. It is related to problems of imbalanced or cost-sensitive classification [57, 103].

• *Semi-Supervised Discretization:* A first attempt to discretize data in semi-supervised classification problems has been devised in [14], showing that it is asymptotically equivalent to the supervised approach.

The research mentioned in this section is out of the scope of this book. We point out that the main objective of this chapter is to give a wide overview of the discretization methods found in the literature and to conduct an experimental comparison of the most relevant discretizers without considering external and advanced factors such as those mentioned above or derived problems from classic supervised classification.

9.3 Properties and Taxonomy

This section presents a taxonomy of discretization methods and the criteria used for building it. First, in Sect. 9.3.1, the main characteristics which will define the categories of the taxonomy will be outlined. Then, in Sect. 9.3.2, we enumerate the discretization methods proposed in the literature and we will consider by using both their complete name and abbreviated name together with the associated reference. Finally, we present the taxonomy.

9.3.1 Common Properties

This section provides a framework for the discussion of the discretizers presented in the next subsection. The issues discussed include several properties involved in the structure of the taxonomy, since they are exclusive to the operation of the discretizer. Other, less critical issues such as parametric properties or stopping conditions will be presented although they are not involved in the taxonomy. Finally, some criteria will also be pointed out in order to compare discretization methods.

9.3.1.1 Main Characteristics of a Discretizer

In [36, 51, 75, 123], various axes have been described in order to make a categorization of discretization methods. We review and explain them in this section, emphasizing the main aspects and relations found among them and unifying the notation. The taxonomy presented will be based on these characteristics:

• *Static versus Dynamic:* This characteristic refers to the moment and independence which the discretizer operates in relation to the learner. A dynamic discretizer acts when the learner is building the model, thus they can only access partial information (local property, see later) embedded in the learner itself, yielding compact and accurate results in conjuntion with the associated learner. Otherwise, a static discretizer proceeds prior to the learning task and it is independent from the learning algorithm [75]. Almost all known discretizers are static, due to the fact that most of the dynamic discretizers are really subparts or stages of DM algorithms when dealing with numerical data [13]. Some examples of well-known dynamic techniques are ID3 discretizer [92] and ITFP [6].

- *Univariate versus Multivariate:* Multivariate techniques, also known as 2D discretization [81], simultaneously consider all attributes to define the initial set of cut points or to decide the best cut point altogether. They can also discretize one attribute at a time when studying the interactions with other attributes, exploiting high order relationships. By contrast, univariate discretizers only work with a single attribute at a time, once an order among attributes has been established, and the resulting discretization scheme in each attribute remains unchanged in later stages. Interest has recently arisen in developing multivariate discretizers since they are very influential in deductive learning [10, 49] and in complex classification problems where high interactions among multiple attributes exist, which univariate discretizers might obviate [42, 121].
- *Supervised versus Unsupervised:* Unsupervised discretizers do not consider the class label whereas supervised ones do. The manner in which the latter consider the class attribute depends on the interaction between input attributes and class labels, and the heuristic measures used to determine the best cut points (entropy, interdependence, etc.). Most discretizers proposed in the literature are supervised and theoretically using class information, should automatically determine the best number of intervals for each attribute. If a discretizer is unsupervised, it does not mean that it cannot be applied over supervised tasks. However, a supervised discretizer can only be applied over supervised DM problems. Representative unsupervised discretizers are EqualWidth and EqualFrequency [73], PKID and FFD [122] and MVD [10].
- *Splitting versus Merging:* This refers to the procedure used to create or define new intervals. Splitting methods establish a cut point among all the possible boundary points and divide the domain into two intervals. By contrast, merging methods start with a pre-defined partition and remove a candidate cut point to mix both adjacent intervals. These properties are highly related to *Top-Down* and *Bottom-up* respectively (explained in the next section). The idea behind them is very similar, except that top-down or bottom-up discretizers assume that the process is incremental (described later), according to a hierarchical discretization construction. In fact, there can be discretizers whose operation is based on splitting or merging more than one interval at a time [72, 96]. Also, some discretizers can be considered *hybrid* due to the fact that they can alternate splits with merges in running time [24, 43].
- *Global versus Local:* To make a decision, a discretizer can either require all available data in the attribute or use only partial information. A discretizer is said to be local when it only makes the partition decision based on local information. Examples of widely used local techniques are MDLP [41] and ID3 [92]. Few discretizers are local, except some based on top-down partition and all the dynamic techniques. In a top-down process, some algorithms follow the divide-and-conquer scheme and when a split is found, the data is recursively divided, restricting access

to partial data. Regarding dynamic discretizers, they find the cut points in internal operations of a DM algorithm, so they never gain access to the full data set.

- *Direct versus Incremental:* Direct discretizers divide the range into k intervals simultaneously, requiring an additional criterion to determine the value of k. They do not only include one-step discretization methods, but also discretizers which perform several stages in their operation, selecting more than a single cut point at every step. By contrast, incremental methods begin with a simple discretization and pass through an improvement process, requiring an additional criterion to know when to stop it. At each step, they find the best candidate boundary to be used as a cut point and afterwards the rest of the decisions are made accordingly. Incremental discretizers are also known as hierarchical discretizers [9]. Both types of discretizers are widespread in the literature, although there is usually a more defined relationship between incremental and supervised ones.
- *Evaluation Measure:* This is the metric used by the discretizer to compare two candidate schemes and decide which is more suitable to be used. We consider five main families of evaluation measures:
  - *Information:* This family includes *entropy* as the most used evaluation measure in discretization (MDLP [41], ID3 [92], FUSINTER [126]) and other derived information theory measures such as the *Gini index* [66].
  - *Statistical:* Statistical evaluation involves the measurement of dependency/correlation among attributes (Zeta [58], ChiMerge [68], Chi2 [76]), probability and bayesian properties [119] (MODL [16]), interdependency [70], contingency coefficient [106], etc.
  - *Rough Sets:* This group is composed of methods that evaluate the discretization schemes by using rough set measures and properties [86], such as lower and upper approximations, class separability, etc.
  - *Wrapper:* This collection comprises methods that rely on the error provided by a classifier that is run for each evaluation. The classifier can be a very simple one, such as a majority class voting classifier (Valley [108]) or general classifiers such as Naïve Bayes (NBIterative [87]).
  - *Binning:* This category refers to the absence of an evaluation measure. It is the simplest method to discretize an attribute by creating a specified number of bins. Each bin is defined as a priori and allocates a specified number of values per attribute. Widely used binning methods are EqualWidth and EqualFrequency.

9.3.1.2 Other Properties

We can discuss other properties related to discretization which also influence the operation and results obtained by a discretizer, but to a lower degree than the characteristics explained above. Furthermore, some of them present a large variety of categorizations and may harm the interpretability of the taxonomy.

- *Parametric versus Non-Parametric:* This property refers to the automatic determination of the number of intervals for each attribute by the discretizer. A nonparametric discretizer computes the appropriate number of intervals for each attribute considering a trade-off between the loss of information or consistency and obtaining the lowest number of them. A parametric discretizer requires a maximum number of intervals desired to be fixed by the user. Examples of non-parametric discretizers are MDLP [41] and CAIM [70]. Examples of parametric ones are ChiMerge [68] and CADD [24].

- *Top-Down versus Bottom Up:* This property is only observed in incremental discretizers. Top-Down methods begin with an empty discretization. Its improvement process is simply to add a new cutpoint to the discretization. On the other hand, Bottom-Up methods begin with a discretization that contains all the possible cutpoints. Its improvement process consists of iteratively merging two intervals, removing a cut point. A classic Top-Down method is MDLP [41] and a well-known Bottom-Up method is ChiMerge [68].
- *Stopping Condition:* This is related to the mechanism used to stop the discretization process and must be specified in nonparametric approaches. Well-known stopping criteria are the Minimum Description Length measure [41], confidence thresholds [68], or inconsistency ratios [26].
- *Disjoint versus Non-Disjoint:* Disjoint methods discretize the value range of the attribute into disassociated intervals, without overlapping, whereas non-disjoint methods dicsretize the value range into intervals that can overlap. The methods reviewed in this chapter are disjoint, while fuzzy discretization is usually non-disjoint [62].
- *Ordinal versus Nominal:* Ordinal discretization transforms quantitative data into ordinal qualitative data whereas nominal discretization transforms it into nominal qualitative data, discarding the information about order. Ordinal discretizers are less common, and not usually considered classic discretizers [80].

9.3.1.3 Criteria to Compare Discretization Methods

When comparing discretization methods, there are a number of criteria that can be used to evaluate the relative strengths and weaknesses of each algorithm. These include the number of intervals, inconsistency, predictive classification rate and time requirements

- *Number of Intervals:* A desirable feature for practical discretization is that discretized attributes have as few values as possible, since a large number of intervals may make the learning slow and ineffective [19].
- *Inconsistency:* A supervision-based measure used to compute the number of unavoidable errors produced in the data set. An unavoidable error is one associated to two examples with the same values for input attributes and different class labels. In general, data sets with continuous attributes are consistent, but when a discretization scheme is applied over the data, an inconsistent data set may be obtained. The desired inconsistency level that a discretizer should obtain is 0.0.

- *Predictive Classification Rate:* A successful algorithm will often be able to discretize the training set without significantly reducing the prediction capability of learners in test data which are prepared to treat numerical data.
- *Time requirements:* A static discretization process is carried out just once on a training set, so it does not seem to be a very important evaluation method. However, if the discretization phase takes too long it can become impractical for real applications. In dynamic discretization, the operation is repeated as many times as the learner requires, so it should be performed efficiently.

9.3.2 Methods and Taxonomy

At the time of writting, more than 80 discretization methods have been proposed in the literature. This section is devoted to enumerating and designating them according to a standard followed in this chapter. We have used 30 discretizers in the experimental study, those that we have identified as the most relevant ones. For more details on their descriptions, the reader can visit the URL associated to the KEEL project.[1] Additionaly, implementations of these algorithms in Java can be found in KEEL software [3, 4].

Table 9.1 presents an enumeration of discretizers reviewed in this chapter. The complete name, abbreviation and reference are provided for each one. This chapter does not collect the descriptions of the discretizers. Instead, we recommend that readers consult the original references to understand the complete operation of the discretizers of interest. Discretizers used in the experimental study are depicted in bold. The ID3 discretizer used in the study is a static version of the well-known discretizer embedded in C4.5.

The properties studied above can be used to categorize the discretizers proposed in the literature. The seven characteristics studied allows us to present the taxonomy of discretization methods in an established order. All techniques enumerated in Table 9.1 are collected in the taxonomy drawn in Fig. 9.2. It illustrates the categorization following a hierarchy based on this order: static/dynamic, univariate/multivariate, supervised/unsupervised, splitting/merging/hybrid, global/local, direct/incremental and evaluation measure. The rationale behind the choice of this order is to achieve a clear representation of the taxonomy.

The proposed taxonomy assists us in the organization of many discretization methods so that we can classify them into categories and analyze their behavior. Also, we can highlight other aspects in which the taxonomy can be useful. For example, it provides a snapshot of existing methods and relations or similarities among them. It also depicts the size of the families, the work done in each one and what is currently missing. Finally, it provides a general overview of the state-of-the-art methods in discretization for researchers/practitioners who are beginning in this field or need to discretize data in real applications.

---

[1] http://www.keel.es.

Table 9.1 Discretizers

Complete name	Abbr. name	Reference	Complete name	Abbr. name	Reference
Equal Width Discretizer	**EqualWidth**	[115]	*No name specified*	Butterworth04	[18]
Equal Frequency Discretizer	**EqualFrequency**	[115]	*No name specified*	Zhang04	[124]
No name specified	Chou91	[27]	**Khiops**	**Khiops**	[15]
Adaptive Quantizer	AQ	[21]	**Class-Attribute Interdependence Maximization**	**CAIM**	[70]
Discretizer 2	D2	[19]	**Extended Chi2**	**Extended Chi2**	[101]
ChiMerge	**ChiMerge**	[68]	**Heterogeneity Discretizer**	**Heter-Disc**	[78]
One-Rule Discretizer	**1R**	[59]	**Unsupervised Correlation Preserving Discretizer**	**UCPD**	[81]
Iterative Dichotomizer 3 Discretizer	**ID3**	[92]	*No name specified*	Multi-MDL	[42]
Minimum Description Length Principle	**MDLP**	[41]	Difference Similitude Set Theory Discretizer	DSST	[116]
Valley	Valley	[108, 109]	Multivariate Interdependent Discretizer	MIDCA	[22]
Class-Attribute Dependent Discretizer	**CADD**	[24]	**MODL**	**MODL**	[16]
ReliefF Discretizer	ReliefF	[69]	Information Theoretic Fuzzy Partitioning	ITFP	[6]
Class-driven Statistical Discretizer	StatDisc	[94]	*No name specified*	Wu06	[118]
No name specified	NBIterative	[87]	Fast Independent Component Analysis	FastICA	[67]
Boolean Reasoning Discretizer	BRDisc	[86]	Linear Program Relaxation	LP-Relaxation	[37]
Minimum Description Length Discretizer	MDL-Disc	[89]	**Hellinger-Based Discretizer**	**HellingerBD**	[72]
Bayesian Discretizer	**Bayesian**	[119]	**Distribution Index-Based Discretizer**	**DIBD**	[117]
No name specified	Friedman96	[45]	Wrapper Estimation of Distribution Algorithm	WEDA	[43]
Cluster Analysis Discretizer	**ClusterAnalysis**	[26]	Clustering + Rought Sets Discretizer	Cluster-RS-Disc	[100]
Zeta	**Zeta**	[58]	**Interval Distance Discretizer**	**IDD**	[96]
Distance-based Discretizer	**Distance**	[20]	**Class-Attribute Contingency Coefficient**	**CACC**	[106]
Finite Mixture Model Discretizer	FMM	[102]	Rectified Chi2	Rectified Chi2	[91]

(continued)

Table 9.1 (continued)

Chi2	**Chi2**	[76]	**Ameva**	**Ameva**	[53]
No name specified	FischerExt	[127]	Unification	Unification	[66]
Contextual Merit Numerical Feature Discretizer	CM-NFD	[60]	Multiple Scanning Discretizer	MultipleScan	[54]
Concurrent Merger	ConMerge	[110]	Optimal Flexible Frequency Discretizer	OFFD	[111]
Knowledge EXplorer Discretizer	KEX-Disc	[11]	**Proportional Discretizer**	**PKID**	[122]
LVQ-based Discretization	LVQ-Disc	[88]	**Fixed Frequency Discretizer**	**FFD**	[122]
No name specified	Multi-Bayesian	[83]	Discretization Class intervals Reduce	DCR	[90]
No name specified	A*	[47]	MVD-CG	MVD-CG	[112]
FUSINTER	**FUSINTER**	[126]	Approximate Equal Frequency Discretizer	AEFD	[65]
Cluster-based Discretizer	Cluster-Disc	[82]	*No name specified*	Jiang09	[65]
Entropy-based Discretization According to Distribution of Boundary points	EDA-DB	[5]	Random Forest Discretizer	RFDisc	[12]
No name specified	Clarke00	[30]	Supervised Multivariate Discretizer	SMD	[64]
Relative Unsupervised Discretizer	RUDE	[79]	Clustering Based Discretization	CBD	[49]
Multivariate Discretization	**MVD**	[10]	Improved MDLP	Improved MDLP	[74]
Modified Learning from Examples Module	MODLEM	[55]	Imfor-Disc	Imfor-Disc	[125]
Modified Chi2	**Modified Chi2**	[105]	Clustering ME-MDL	Cluster ME-MDL	[56]
HyperCluster Finder	HCF	[84]	Effective Bottom-up Discretizer	EBDA	[98]
Entropy-based Discretization with Inconsistency Checking	EDIC	[73]	Contextual Discretizer	Contextual-Disc	[85]
Unparametrized Supervised Discretizer	**USD**	[52]	**Hypercube Division Discretizer**	**HDD**	[121]
Rough Set Discretizer	RS-Disc	[34]	Range Coefficient of Dispersion and Skewness	DRDS	[7]
Rough Set Genetic Algorithm Discretizer	RS-GA-Disc	[23]	Entropy-based Discretization using Scope of Classes	EDISC	[99]
Genetic Algorithm Discretizer	GA-Disc	[33]	Universal Discretizer	UniDis	[97]
Self Organizing Map Discretizer	SOM-Disc	[107]	Maximum AUC-based discretizer	MAD	[71]
Optimal Class-Dependent Discretizer	OCDD	[77]			

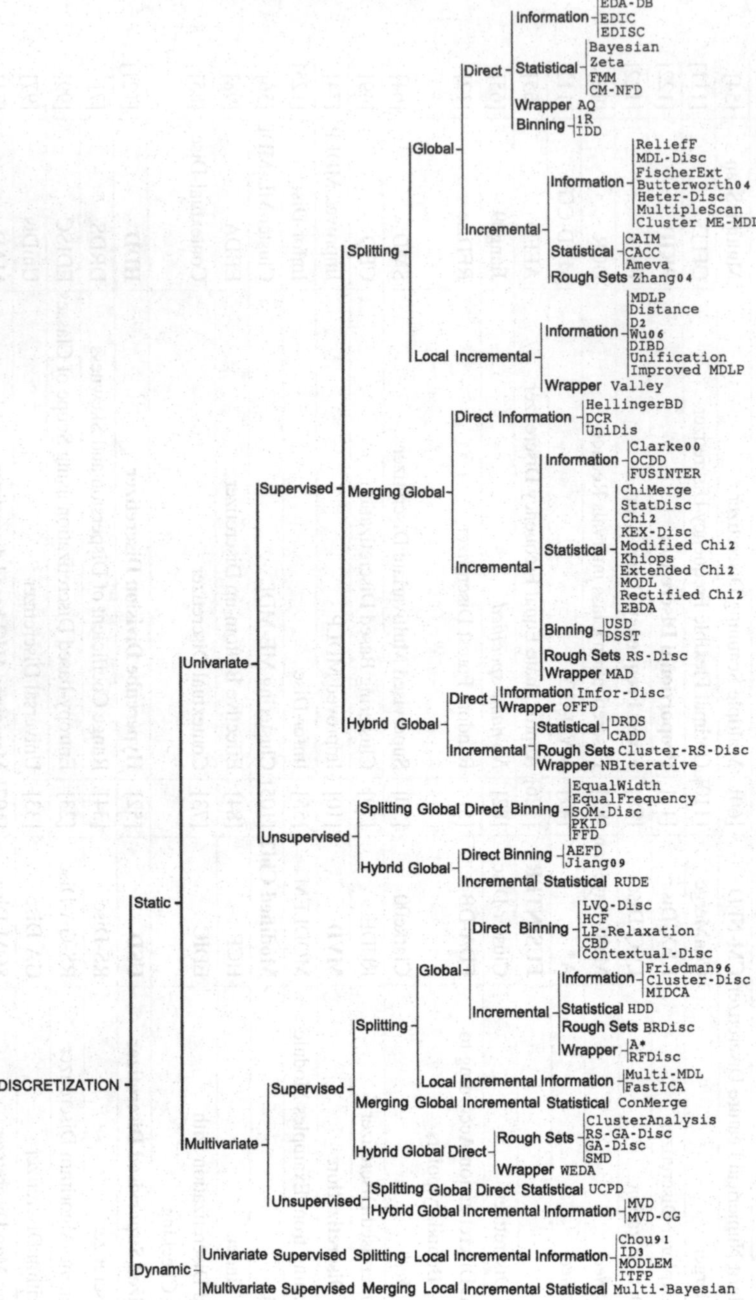

Fig. 9.2 Discretization taxonomy

9.3.3 Description of the Most Representative Discretization Methods

This section is devoted to provide an in-depth description of the most representative methods according to the previous taxonomy, the degree of usage in recent years in the specialized literature and the results are reported in the following section of this chapter, related to the experimental comparative analysis. We will discuss 10 discretizers in more detail: Equal Width/Frequency, MDLP, ChiMerge, Distance, Chi2, PKID, FFD, FUSINTER, CAIM and Modified Chi2. They will be distributed according to the splitting or merging criterion, which can be explained separately to the rest of mechanisms of each discretizer because it is usually a shared process.

9.3.3.1 Splitting Methods

We start with a generalized pseudocode for splitting discretization methods.

Algorithm 1 Splitting Algorithm

```
Require: S = Sorted values of attribute A
  procedure SPLITTING(S)
     if StoppingCriterion() == true then
        Return
     end if
     T = GetBestSplitPoint(S)
     S1 = GetLeftPart(S,T)
     S2 = GetRightPart(S,T)
     Splitting(S1)
     Splitting(S2)
  end procedure
```

The splitting algorithm above consists of all four steps in the discretization scheme, *sort* the feature values, (2) *search* for an appropriate cut point, (3) *split* the range of continuous values according to the cut point, and (4) *stop* when a stopping criterion satisfies, otherwise go to (2). In the following, we include the description of the most representative discretizers based on splitting criterion.

Equal Width or Frequency [75]
They belong to the simplest family of methods that discretize an attribute by creating a specified number of bins, which are created by equal width or equal frequency. This family is known as *Binning* methods. The arity m must be specified at the beginning of both discretizers and determine the number of bins. Each bin is associated with a different discrete value. In *equal width*, the continuous range of a feature is divided into intervals that have an equal width and each interval represents a bin. The arity can be calculated by the relationship between the chosen width for each interval and the total length of the attribute range. In *equal frequency*, an equal number of

continuous values are placed in each bin. Thus, the width of each interval is computed by dividing the length of the attribute range by the desired arity.

There is no need for a stopping criterion as the number of bins is computed directly at the beginning of the process. In practice, both methods are equivalent as they only depend on the number of bins desired, regardless of the calculation method.

MDLP [41]
This discretizer uses the entropy measure to evaluate candidate cut points. Entropy is one of the most commonly used discretization measures in the literature. The entropy of a sample variable X is

$$H(X) = -\sum_x p_x \log p_x$$

where x represents a value of X and p_x its estimated probability of occurring. It corresponds to the average amount of information per event where information of an event is defines as:

$$I(x) = -\log p_x$$

Information is high for lower probable events and low otherwise. This discretizer uses the *Information Gain* of a cut point, which is defined as

$$G(A, T; S) = H(S) - H(A, T; S) = H(S) - \frac{|S_1|}{N}H(S_1) - \frac{|S_2|}{N}H(S_2)$$

where A is the attribute in question, T is a candidate cut point and S is the set of N examples. So, S_i is a partitioned subset of examples produced by T.

The MDLP discretizer applies the *Minimum Description Length Principle* to decide the acceptation or rejection for each cut point and to govern the stopping criterion. It is defined in information theory to be the minimum number of bits required to uniquely specify an object out of the universe of all objects. It computes a final cost of coding and takes part in the decision making of the discretizer.

In summary, the MDLP criterion is that the partition induced by a cut point T for a set S of N examples is accepted iff

$$G(A, T; S) > \frac{\log_2(N-1)}{N} + \frac{\delta(A, T; S)}{N}$$

where $\delta(A, T; S) = \log_2(3^c - 2) - [c \cdot H(S) - c_1 \cdot H(S_1) - c_2 \cdot H(S_2)]$. We recall that c stands for the number of classes of the data set in supervised learning.

The stopping criterion is given in the MDLP itself, due to the fact that $\delta(A, T; S)$ acts as a threshold to stop accepting new partitions.

Distance [20]
This method introduces a distance measure called *Mantaras Distance* to evaluate the cut points. Let us consider two partitions S_a and S_b on a range of continuous values,

each containing c_a and C_b number of classes. The Mantaras distance between two partitions due to a single cut point is given below.

$$Dist(S_a, S_b) = \frac{I(S_a|S_b) + I(S_b|S_a)}{I(S_a \cap S_b)}$$

Since,

$$I(S_b|S_a) = I(S_b \cap S_a) - I(S_a)$$

$$Dist(S_a, S_b) = 2 - \frac{I(S_a) + I(S_b)}{I(S_a \cap S_b)}$$

where,

$$I(S_a) = -\sum_{i=1}^{c_a} S_i \log_2 S_i$$

$$I(S_b) = -\sum_{j=1}^{c_b} S_j \log_2 S_j$$

$$I(S_a \cap S_b) = -\sum_{i=1}^{c_a}\sum_{j=1}^{c_b} S_{ij} \log_2 S_{ij}$$

$$S_i = \frac{|C_i|}{N}$$

$$|C_i| = \text{total count of class } i$$

$$N = \text{total number of instances}$$

$$S_{ij} = S_i \times S_j$$

It chooses the cut point that minimizes the distance. As a stopping criterion, it uses the minimum description length discussed previously to determine whether more cut points should be added.

PKID [122]
In order to maintain a low bias and a low variance in a learning scheme, it is recommendable to increase both the interval frequency and the number of intervals as the amount of training data increases too. A good way to achieve this is to set interval frequency and interval number equally proportional to the amount of training data. This the main purpose of *proportional discretization* (PKID).

When discretizing a continuous attribute for which there are N instances, supposing that the desired interval frequency is s and the desired interval number is t, PKID calculates s and t by the following expressions:

$$s \times t = n$$
$$s = t$$

Thus, this discretizer is an equal frequency (or equal width) discretizer where both the interval frequency and number of intervals have the same quantity and they only depend on the number of instances in training data.

FFD [122]
FFD stands for *fixed frequency discretization* and was proposed for managing bias and variance especially in naive-bayes based classifiers. To discretize a continuous attribute, FFD sets a *sufficient interval frequency*, f. Then it discretizes the ascendingly sorted values into intervals of frequency f. Thus each interval has approximately the same number f of training instances with adjacent values. Incorporating f, FFD aims to ensure that the interval frequency is sufficient so that there are enough training instances in each interval to reliably estimate the Naïve Bayes probabilities.

There may be confusion when distinguishing equal frequency discretization from FFD. The former one fixes the interval number, thus it arbitrarily chooses the interval number and then discretizes a continuous attribute into intervals such that each interval has the same number of training instances. On the other hand, the later method, FFD, fixes the interval frequency by the value f. It then sets cut points so that each interval contains f training instances, controlling the discretization variance.

CAIM [70]
CAIM stands for Class-Attribute Interdependency Maximization criterion, which measures the dependency between the class variable C and the discretized variable D for attibute A. The method requires the computation of the quanta matrix [24], which, in summary, collects a snapshot of the number of real values of A within each interval and for each class of the corresponding example. The criterion is calculated as:

$$CAIM(C, D, A) = \frac{\sum_{r=1}^{n} \frac{\max_r^2}{M_{+r}}}{m}$$

where m is the number of intervals, r iterates through all intervals, i.e. $r = 1, 2, \ldots, m$, $\max_r$ is the maximum value among all q_{ir} values (maximum value within the rth column of the quanta matrix), M_{+r} is the total number of continuous values of attribute A that are within the interval $(d_{r-1}, d_r]$.

According to the authors, CAIM has the following properties:

- The larger the value of CAIM, the higher the interdependence between the class labels and the discrete intervals.
- It generates discretization schemes where each interval has all of its values grouped within a single class label.
- It has taken into account the negative impact that values belonging to classes, other than the class with the maximum number of values within an interval, have on the discretization scheme.

- The algorithm favors discretization schemes with a smaller number of intervals.

The stopping criterion is satisfied when the CAIM value in a new iteration is equal to or less than the CAIM value of the past iteration or if the number of intervals is less than the number of classes of the problem.

9.3.3.2 Merging Methods

We start with a generalized pseudocode for merging discretization methods.

Algorithm 2 Merging Algorithm

```
Require: S = Sorted values of attribute A
  procedure MERGING(S)
     if StoppingCriterion() == true then
        Return
     end if
     T = GetBestAdjacentIntervals(S)
     S = MergeAdjacentIntervals(S,T)
     Merging(S)
  end procedure
```

The merging algorithm above consists of four steps in the discretization process: (1) *sort* the feature values, (2) *search* for the best two neighboring intervals, (3) *merge* the pair into one interval, and (4) *stop* when a stopping criterion satisfies, otherwise go to (2). In the following, we include the description of the most representative discretizers based on merging criteria.

ChiMerge [68]
First of all, we introduce the χ^2 measure as other important evaluation metric in discretization. χ^2 is a statistical measure that conducts a significance test on the relationship between the values of an attribute and the class. The rationale behind this is that in accurate discretization, the relative class frequencies should be fairly consistent within an interval but two adjacent intervals should not have similar relative class frequency. The χ^2 statistic determines the similarity of adjacent intervals based on some significance level. Actually, it tests the hypothesis that two adjacent intervals of an attribute are independent of the class. If they are independent, they should be merged; otherwise they should remain separate. χ^2 is computed as:

$$\chi^2 = \sum_{i=1}^{2} \sum_{j=1}^{c} \frac{(N_{ij} - E_{ij})^2}{E_{ij}}$$

where:

$$
\begin{aligned}
c &= \text{number of classes} \\
N_{ij} &= \text{number of distinct values in the } i\text{th interval, } j\text{th class} \\
R_i &= \text{number of examples in } i\text{th interval} = \sum_{j=1}^{c} N_{ij} \\
C_j &= \text{number of examples in } j\text{th class} = \sum_{i=1}^{m} N_{ij} \\
N &= \text{total number of examples} = \sum_{j=1}^{c} C_j \\
E_{ij} &= \text{expected frequency of } N_{ij} = (R_i \times C_j)/N
\end{aligned}
$$

ChiMerge is a supervised, bottom-up discretizer. At the beginning, each distinct value of the attribute is considered to be one interval. χ^2 tests are performed for every pair of adjacent intervals. Those adjacent intervals with the least χ^2 value are merged until the chosen stopping criterion is satisfied. The significance level for χ^2 is an input parameter that determines the threshold for the stopping criterion. Another parameter used is the called *max-interval* which can be included to avoid the excessive number of intervals from being created. The recommended value for the significance level should be included between the range from 0.90 to 0.99. The *max-interval* parameter should be set to 10 or 15.

Chi2 [76]
It can be explained as an automated version of ChiMerge. Here, the statistical significance level keeps changing to merge more and more adjacent intervals as long as an inconsistency criterion is satisfied. We understand inconsistency to be two instances that match but belong to different classes. It is even possible to completely remove an attribute because the inconsistency property does not appear during the process of discretizing an attribute, acting as a feature selector. Like ChiMerge, χ^2 statistic is used to discretize the continuous attributes until some inconsistencies are found in the data.

The stopping criterion is achieved when there are inconsistencies in the data considering a limit of zero or δ inconsistency level as default.

Modified Chi2 [105]
In the original Chi2 algorithm, the stopping criterion was defined as the point at which the inconsistency rate exceeded a predefined rate δ. The δ value could be given after some tests on the training data for different data sets. The modification proposed was to use the level of consistency checking coined from Rough Sets Theory. Thus, this level of consistency replaces the basic inconsistency checking, ensuring that the fidelity of the training data could be maintained to be the same after discretization and making the process completely automatic.

Another fixed problem is related to the merging criterion used in Chi2, which was not very accurate, leading to overmerging. The original Chi2 algorithm computes the χ^2 value using an initially predefined degree of freedom (number of classes minus one). From the view of statistics, this was inaccurate because the degree of freedom may change according to the two adjacent intervals to be merged. This fact may change the order of merging and benefit the inconsistency after the discretization.

FUSINTER [126]
This method uses the same strategy as the ChiMerge method, but rather than trying to merge adjacent intervals locally, FUSINTER tries to find the partition which optimizes the measure. Next, we provide a short description of the FUSINTER algorithm.

- Obtain the boundary cut points after an increasing sorting of the values and the formation of intervals with run of examples of the same class. Superposition of several classes into an unique cut point is also allowed.
- Construct a matrix with a similar structure like a quanta matrix.
- Find two adjacent intervals whose merging would improve the value of the criterion and check if they can be merged using a differential criterion.
- Repeat until no improvement is possible or only an interval is achieved.

Two criteria can be used for deciding the merging of intervals. The first is the Shannon's entropy and the second is the quadratic entropy.

9.4 Experimental Comparative Analysis

This section presents the experimental framework and the results collected and discussions on them. Sect. 9.4.1 will describe the complete experimental set up. Then, we offer the study and analysis of the results obtained over the data sets used in Sect. 9.4.2.

9.4.1 Experimental Set up

The goal of this section is to show all the properties and issues related to the experimental study. We specify the data sets, validation procedure, classifiers used, parameters of the classifiers and discretizers, and performance metrics. Data sets and statistical tests used to contrast were described in the Chap. 2 of this book. Here, we will only specify the name of the data sets used. The performance of discretization algorithms is analyzed by using 40 data sets taken from the UCI ML Database Repository [8] and KEEL data set repository [3]. They are enumerated in Table 9.2.

In this study, six classifiers have been used in order to find differences in performance among the discretizers. The classifiers are:

Table 9.2 Enumeration of data sets used in the experimental study

Data set	Data set
Abalone	Appendicitis
Australian	Autos
Balance	Banana
Bands	Bupa
Cleveland	Contraceptive
Crx	Dermatology
Ecoli	Flare
Glass	Haberman
Hayes	Heart
Hepatitis	Iris
Mammographic	Movement
Newthyroid	Pageblocks
Penbased	Phoneme
Pima	Saheart
Satimage	Segment
Sonar	Spambase
Specfheart	Tae
Titanic	Vehicle
Vowel	Wine
Wisconsin	Yeast

- *C4.5* [92]: A well-known decision tree, considered one of the top 10 DM algorithms [120].
- *DataSqueezer* [28]: This learner belongs to the family of inductive rule extraction. In spite of its relative simplicity, DataSqueezer is a very effective learner. The rules generated by the algorithm are compact and comprehensible, but accuracy is to some extent degraded in order to achieve this goal.
- *KNN*: One of the simplest and most effective methods based on similarities among a set of objects. It is also considered one of the top 10 DM algorithms [120] and it can handle nominal attributes using proper distance functions such as HVDM [114]. It belongs to the lazy learning family [2, 48].
- *Naïve Bayes*: This is another of the top 10 DM algorithms [120]. Its aim is to construct a rule which will allow us to assign future objects to a class, assuming independence of attributes when probabilities are established.
- *PUBLIC* [93]: It is an advanced decision tree that integrates the pruning phase with the building stage of the tree in order to avoid the expansion of branches that would be pruned afterwards.
- *Ripper* [32]: This is a widely used rule induction method based on a *separate and conquer* strategy. It incorporates diverse mechanisms to avoid overfitting and to handle numeric and nominal attributes simultaneously. The models obtained are in the form of decision lists.

Table 9.3 Parameters of the discretizers and classifiers

Method	Parameters
C4.5	Pruned tree, confidence = 0.25, 2 examples per leaf
DataSqueezer	Pruning and generalization threshold = 0.05
KNN	$K = 3$, HVDM distance
PUBLIC	25 nodes between prune
Ripper	$k = 2$, grow set = 0.66
1R	6 examples of the same class per interval
CADD	Confidence threshold = 0.01
Chi2	Inconsistency threshold = 0.02
ChiMerge	Confidence threshold = 0.05
FDD	Frequency size = 30
FUSINTER	$\alpha = 0.975$, $\lambda = 1$
HDD	Coefficient = 0.8
IDD	Neighborhood = 3, windows size = 3, nominal distance
MODL	Optimized process type
UCPD	Intervals = [3, 6], KNN map type, neighborhood = 6,
	Minimum support = 25, merged threshold = 0.5,
	Scaling factor = 0.5, use discrete

The data sets considered are partitioned using the 10-FCV procedure. The parameters of the discretizers and classifiers are those recommended by their respective authors. They are specified in Table 9.3 for those methods which require them. We assume that the choice of the values of parameters is optimally chosen by their own authors. Nevertheless, in discretizers that require the input of the number of intervals as a parameter, we use a rule of thumb which is dependent on the number of instances in the data set. It consists in dividing the number of instances by 100 and taking the maximum value between this result and the number of classes. All discretizers and classifiers are run one time in each partition because they are non-stochastic.

Two performance measures are widely used because of their simplicity and successful application when multi-class classification problems are dealt with. We refer to accuracy and Cohen's kappa [31] measures, which will be adopted to measure the efficacy discretizers in terms of the generalization classification rate. The explanation of Cohen's kappa was given in Chap. 2.

The empirical study involves 30 discretization methods from those listed in Table 9.1. We want to outline that the implementations are only based on the descriptions and specifications given by the respective authors in their papers.

Table 9.4 Average results collected from intrinsic properties of the discretizers: number of intervals obtained and inconsistency rates in training and test data

Number int.		Incons. train		Incons. tst	
Heter-Disc	8.3125	ID3	0.0504	ID3	0.0349
MVD	18.4575	PKID	0.0581	PKID	0.0358
Distance	23.2125	Modified Chi2	0.0693	FFD	0.0377
UCPD	35.0225	FFD	0.0693	HDD	0.0405
MDLP	36.6600	HDD	0.0755	Modified Chi2	0.0409
Chi2	46.6350	USD	0.0874	USD	0.0512
FUSINTER	59.9850	ClusterAnalysis	0.0958	Khiops	0.0599
DIBD	64.4025	Khiops	0.1157	ClusterAnalysis	0.0623
CADD	67.7100	EqualWidth	0.1222	EqualWidth	0.0627
ChiMerge	69.5625	EqualFrequency	0.1355	EqualFrequency	0.0652
CAIM	72.5125	Chi2	0.1360	Chi2	0.0653
Zeta	75.9325	Bayesian	0.1642	FUSINTER	0.0854
Ameva	78.8425	MODL	0.1716	MODL	0.0970
Khiops	130.3000	FUSINTER	0.1735	HellingerBD	0.1054
1R	162.1925	HellingerBD	0.1975	Bayesian	0.1139
EqualWidth	171.7200	IDD	0.2061	UCPD	0.1383
Extended Chi2	205.2650	ChiMerge	0.2504	ChiMerge	0.1432
HellingerBD	244.6925	UCPD	0.2605	IDD	0.1570
EqualFrequency	267.7250	CAIM	0.2810	CAIM	0.1589
PKID	295.9550	Extended Chi2	0.3048	Extended Chi2	0.1762
MODL	335.8700	Ameva	0.3050	Ameva	0.1932
FFD	342.6050	1R	0.3112	CACC	0.2047
IDD	349.1250	CACC	0.3118	1R	0.2441
Modified Chi2	353.6000	MDLP	0.3783	Zeta	0.2454
CACC	505.5775	Zeta	0.3913	MDLP	0.2501
ClusterAnalysis	1116.1800	MVD	0.4237	DIBD	0.2757
USD	1276.1775	Distance	0.4274	Distance	0.2987
Bayesian	1336.0175	DIBD	0.4367	MVD	0.3171
ID3	1858.3000	CADD	0.6532	CADD	0.5688
HDD	2202.5275	Heter-Disc	0.6749	Heter-Disc	0.5708

9.4.2 Analysis and Empirical Results

Table 9.4 presents the average results corresponding to the number of intervals and inconsistency rate in training and test data by all the discretizers over the 40 data sets. Similarly, Tables 9.5 and 9.6 collect the average results associated to accuracy and kappa measures for each classifier considered. For each metric, the discretizers are ordered from the best to the worst. In Tables 9.5 and 9.6, we highlight those

Table 9.5 Average results of accuracy considering the six classifiers

C4.5		*DataSqueezer*		*KNN*		*Naïve Bayes*		*PUBLIC*		*Ripper*	
FUSINTER	0.7588	Distance	0.5666	PKID	0.7699	PKID	0.7587	FUSINTER	0.7448	Modified Chi2	0.7241
ChiMerge	0.7494	CAIM	0.5547	FFD	0.7594	Modified Chi2	0.7578	CAIM	0.7420	Chi2	0.7196
Zeta	0.7488	Ameva	0.5518	Modified Chi2	0.7573	FUSINTER	0.7576	ChiMerge	0.7390	PKID	0.7097
CAIM	0.7484	MDLP	0.5475	EqualFrequency	0.7557	ChiMerge	0.7543	MDLP	0.7334	MODL	0.7089
UCPD	0.7447	Zeta	0.5475	Khiops	0.7512	FFD	0.7535	Distance	0.7305	FUSINTER	0.7078
Distance	0.7446	ChiMerge	0.5472	EqualWidth	0.7472	CAIM	0.7535	Zeta	0.7301	Khiops	0.6999
MDLP	0.7444	CACC	0.5430	FUSINTER	0.7440	EqualWidth	0.7517	Chi2	0.7278	FFD	0.6970
Chi2	0.7442	Heter-Disc	0.5374	ChiMerge	0.7389	Zeta	0.7507	UCPD	0.7254	EqualWidth	0.6899
Modified Chi2	0.7396	DIBD	0.5322	CAIM	0.7381	EqualFrequency	0.7491	Modified Chi2	0.7250	EqualFrequency	0.6890
Ameva	0.7351	UCPD	0.5172	MODL	0.7372	MODL	0.7479	Khiops	0.7200	CAIM	0.6870
Khiops	0.7312	MVD	0.5147	HellingerBD	0.7327	Chi2	0.7476	Ameva	0.7168	HellingerBD	0.6816
MODL	0.7310	FUSINTER	0.5126	Chi2	0.7267	Khiops	0.7455	HellingerBD	0.7119	USD	0.6807
EqualFrequency	0.7304	Bayesian	0.4915	USD	0.7228	USD	0.7428	EqualFrequency	0.7110	ChiMerge	0.6804
EqualWidth	0.7252	Extended Chi2	0.4913	Ameva	0.7220	ID3	0.7381	MODL	0.7103	ID3	0.6787

(continued)

Table 9.5 (continued)

HellingerBD	0.7240	Chi2	0.4874	ID3	0.7172	Ameva	0.7375	CACC	0.7069	Zeta	0.6786
CACC	0.7203	HellingerBD	0.4868	ClusterAnalysis	0.7132	Distance	0.7372	DIBD	0.7002	HDD	0.6700
Extended Chi2	0.7172	MODL	0.4812	Zeta	0.7126	MDLP	0.7369	EqualWidth	0.6998	Ameva	0.6665
DIBD	0.7141	CADD	0.4780	HDD	0.7104	ClusterAnalysis	0.7363	Extended Chi2	0.6974	UCPD	0.6651
FFD	0.7091	EqualFrequency	0.4711	UCPD	0.7090	HellingerBD	0.7363	HDD	0.6789	CACC	0.6562
PKID	0.7079	1R	0.4702	MDLP	0.7002	HDD	0.7360	FFD	0.6770	Extended Chi2	0.6545
HDD	0.6941	EqualWidth	0.4680	Distance	0.6888	UCPD	0.7227	PKID	0.6758	Bayesian	0.6521
USD	0.6835	IDD	0.4679	IDD	0.6860	Extended Chi2	0.7180	USD	0.6698	ClusterAnalysis	0.6464
ClusterAnalysis	0.6813	USD	0.4651	Bayesian	0.6844	CACC	0.7176	Bayesian	0.6551	MDLP	0.6439
ID3	0.6720	Khiops	0.4567	CACC	0.6813	Bayesian	0.7167	ClusterAnalysis	0.6477	Distance	0.6402
1R	0.6695	Modified Chi2	0.4526	DIBD	0.6731	DIBD	0.7036	ID3	0.6406	IDD	0.6219
Bayesian	0.6675	HDD	0.4308	1R	0.6721	IDD	0.6966	MVD	0.6401	Heter-Disc	0.6084
IDD	0.6606	ClusterAnalysis	0.4282	Extended Chi2	0.6695	1R	0.6774	IDD	0.6352	1R	0.6058
MVD	0.6499	PKID	0.3942	MVD	0.6062	MVD	0.6501	1R	0.6332	DIBD	0.5953
Heter-Disc	0.6443	ID3	0.3896	Heter-Disc	0.5524	Heter-Disc	0.6307	Heter-Disc	0.6317	MVD	0.5921
CADD	0.5689	FFD	0.3848	CADD	0.5064	CADD	0.5669	CADD	0.5584	CADD	0.4130

Table 9.6 Average results of kappa considering the six classifiers

C4.5		*DataSqueezer*		*KNN*		*Naïve Bayes*		*PUBLIC*		*Ripper*	
FUSINTER	0.5550	CACC	0.2719	PKID	0.5784	PKID	0.5762	CAIM	0.5279	Modified Chi2	0.5180
ChiMerge	0.5433	Ameva	0.2712	FFD	0.5617	Modified Chi2	0.5742	FUSINTER	0.5204	Chi2	0.5163
CAIM	0.5427	CAIM	0.2618	Modified Chi2	0.5492	FUSINTER	0.5737	ChiMerge	0.5158	MODL	0.5123
Zeta	0.5379	ChiMerge	0.2501	Khiops	0.5457	FFD	0.5710	MDLP	0.5118	FUSINTER	0.5073
MDLP	0.5305	FUSINTER	0.2421	EqualFrequency	0.5438	ChiMerge	0.5650	Distance	0.5074	Khiops	0.4939
UCPD	0.5299	UCPD	0.2324	EqualWidth	0.5338	Chi2	0.5620	Zeta	0.5010	PKID	0.4915
Ameva	0.5297	Zeta	0.2189	CAIM	0.5260	CAIM	0.5616	Ameva	0.4986	EqualFrequency	0.4892
Chi2	0.5290	USD	0.2174	FUSINTER	0.5242	EqualWidth	0.5593	Chi2	0.4899	ChiMerge	0.4878
Distance	0.5288	Distance	0.2099	ChiMerge	0.5232	Khiops	0.5570	UCPD	0.4888	EqualWidth	0.4875
Modified Chi2	0.5163	Khiops	0.2038	MODL	0.5205	EqualFrequency	0.5564	Khiops	0.4846	CAIM	0.4870
MODL	0.5131	HDD	0.2030	HellingerBD	0.5111	MODL	0.5564	CACC	0.4746	Ameva	0.4810
EqualFrequency	0.5108	EqualFrequency	0.2016	Chi2	0.5100	USD	0.5458	HellingerBD	0.4736	FFD	0.4809
Khiops	0.5078	HellingerBD	0.1965	Ameva	0.5041	Zeta	0.5457	Modified Chi2	0.4697	Zeta	0.4769
HellingerBD	0.4984	Bayesian	0.1941	USD	0.4943	Ameva	0.5456	MODL	0.4620	HellingerBD	0.4729
CACC	0.4961	MODL	0.1918	HDD	0.4878	ID3	0.5403	EqualFrequency	0.4535	USD	0.4560
EqualWidth	0.4909	MDLP	0.1875	ClusterAnalysis	0.4863	HDD	0.5394	DIBD	0.4431	UCPD	0.4552

(continued)

Table 9.6 (continued)

Extended Chi2	0.4766	PKID	0.1846	Zeta	0.4831	MDLP	0.5389	EqualWidth	0.4386	CACC	0.4504
DIBD	0.4759	ID3	0.1818	ID3	0.4769	Distance	0.5368	Extended Chi2	0.4358	MDLP	0.4449
FFD	0.4605	EqualWidth	0.1801	UCPD	0.4763	HellingerBD	0.5353	HDD	0.4048	Distance	0.4429
PKID	0.4526	Modified Chi2	0.1788	MDLP	0.4656	ClusterAnalysis	0.5252	FFD	0.3969	HDD	0.4403
HDD	0.4287	DIBD	0.1778	Distance	0.4470	UCPD	0.5194	PKID	0.3883	ID3	0.4359
USD	0.4282	Chi2	0.1743	CACC	0.4367	CACC	0.5128	USD	0.3845	Extended Chi2	0.4290
ClusterAnalysis	0.4044	IDD	0.1648	IDD	0.4329	Extended Chi2	0.4910	MVD	0.3461	ClusterAnalysis	0.4252
ID3	0.3803	FFD	0.1635	Extended Chi2	0.4226	Bayesian	0.4757	ClusterAnalysis	0.3453	Bayesian	0.3987
IDD	0.3803	ClusterAnalysis	0.1613	Bayesian	0.4201	DIBD	0.4731	Bayesian	0.3419	DIBD	0.3759
MVD	0.3759	Extended Chi2	0.1465	DIBD	0.4167	IDD	0.4618	ID3	0.3241	IDD	0.3650
Bayesian	0.3716	MVD	0.1312	1R	0.3940	1R	0.3980	IDD	0.3066	MVD	0.3446
1R	0.3574	1R	0.1147	MVD	0.3429	MVD	0.3977	1R	0.3004	1R	0.3371
Heter-Disc	0.2709	Heter-Disc	0.1024	Heter-Disc	0.2172	Heter-Disc	0.2583	Heter-Disc	0.2570	Heter-Disc	0.2402
CADD	0.1524	CADD	0.0260	CADD	0.1669	CADD	0.1729	CADD	0.1489	CADD	0.1602

Table 9.7 Wilcoxon test results in number of intervals and inconsistencies

	N. Intervals		Incons. Tra		Incons. Tst	
	+	±	+	±	+	±
1R	10	21	3	17	2	20
Ameva	13	21	6	16	4	21
Bayesian	2	4	10	29	7	29
CACC	7	22	4	17	4	21
CADD	21	28	0	1	0	1
CAIM	14	23	6	19	6	20
Chi2	15	26	9	20	9	20
ChiMerge	15	23	6	20	6	23
ClusterAnalysis	1	4	15	29	9	29
DIBD	21	27	2	7	2	8
Distance	26	28	2	6	2	6
EqualFrequency	7	12	12	26	11	29
EqualWidth	11	18	16	26	13	29
Extended Chi2	14	27	2	14	2	18
FFD	5	8	21	29	16	29
FUSINTER	14	22	11	23	8	29
HDD	0	2	18	29	14	29
HellingerBD	9	15	8	21	7	26
Heter-Disc	29	29	0	1	0	1
ID3	0	1	23	29	16	29
IDD	5	11	8	28	6	29
Khiops	9	15	15	27	12	29
MDLP	22	27	3	9	3	11
Modified Chi2	7	13	17	26	15	29
MODL	5	14	12	24	7	29
MVD	23	28	2	13	2	13
PKID	5	8	22	29	16	29
UCPD	17	25	6	17	5	20
USD	2	4	18	29	15	29
Zeta	12	23	3	9	3	13

discretizers whose performance is within 5 % of the range between the best and the worst method in each measure, that is, $value_{best} - (0.05 \cdot (value_{best} - value_{worst}))$. They should be considered as outstanding methods in each category, regardless of their specific position in the table.

The Wilcoxon test [35, 50, 113] is adopted in this study considering a level of significance equal to $\alpha = 0.05$. Tables 9.7, 9.8 and 9.9 show a summary of all possible

Table 9.8 Wilcoxon test results in accuracy

	C4.5		Data Squeezer		KNN		Naïve Bayes		PUBLIC		Ripper	
	+	±	+	±	+	±	+	±	+	±	+	±
1R	1	12	3	23	2	19	1	9	1	12	1	11
Ameva	14	29	17	29	8	26	9	29	13	29	9	29
Bayesian	1	9	5	26	2	12	2	11	0	11	2	17
CACC	9	28	16	29	2	18	5	28	9	29	4	26
CADD	0	1	1	22	0	1	0	1	0	6	0	0
CAIM	16	29	16	29	11	28	10	29	16	29	11	28
Chi2	13	29	4	26	6	27	9	29	11	29	19	29
ChiMerge	17	29	18	29	13	28	10	29	17	29	9	28
ClusterAnalysis	1	10	0	12	5	24	6	27	1	11	2	20
DIBD	6	21	8	29	2	9	2	8	9	23	1	5
Distance	13	29	16	29	2	17	7	26	13	28	2	13
EqualFrequency	10	27	3	21	18	29	9	29	10	26	11	27
EqualWidth	7	20	2	18	11	28	8	29	6	20	9	27
Extended Chi2	9	27	4	26	3	19	3	17	6	25	2	25
FFD	5	15	0	5	20	28	8	29	1	13	10	27
FUSINTER	21	29	9	29	12	28	15	29	20	29	11	29
HDD	1	18	0	14	4	23	5	28	0	24	7	26
HellingerBD	10	27	4	22	7	26	7	28	10	26	6	26
Heter-Disc	0	9	9	29	0	2	0	3	0	11	1	10
ID3	1	10	0	5	5	22	4	28	0	11	5	26
IDD	1	10	3	23	4	21	2	14	0	12	1	16
Khiops	12	27	3	18	18	29	9	29	9	27	11	29
MDLP	14	29	14	29	3	22	8	29	15	29	2	16
Modified Chi2	11	27	3	21	17	28	10	29	9	29	23	29
MODL	12	28	5	23	14	28	9	29	10	28	17	29
MVD	1	15	5	29	1	8	1	7	0	19	1	13
PKID	5	15	0	6	27	29	9	29	1	13	15	29
UCPD	14	29	7	26	4	17	2	15	14	28	3	19
USD	1	13	3	19	6	23	6	29	1	19	7	25
Zeta	14	29	17	29	4	20	9	29	14	29	7	27

comparisons involved in the Wilcoxon test among all discretizers and measures, for number of intervals and inconsistency rate, accuracy and kappa respectively. Again, the individual comparisons between all possible discretizers are exhibited in the aforementioned URL, where a detailed report of statistical results can be found for each measure and classifier. Tables 9.7, 9.8 and 9.9 summarize, for each method in the rows, the number of discretizers outperformed by using the Wilcoxon test under the

Table 9.9 Wilcoxon test results in kappa

	C4.5		Data Squeezer		KNN		Naïve Bayes		PUBLIC		Ripper	
	+	±	+	±	+	±	+	±	+	±	+	±
1R	1	11	0	15	2	16	2	8	1	13	1	11
Ameva	15	29	24	29	11	26	11	29	16	29	9	29
Bayesian	1	8	1	24	2	10	2	8	1	11	2	17
CACC	11	28	25	29	3	16	7	25	13	29	4	26
CADD	0	1	0	3	0	1	0	1	0	2	0	0
CAIM	17	29	22	29	13	28	11	29	21	29	11	28
Chi2	14	29	2	24	11	27	10	29	13	29	19	29
ChiMerge	19	29	22	29	13	28	11	29	18	29	9	28
ClusterAnalysis	2	10	2	21	5	23	6	22	1	11	2	20
DIBD	8	20	1	24	2	10	2	7	7	18	1	5
Distance	16	29	1	26	2	16	7	28	16	29	2	13
EqualFrequency	11	25	3	25	18	29	10	29	10	23	11	27
EqualWidth	7	20	2	23	14	27	8	28	6	18	9	27
Extended Chi2	10	27	1	20	2	17	3	16	6	23	2	25
FFD	6	14	1	19	23	28	12	29	2	14	10	27
FUSINTER	21	29	16	29	14	28	18	29	19	29	11	29
HDD	2	17	5	25	5	22	6	25	1	22	7	26
HellingerBD	11	23	4	23	9	26	7	21	11	24	6	26
Heter-Disc	0	6	0	12	0	2	0	2	0	8	1	10
ID3	1	8	2	22	4	20	6	26	0	10	5	26
IDD	1	9	1	23	2	18	2	15	1	11	1	16
Khiops	11	24	5	24	18	29	10	29	13	25	11	29
MDLP	16	29	1	24	6	22	8	29	19	29	2	16
Modified Chi2	12	27	1	21	17	27	14	29	9	28	23	29
MODL	12	28	4	24	14	27	12	29	11	28	17	29
MVD	1	12	0	19	1	10	1	6	1	16	1	13
PKID	5	14	2	23	27	29	14	29	2	14	15	29
UCPD	14	29	15	28	4	16	4	16	13	25	3	19
USD	4	13	9	25	6	23	6	25	3	15	7	25
Zeta	15	29	9	27	3	18	6	27	16	29	7	27

column represented by the '+' symbol. The column with the '±' symbol indicates the number of wins and ties obtained by the method in the row. The maximum value for each column is highlighted by a shaded cell.

Once the results are presented in the mentioned tables and graphics, we can stress some interesting properties observed from them, and we can point out the best performing discretizers:

- Regarding the number of intervals, the discretizers which divide the numerical attributes in fewer intervals are *Heter-Disc*, *MVD* and *Distance*, whereas discretizers which require a large number of cut points are *HDD*, *ID3* and *Bayesian*. The Wilcoxon test confirms that *Heter-Disc* is the discretizer that obtains the least intervals outperforming the rest.
- The inconsistency rate both in training data and test data follows a similar trend for all discretizers, considering that the inconsistency obtained in test data is always lower than in training data. *ID3* is the discretizer that obtains the lowest average inconsistency rate in training and test data, albeit the Wilcoxon test cannot find significant differences between it and the other two discretizers: *FFD* and *PKID*. We can observe a close relationship between the number of intervals produced and the inconsistency rate, where discretizers that compute fewer cut points are usually those which have a high inconsistency rate. They risk the consistency of the data in order to simplify the result, although the consistency is not usually correlated with the accuracy, as we will see below.
- In decision trees (*C4.5* and *PUBLIC*), a subset of discretizers can be stressed as the best performing ones. Considering average accuracy, *FUSINTER*, *ChiMerge* and *CAIM* stand out from the rest. Considering average kappa, *Zeta* and *MDLP* are also added to this subset. The Wilcoxon test confirms this result and adds another discretizer, *Distance*, which outperforms 16 of the 29 methods. All methods emphasized are supervised, incremental (except *Zeta*) and use statistical and information measures as evaluators. Splitting/Merging and Local/Global properties have no effect on decision trees.
- Considering rule induction (*DataSqueezer* and *Ripper*), the best performing discretizers are *Distance*, *Modified Chi2*, *Chi2*, *PKID* and *MODL* in average accuracy and *CACC*, *Ameva*, *CAIM* and *FUSINTER* in average kappa. In this case, the results are very irregular due to the fact that the Wilcoxon test emphasizes the *ChiMerge* as the best performing discretizer for *DataSqueezer* instead of *Distance* and incorporates *Zeta* in the subset. With *Ripper*, the Wilcoxon test confirms the results obtained by averaging accuracy and kappa. It is difficult to discern a common set of properties that define the best performing discretizers due to the fact that rule induction methods differ in their operation to a greater extent than decision trees. However, we can say that, in the subset of best methods, incremental and supervised discretizers predominate in the statistical evaluation.
- Lazy and bayesian learning can be analyzed together, due to the fact that the HVDM distance used in KNN is highly related to the computation of bayesian probabilities considering attribute independence [114]. With respect to lazy and bayesian learning, *KNN* and *Naïve Bayes*, the subset of remarkable discretizers is formed by *PKID*, *FFD*, *Modified Chi2*, *FUSINTER*, *ChiMerge*, *CAIM*, *EqualWidth* and *Zeta*, when average accuracy is used; and *Chi2*, *Khiops*, *EqualFrequency* and *MODL* must be added when average kappa is considered. The statistcal report by Wilcoxon informs us of the existence of two outstanding methods: *PKID* for *KNN*, which outperforms 27/29 and *FUSINTER* for *Naïve Bayes*. Here, supervised and unsupervised, direct and incremental, binning and statistical/information evalua-

tion are characteristics present in the best performing methods. However, we can see that all of them are global, thus identifying a trend towards binning methods.

- In general, accuracy and kappa performance registered by discretizers do not differ too much. The behavior in both evaluation metrics are quite similar, taking into account that the differences in kappa are usually lower due to the compensation of random success offered by it. Surprisingly, in *DataSqueezer*, accuracy and kappa offer the greatest differences in behavior, but they are motivated by the fact that this method focuses on obtaining simple rule sets, leaving precision in the background.
- It is obvious that there is a direct dependence between discretization and the classifier used. We have pointed out that a similar behavior in decision trees and lazy/bayesian learning can be detected, whereas in rule induction learning, the operation of the algorithm conditions the effectiveness of the discretizer. Knowing a subset of suitable discretizers for each type of discretizer is a good starting point to understand and propose improvements in the area.
- Another interesting observation can be made about the relationship between accuracy and the number of intervals yielded by a discretizer. A discretizer that computes few cut points does not have to obtain poor results in accuracy and vice versa.
- Finally, we can stress a subset of global best discretizers considering a trade-off between the number of intervals and accuracy obtained. In this subset, we can include *FUSINTER*, *Distance*, *Chi2*, *MDLP* and *UCPD*.

On the other hand, an analysis centered on the 30 discretizers studied is given as follows:

- Many classic discretizers are usually the best performing ones. This is the case of *ChiMerge*, *MDLP*, *Zeta*, *Distance* and *Chi2*.
- Other classic discretizers are not as good as they should be, considering that they have been improved over the years: *EqualWidth*, *EqualFrequency*, *1R*, *ID3* (the static version is much worse than the dynamic inserted in C4.5 operation), *CADD*, *Bayesian* and *ClusterAnalysis*.
- Slight modifications of classic methods have greatly enhanced their results, such as, for example, *FUSINTER*, *Modified Chi2*, *PKID* and *FFD*; but in other cases, the extensions have diminished their performance: *USD*, *Extended Chi2*.
- Promising techniques that have been evaluated under unfavorable circumstances are *MVD* and *UCP*, which are unsupervised methods useful for application to other DM problems apart from classification.
- Recent proposed methods that have been demonstrated to be competitive compared with classic methods and even outperforming them in some scenarios are *Khiops*, *CAIM*, *MODL*, *Ameva* and *CACC*. However, recent proposals that have reported bad results in general are *Heter-Disc*, *HellingerBD*, *DIBD*, *IDD* and *HDD*.
- Finally, this study involves a higher number of data sets than the quantity considered in previous works and the conclusions achieved are impartial towards an specific discretizer. However, we have to stress some coincidences with the conclusions of these previous works. For example in [105], the authors propose

an improved version of *Chi2* in terms of accuracy, removing the user parameter choice. We check and measure the actual improvement. In [122], the authors develop an intense theoretical and analytical study concerning *Naïve Bayes* and propose *PKID* and *FFD* according to their conclusions. Thus, we corroborate that *PKID* is the best suitable method for *Naïve Bayes* and even for *KNN*. Finally, we may note that *CAIM* is one of the simplest discretizers and its effectiveness has also been shown in this study.

References

1. Agrawal, R., Srikant, R.: Fast algorithms for mining association rules. In: Proceedings of the 20th Very Large Data Bases conference (VLDB), pp. 487–499 (1994)
2. Aha, D.W. (ed.): Lazy Learning. Springer, New York (2010)
3. Alcalá-Fdez, J., Fernández, A., Luengo, J., Derrac, J., García, S., Sánchez, L., Herrera, F.: KEEL data-mining software tool: data set repository, integration of algorithms and experimental analysis framework. J. Multiple-Valued Logic Soft Comput. **17**(2–3), 255–287 (2011)
4. Alcalá-Fdez, J., Sánchez, L., García, S., del Jesus, M.J., Ventura, S., Garrell, J.M., Otero, J., Romero, C., Bacardit, J., Rivas, V.M., Fernández, J.C., Herrera, F.: KEEL: a software tool to assess evolutionary algorithms for data mining problems. Soft Comput. **13**(3), 307–318 (2009)
5. An, A., Cercone, N.: Discretization of Continuous Attributes for Learning Classification Rules. In: Proceedings of the Third Conference on Methodologies for Knowledge Discovery and Data Mining, pp. 509–514 (1999)
6. Au, W.H., Chan, K.C.C., Wong, A.K.C.: A fuzzy approach to partitioning continuous attributes for classification. IEEE Trans. Knowl. Data Eng. **18**(5), 715–719 (2006)
7. Augasta, M.G., Kathirvalavakumar, T.: A new discretization algorithm based on range coefficient of dispersion and skewness for neural networks classifier. Appl. Soft Comput. **12**(2), 619–625 (2012)
8. Bache, K., Lichman, M.: UCI machine learning repository (2013). http://archive.ics.uci.edu/ml
9. Bakar, A.A., Othman, Z.A., Shuib, N.L.M.: Building a new taxonomy for data discretization techniques. In: Proceedings on Conference on Data Mining and Optimization (DMO), pp. 132–140 (2009)
10. Bay, S.D.: Multivariate discretization for set mining. Knowl. Inf. Syst. **3**, 491–512 (2001)
11. Berka, P., Bruha, I.: Empirical comparison of various discretization procedures. Int. J. Pattern Recognit. Artif. Intell. **12**(7), 1017–1032 (1998)
12. Berrado, A., Runger, G.C.: Supervised multivariate discretization in mixed data with random forests. In: ACS/IEEE International Conference on Computer Systems and Applications (ICCSA), pp. 211–217 (2009)
13. Berzal, F., Cubero, J.C., Marín, N., Sánchez, D.: Building multi-way decision trees with numerical attributes. Inform. Sci. **165**, 73–90 (2004)
14. Bondu, A., Boulle, M., Lemaire, V.: A non-parametric semi-supervised discretization method. Knowl. Inf. Syst. **24**, 35–57 (2010)
15. Boulle, M.: Khiops: a statistical discretization method of continuous attributes. Mach. Learn. **55**, 53–69 (2004)
16. Boullé, M.: MODL: a bayes optimal discretization method for continuous attributes. Mach. Learn. **65**(1), 131–165 (2006)
17. Breiman, L., Friedman, J., Stone, C.J., Olshen, R.A.: Classification and Regression Trees. Chapman and Hall/CRC, New York (1984)
18. Butterworth, R., Simovici, D.A., Santos, G.S., Ohno-Machado, L.: A greedy algorithm for supervised discretization. J. Biomed. Inform. **37**, 285–292 (2004)

19. Catlett, J.: On changing continuous attributes into ordered discrete attributes. In European Working Session on Learning (EWSL), Lecture Notes on Computer Science, vol. 482, pp. 164–178. Springer (1991)
20. Cerquides, J., Mantaras, R.L.D.: Proposal and empirical comparison of a parallelizable distance-based discretization method. In: Proceedings of the Third International Conference on Knowledge Discovery and Data Mining (KDD), pp. 139–142 (1997)
21. Chan, C., Batur, C., Srinivasan, A.: Determination of quantization intervals in rule based model for dynamic systems. In: Proceedings of the Conference on Systems and Man and and Cybernetics, pp. 1719–1723 (1991)
22. Chao, S., Li, Y.: Multivariate interdependent discretization for continuous attribute. Proc. Third Int. Conf. Inf. Technol. Appl. (ICITA) **2**, 167–172 (2005)
23. Chen, C.W., Li, Z.G., Qiao, S.Y., Wen, S.P.: Study on discretization in rough set based on genetic algorithm. In: Proceedings of the Second International Conference on Machine Learning and Cybernetics (ICMLC), pp. 1430–1434 (2003)
24. Ching, J.Y., Wong, A.K.C., Chan, K.C.C.: Class-dependent discretization for inductive learning from continuous and mixed-mode data. IEEE Trans. Pattern Anal. Mach. Intell. **17**, 641–651 (1995)
25. Chlebus, B., Nguyen, S.H.: On finding optimal discretizations for two attributes. Lect. Notes Artif. Intell. **1424**, 537–544 (1998)
26. Chmielewski, M.R., Grzymala-Busse, J.W.: Global discretization of continuous attributes as preprocessing for machine learning. Int. J. Approximate Reasoning **15**(4), 319–331 (1996)
27. Chou, P.A.: Optimal partitioning for classification and regression trees. IEEE Trans. Pattern Anal. Mach. Intell. **13**, 340–354 (1991)
28. Cios, K.J., Kurgan, L.A., Dick, S.: Highly scalable and robust rule learner: performance evaluation and comparison. IEEE Trans. Syst. Man Cybern. Part B **36**, 32–53 (2006)
29. Cios, K.J., Pedrycz, W., Swiniarski, R.W., Kurgan, L.A.: Data Mining: A Knowledge Discovery Approach. Springer, New York (2007)
30. Clarke, E.J., Barton, B.A.: Entropy and MDL discretization of continuous variables for bayesian belief networks. Int. J. Intell. Syst. **15**, 61–92 (2000)
31. Cohen, J.A.: Coefficient of agreement for nominal scales. Educ. Psychol. Measur. **20**, 37–46 (1960)
32. Cohen, W.W.: Fast Effective Rule Induction. In: Proceedings of the Twelfth International Conference on Machine Learning (ICML), pp. 115–123 (1995)
33. Dai, J.H.: A genetic algorithm for discretization of decision systems. In: Proceedings of the Third International Conference on Machine Learning and Cybernetics (ICMLC), pp. 1319–1323 (2004)
34. Dai, J.H., Li, Y.X.: Study on discretization based on rough set theory. In: Proceedings of the First International Conference on Machine Learning and Cybernetics (ICMLC), pp. 1371–1373 (2002)
35. Demšar, J.: Statistical comparisons of classifiers over multiple data sets. J. Mach. Learn. Res. **7**, 1–30 (2006)
36. Dougherty, J., Kohavi, R., Sahami, M.: Supervised and unsupervised discretization of continuous features. In: Proceedings of the Twelfth International Conference on Machine Learning (ICML), pp. 194–202 (1995)
37. Elomaa, T., Kujala, J., Rousu, J.: Practical approximation of optimal multivariate discretization. In: Proceedings of the 16th International Symposium on Methodologies for Intelligent Systems (ISMIS), pp. 612–621 (2006)
38. Elomaa, T., Rousu, J.: General and efficient multisplitting of numerical attributes. Mach. Learn. **36**, 201–244 (1999)
39. Elomaa, T., Rousu, J.: Necessary and sufficient pre-processing in numerical range discretization. Knowl. Inf. Syst. **5**, 162–182 (2003)
40. Elomaa, T., Rousu, J.: Efficient multisplitting revisited: Optima-preserving elimination of partition candidates. Data Min. Knowl. Disc. **8**, 97–126 (2004)

41. Fayyad, U.M., Irani, K.B.: Multi-interval discretization of continuous-valued attributes for classification learning. In: Proceedings of the 13th International Joint Conference on Artificial Intelligence (IJCAI), pp. 1022–1029 (1993)
42. Ferrandiz, S., Boullé, M.: Multivariate discretization by recursive supervised bipartition of graph. In: Proceedings of the 4th Conference on Machine Learning and Data Mining (MLDM), pp. 253–264 (2005)
43. Flores, J.L., Inza, I., Larrañaga, P.: Larra: Wrapper discretization by means of estimation of distribution algorithms. Intell. Data Anal. **11**(5), 525–545 (2007)
44. Flores, M.J., Gámez, J.A., Martínez, A.M., Puerta, J.M.: Handling numeric attributes when comparing bayesian network classifiers: does the discretization method matter? Appl. Intell. **34**, 372–385 (2011)
45. Friedman, N., Goldszmidt, M.: Discretizing continuous attributes while learning bayesian networks. In: Proceedings of the 13th International Conference on Machine Learning (ICML), pp. 157–165 (1996)
46. Gaddam, S.R., Phoha, V.V., Balagani, K.S.: K-Means+ID3: a novel method for supervised anomaly detection by cascading k-means clustering and ID3 decision tree learning methods. IEEE Trans. Knowl. Data Eng. **19**, 345–354 (2007)
47. Gama, J., Torgo, L., Soares, C.: Dynamic discretization of continuous attributes. In: Proceedings of the 6th Ibero-American Conference on AI: Progress in Artificial Intelligence, IBERAMIA, pp. 160–169 (1998)
48. Garcia, E.K., Feldman, S., Gupta, M.R., Srivastava, S.: Completely lazy learning. IEEE Trans. Knowl. Data Eng. **22**, 1274–1285 (2010)
49. García, M.N.M., Lucas, J.P., Batista, V.F.L., Martín, M.J.P.: Multivariate discretization for associative classification in a sparse data application domain. In: Proceedings of the 5th International Conference on Hybrid Artificial Intelligent Systems (HAIS), pp. 104–111 (2010)
50. García, S., Herrera, F.: An extension on "statistical comparisons of classifiers over multiple data sets" for all pairwise comparisons. J. Mach. Learn. Res. **9**, 2677–2694 (2008)
51. García, S., Luengo, J., Sáez, J.A., López, V., Herrera, F.: A survey of discretization techniques: taxonomy and empirical analysis in supervised learning. IEEE Trans. Knowl. Data Eng. **25**(4), 734–750 (2013)
52. Giráldez, R., Aguilar-Ruiz, J., Riquelme, J., Ferrer-Troyano, F., Rodríguez-Baena, D.: Discretization oriented to decision rules generation. Frontiers Artif. Intell. Appl. **82**, 275–279 (2002)
53. González-Abril, L., Cuberos, F.J., Velasco, F., Ortega, J.A.: AMEVA: an autonomous discretization algorithm. Expert Syst. Appl. **36**, 5327–5332 (2009)
54. Grzymala-Busse, J.W.: A multiple scanning strategy for entropy based discretization. In: Proceedings of the 18th International Symposium on Foundations of Intelligent Systems, ISMIS, pp. 25–34 (2009)
55. Grzymala-Busse, J.W., Stefanowski, J.: Three discretization methods for rule induction. Int. J. Intell. Syst. **16**(1), 29–38 (2001)
56. Gupta, A., Mehrotra, K.G., Mohan, C.: A clustering-based discretization for supervised learning. Stat. Probab. Lett. **80**(9–10), 816–824 (2010)
57. He, H., Garcia, E.A.: Learning from imbalanced data. IEEE Trans. Knowl. Data Eng. **21**(9), 1263–1284 (2009)
58. Ho, K.M., Scott, P.D.: Zeta: A global method for discretization of continuous variables. In: Proceedings of the Third International Conference on Knowledge Discovery and Data Mining (KDD), pp. 191–194 (1997)
59. Holte, R.C.: Very simple classification rules perform well on most commonly used datasets. Mach. Learn. **11**, 63–90 (1993)
60. Hong, S.J.: Use of contextual information for feature ranking and discretization. IEEE Trans. Knowl. Data Eng. **9**, 718–730 (1997)
61. Hu, H.W., Chen, Y.L., Tang, K.: A dynamic discretization approach for constructing decision trees with a continuous label. IEEE Trans. Knowl. Data Eng. **21**(11), 1505–1514 (2009)

62. Ishibuchi, H., Yamamoto, T., Nakashima, T.: Fuzzy data mining: Effect of fuzzy discretization. In: IEEE International Conference on Data Mining (ICDM), pp. 241–248 (2001)
63. Janssens, D., Brijs, T., Vanhoof, K., Wets, G.: Evaluating the performance of cost-based discretization versus entropy- and error-based discretization. Comput. Oper. Res. **33**(11), 3107–3123 (2006)
64. Jiang, F., Zhao, Z., Ge, Y.: A supervised and multivariate discretization algorithm for rough sets. In: Proceedings of the 5th international conference on Rough set and knowledge technology, RSKT, pp. 596–603 (2010)
65. Jiang, S., Yu, W.: A local density approach for unsupervised feature discretization. In: Proceedings of the 5th International Conference on Advanced Data Mining and Applications, ADMA, pp. 512–519 (2009)
66. Jin, R., Breitbart, Y., Muoh, C.: Data discretization unification. Knowl. Inf. Syst. **19**, 1–29 (2009)
67. Kang, Y., Wang, S., Liu, X., Lai, H., Wang, H., Miao, B.: An ICA-based multivariate discretization algorithm. In: Proceedings of the First International Conference on Knowledge Science, Engineering and Management (KSEM), pp. 556–562 (2006)
68. Kerber, R.: Chimerge: Discretization of numeric attributes. In: National Conference on Artifical Intelligence American Association for Artificial Intelligence (AAAI), pp. 123–128 (1992)
69. Kononenko, I., Sikonja, M.R.: Discretization of continuous attributes using relieff. In: Proceedings of Elektrotehnika in Racunalnika Konferenca (ERK) (1995)
70. Kurgan, L.A., Cios, K.J.: CAIM discretization algorithm. IEEE Trans. Knowl. Data Eng. **16**(2), 145–153 (2004)
71. Kurtcephe, M., Güvenir, H.A.: A discretization method based on maximizing the area under receiver operating characteristic curve. Int. J. Pattern Recognit. Artif. Intell. **27**(1), 8 (2013)
72. Lee, C.H.: A hellinger-based discretization method for numeric attributes in classification learning. Knowl. Based Syst. **20**, 419–425 (2007)
73. Li, R.P., Wang, Z.O.: An entropy-based discretization method for classification rules with inconsistency checking. In: Proceedings of the First International Conference on Machine Learning and Cybernetics (ICMLC), pp. 243–246 (2002)
74. Li, W.L., Yu, R.H., Wang, X.Z.: Discretization of continuous-valued attributes in decision tree generation. In: Proocedings of the Second International Conference on Machine Learning and Cybernetics (ICMLC), pp. 194–198 (2010)
75. Liu, H., Hussain, F., Tan, C.L., Dash, M.: Discretization: an enabling technique. Data Min. Knowl. Disc. **6**(4), 393–423 (2002)
76. Liu, H., Setiono, R.: Feature selection via discretization. IEEE Trans. Knowl. Data Eng. **9**, 642–645 (1997)
77. Liu, L., Wong, A.K.C., Wang, Y.: A global optimal algorithm for class-dependent discretization of continuous data. Intell. Data Anal. **8**, 151–170 (2004)
78. Liu, X., Wang, H.: A discretization algorithm based on a heterogeneity criterion. IEEE Trans. Knowl. Data Eng. **17**, 1166–1173 (2005)
79. Ludl, M.C., Widmer, G.: Relative unsupervised discretization for association rule mining, In: Proceedings of the 4th European Conference on Principles of Data Mining and Knowledge Discovery, The Fourth European Conference on Principles and Practice of Knowledge Discovery in Databases (PKDD), pp. 148–158 (2000)
80. Macskassy, S.A., Hirsh, H., Banerjee, A., Dayanik, A.A.: Using text classifiers for numerical classification. In: Proceedings of the 17th International Joint Conference on Artificial Intelligence, Vol. 2 (IJCAI), pp. 885–890 (2001)
81. Mehta, S., Parthasarathy, S., Yang, H.: Toward unsupervised correlation preserving discretization. IEEE Trans. Knowl. Data Eng. **17**, 1174–1185 (2005)
82. Monti, S., Cooper, G.: A latent variable model for multivariate discretization. In: Proceedings of the Seventh International Workshop on AI & Statistics (Uncertainty) (1999)
83. Monti, S., Cooper, G.F.: A multivariate discretization method for learning bayesian networks from mixed data. In: Proceedings on Uncertainty in Artificial Intelligence (UAI), pp. 404–413 (1998)

84. Muhlenbach, F., Rakotomalala, R.: Multivariate supervised discretization, a neighborhood graph approach. In: Proceedings of the 2002 IEEE International Conference on Data Mining, ICDM, pp. 314–320 (2002)
85. Nemmiche-Alachaher, L.: Contextual approach to data discretization. In: Proceedings of the International Multi-Conference on Computing in the Global Information Technology (ICCGI), pp. 35–40 (2010)
86. Nguyen, S.H., Skowron, A.: Quantization of real value attributes - rough set and boolean reasoning approach. In: Proceedings of the Second Joint Annual Conference on Information Sciences (JCIS), pp. 34–37 (1995)
87. Pazzani, M.J.: An iterative improvement approach for the discretization of numeric attributes in bayesian classifiers. In: Proceedings of the First International Conference on Knowledge Discovery and Data Mining (KDD), pp. 228–233 (1995)
88. Perner, P., Trautzsch, S.: Multi-interval discretization methods for decision tree learning. In: Advances in Pattern Recognition, Joint IAPR International Workshops SSPR 98 and SPR 98, pp. 475–482 (1998)
89. Pfahringer, B.: Compression-based discretization of continuous attributes. In: Proceedings of the 12th International Conference on Machine Learning (ICML), pp. 456–463 (1995)
90. Pongaksorn, P., Rakthanmanon, T., Waiyamai, K.: DCR: Discretization using class information to reduce number of intervals. In: Proceedings of the International Conference on Quality issues, measures of interestingness and evaluation of data mining model (QIMIE), pp. 17–28 (2009)
91. Qu, W., Yan, D., Sang, Y., Liang, H., Kitsuregawa, M., Li, K.: A novel chi2 algorithm for discretization of continuous attributes. In: Proceedings of the 10th Asia-Pacific web conference on Progress in WWW research and development, APWeb, pp. 560–571 (2008)
92. Quinlan, J.R.: C4.5: Programs for Machine Learning. Morgan Kaufmann Publishers Inc, San Francisco (1993)
93. Rastogi, R., Shim, K.: PUBLIC: a decision tree classifier that integrates building and pruning. Data Min. Knowl. Disc. **4**, 315–344 (2000)
94. Richeldi, M., Rossotto, M.: Class-driven statistical discretization of continuous attributes. In: Proceedings of the 8th European Conference on Machine Learning (ECML), ECML '95, pp. 335–338 (1995)
95. Roy, A., Pal, S.K.: Fuzzy discretization of feature space for a rough set classifier. Pattern Recognit. Lett. **24**, 895–902 (2003)
96. Ruiz, F.J., Angulo, C., Agell, N.: IDD: a supervised interval distance-based method for discretization. IEEE Trans. Knowl. Data Eng. **20**(9), 1230–1238 (2008)
97. Sang, Y., Jin, Y., Li, K., Qi, H.: UniDis: a universal discretization technique. J. Intell. Inf. Syst. **40**(2), 327–348 (2013)
98. Sang, Y., Li, K., Shen, Y.: EBDA: An effective bottom-up discretization algorithm for continuous attributes. In: Proceedings of the 10th IEEE International Conference on Computer and Information Technology (CIT), pp. 2455–2462 (2010)
99. Shehzad, K.: Edisc: a class-tailored discretization technique for rule-based classification. IEEE Trans. Knowl. Data Eng. **24**(8), 1435–1447 (2012)
100. Singh, G.K., Minz, S.: Discretization using clustering and rough set theory. In: Proceedings of the 17th International Conference on Computer Theory and Applications (ICCTA), pp. 330–336 (2007)
101. Su, C.T., Hsu, J.H.: An extended chi2 algorithm for discretization of real value attributes. IEEE Trans. Knowl. Data Eng. **17**, 437–441 (2005)
102. Subramonian, R., Venkata, R., Chen, J.: A visual interactive framework for attribute discretization. In: Proceedings of the First International Conference on Knowledge Discovery and Data Mining (KDD), pp. 82–88 (1997)
103. Sun, Y., Wong, A.K.C., Kamel, M.S.: Classification of imbalanced data: a review. Int. J. Pattern Recognit. Artif. Intell. **23**(4), 687–719 (2009)
104. Susmaga, R.: Analyzing discretizations of continuous attributes given a monotonic discrimination function. Intell. Data Anal. **1**(1–4), 157–179 (1997)

105. Tay, F.E.H., Shen, L.: A modified chi2 algorithm for discretization. IEEE Trans. Knowl. Data Eng. **14**, 666–670 (2002)
106. Tsai, C.J., Lee, C.I., Yang, W.P.: A discretization algorithm based on class-attribute contingency coefficient. Inf. Sci. **178**, 714–731 (2008)
107. Vannucci, M., Colla, V.: Meaningful discretization of continuous features for association rules mining by means of a SOM. In: Proocedings of the 12th European Symposium on Artificial Neural Networks (ESANN), pp. 489–494 (2004)
108. Ventura, D., Martinez, T.R.: BRACE: A paradigm for the discretization of continuously valued data, In: Proceedings of the Seventh Annual Florida AI Research Symposium (FLAIRS), pp. 117–121 (1994)
109. Ventura, D., Martinez, T.R.: An empirical comparison of discretization methods. In: Proceedings of the 10th International Symposium on Computer and Information Sciences (ISCIS), pp. 443–450 (1995)
110. Wang, K., Liu, B.: Concurrent discretization of multiple attributes. In: Proceedings of the Pacific Rim International Conference on Artificial Intelligence (PRICAI), pp. 250–259 (1998)
111. Wang, S., Min, F., Wang, Z., Cao, T.: OFFD: Optimal flexible frequency discretization for naive bayes classification. In: Proceedings of the 5th International Conference on Advanced Data Mining and Applications, ADMA, pp. 704–712 (2009)
112. Wei, H.: A novel multivariate discretization method for mining association rules. In: 2009 Asia-Pacific Conference on Information Processing (APCIP), pp. 378–381 (2009)
113. Wilcoxon, F.: Individual comparisons by ranking methods. Biometrics **1**, 80–83 (1945)
114. Wilson, D.R., Martinez, T.R.: Reduction techniques for instance-based learning algorithms. Mach. Learn. **38**(3), 257–286 (2000)
115. Wong, A.K.C., Chiu, D.K.Y.: Synthesizing statistical knowledge from incomplete mixed-mode data. IEEE Trans. Pattern Anal. Mach. Intell. **9**, 796–805 (1987)
116. Wu, M., Huang, X.C., Luo, X., Yan, P.L.: Discretization algorithm based on difference-similitude set theory. In: Proceedings of the Fourth International Conference on Machine Learning and Cybernetics (ICMLC), pp. 1752–1755 (2005)
117. Wu, Q., Bell, D.A., Prasad, G., McGinnity, T.M.: A distribution-index-based discretizer for decision-making with symbolic ai approaches. IEEE Trans. Knowl. Data Eng. **19**, 17–28 (2007)
118. Wu, Q., Cai, J., Prasad, G., McGinnity, T.M., Bell, D.A., Guan, J.: A novel discretizer for knowledge discovery approaches based on rough sets. In: Proceedings of the First International Conference on Rough Sets and Knowledge Technology (RSKT), pp. 241–246 (2006)
119. Wu, X.: A bayesian discretizer for real-valued attributes. Comput. J. **39**, 688–691 (1996)
120. Wu, X., Kumar, V. (eds.): The Top Ten Algorithms in Data Mining. Data Mining and Knowledge Discovery. Chapman and Hall/CRC, Taylor and Francis, Boca Raton (2009)
121. Yang, P., Li, J.S., Huang, Y.X.: HDD: a hypercube division-based algorithm for discretisation. Int. J. Syst. Sci. **42**(4), 557–566 (2011)
122. Yang, Y., Webb, G.I.: Discretization for naive-bayes learning: managing discretization bias and variance. Mach. Learn. **74**(1), 39–74 (2009)
123. Yang, Y., Webb, G.I., Wu, X.: Discretization methods. In: Data Mining and Knowledge Discovery Handbook, pp. 101–116 (2010)
124. Zhang, G., Hu, L., Jin, W.: Discretization of continuous attributes in rough set theory and its application. In: Proceedings of the 2004 IEEE Conference on Cybernetics and Intelligent Systems (CIS), pp. 1020–1026 (2004)
125. Zhu, W., Wang, J., Zhang, Y., Jia, L.: A discretization algorithm based on information distance criterion and ant colony optimization algorithm for knowledge extracting on industrial database. In: Proceedings of the 2010 IEEE International Conference on Mechatronics and Automation (ICMA), pp. 1477–1482 (2010)
126. Zighed, D.A., Rabaséda, S., Rakotomalala, R.: FUSINTER: a method for discretization of continuous attributes. Int. J. Uncertainty, Fuzziness Knowl. Based Syst. **6**, 307–326 (1998)
127. Zighed, D.A., Rakotomalala, R., Feschet, F.: Optimal multiple intervals discretization of continuous attributes for supervised learning. In: Proceedings of the First International Conference on Knowledge Discovery and Data Mining (KDD), pp. 295–298 (1997)

Chapter 10
A Data Mining Software Package Including Data Preparation and Reduction: KEEL

Abstract KEEL software is an open source Data Mining tool widely used in research and real life applications. Most of the algorithms described, if not all of them, throughout the book are actually implemented and publicly available in this Data Mining platform. Since KEEL enables the user to create and run single or concatenated preprocessing techniques in the data, such software is carefully introduced in this section, intuitively guiding the reader across the step needed to set up all the data preparations that might be needed. It is also interesting to note that the experimental analyses carried out in this book have been created using KEEL, allowing the consultant to quickly compare and adapt the results presented here. An extensive revision of Data Mining software tools are presented in Sect. 10.1. Among them, we will focus on the open source KEEL platform in Sect. 10.2 providing details of its main features and usage. For the practitioners interest, the most common used data sources are introduced in Sect. 10.3 and the steps needed to integrate any new algorithm in it in Sect. 10.4. Once the results have been obtained, the appropriate comparison guidelines are provided in Sect. 10.5. The most important aspects of the tool are summarized in Sect. 10.6.

10.1 Data Mining Softwares and Toolboxes

As we have indicated in Chap. 1, Data Mining (DM) is the process for automatic discovery of high level knowledge by obtaining information from real world, large and complex data sets [1], and is the core step of a broader process, called KDD. In addition to the DM step, the KDD process includes application of several preprocessing methods aimed at faciliting application of DM algorithms and postprocessing methods for refining and improving the discovered knowledge. The evolution of the available techniques and their wide adoption demands to gather all the steps involved in the KDD process in the least amount of pieces of software as possible for the sake of easier application and comparisons among the results obtained, yet allowing non expert practitioners to have access to KDD techniques.

S. García et al., *Data Preprocessing in Data Mining*,
Intelligent Systems Reference Library 72, DOI 10.1007/978-3-319-10247-4_10

Many DM software tools have been developed in the last few years due to the popularization of DM. Although a lot of them are commercially distributed (some of the leading commercial software are mining suites such as SPSS Clementine,[1] Oracle Data Mining[2] and KnowledgeSTUDIO[3]), only a few were available as open source. Fortunately this tendency has changed and free and open source DM tools have appeared to cover many specialized tasks in the process as well as general tools that include most of the steps of KDD. Among the latter we can highlight Weka [2], Orange [3] or Java-ML [4] as the most well-known of a growing family of open source toolboxes for DM.

Most programming languages have a DM software so any user has the possibility of performing experiments. While Weka, RapidMiner[4] [5], Java-ML and $\alpha Miv\epsilon r$ are written in Java, ADaM[5] and Orange are written in Python. Statistical languages also have their software tools as Rattle [6] for R.

It is also common to find libraries for some popular programming languages that can be added to a particular project. Their aim is not the novel user but an experienced practitioner who wants to add functionality to real-world cases without dealing with a multi-purpose GUI or having to rip off the methods they want. A well-known library written in C++ for fast programs is MLC++,[6] and R has their own statistical analysis package.[7] In Java the MLJ library[8] is available to be integrated in any project with ease.

Apart from the aforementioned toolboxes, the reader can find more alternatives to suit to their needs. Many specialized webpages are devoted to the presentation, promotion and publishing of DM news and software. We recommend visiting the KDnuggets software directory[9] and the-Data-Mine site.[10] In the research field open source tools are playing an increasingly important role as is pointed out in [7]. To this regard the link page of the Knowledge Extraction based on Evolutionary Learning (KEEL) webpage[11] contains an extensive list of open source DM tools and related fields such as metaheuristic optimization.

KEEL [8] is a open source Java software tool which empowers the user to assess the behavior of ML, evolutionary learning and soft computing based techniques for different kinds of DM problems: regression, classification, clustering, pattern mining and so on. This tool can offer several advantages:

[1] http://www.spss.com/clementine.

[2] http://www.oracle.com/technology/products/bi/odm.

[3] http://www.angoss.com/products/studio/index.php.

[4] http://sourceforge.net/projects/rapidminer/.

[5] http://projects.itsc.uah.edu/datamining/adam/.

[6] http://www.sgi.com/tech/mlc/.

[7] http://www.r-project.org/.

[8] http://www.kddresearch.org/Groups/Machine-Learning/MLJ/.

[9] http://www.kdnuggets.com/software.

[10] http://the-data-mine.com/bin/view/Software.

[11] http://sci2s.ugr.es/keel/links.php.

- *It reduces programming work*. It includes libraries of different paradigms as evolutionary learning algorithms based on different paradigms (Pittsburgh, Michigan and IRL), fuzzy learning, lazy learning, ANNs, SVMs models and many more; simplifying the integration of DM algorithms with different pre-processing techniques. It can alleviate the work of programming and enable researchers to focus on the analysis of their new learning models in comparison with the existing ones.
- *It extends the range of possible users applying ML algorithms*. An extensive library of ML techniques together with easy-to-use software considerably reduce the level of knowledge and experience required by researchers in DM. As a result researchers with less knowledge, when using this tool, would be able to successfully apply these algorithms to their problems.
- *It has an unparalleled range of preprocessing methods included for DM*, from discretization algorithms to noisy data filters. Few DM platforms offer the same amount of preprocessing techniques as KEEL does. This fact combined with a well-known data format facilitates the user to treat and include their data in the KEEL work flow and to easily prepare it to be used with their favourite techniques.
- *Cross platform compatibility*. Due to the use of a strict object-oriented approach for the library and software tool, these can be used on any machine with Java. As a result, any researcher can use KEEL on their machine, regardless of the operating system.

10.2 KEEL: Knowledge Extraction Based on Evolutionary Learning

KEEL[12] is a software tool that facilitates the analysis of the behaviour of ML in the different areas of learning and pre-processing tasks, making the management of these techniques easy for the user. The models correspond with the most well-known and employed models in each methodology, such as feature and instance selection [9, 10], decision trees [11], SVMs [12], noise filters [13], lazy learning [14], evolutionary fuzzy rule learning [15], genetic ANNs [16], Learning Classifier Systems [17], and many more.

The current available version of KEEL consists of the following function blocks[13]:

- *Data Management*: This part is made up of a set of tools that can be used to build new data, to export and import data in other formats to or from KEEL format, data edition and visualization, to apply transformations and partitioning to data, etc…
- *Design of Experiments (off-line module)*: The aim of this part is the design of the desired experimentation over the selected data sets and providing for many options in different areas: type of validation, type of learning (classification, regression, unsupervised learning), etc…

[12] http://keel.es.

[13] http://www.keel.es/software/prototypes/version1.0/\/ManualKeel.pdf.

- *Educational Experiments (on-line module)*: With a similar structure to the aforementioned, this permits the design of experiment to be run step-by-step in order to display the learning process of a certain model by using the software tool for educational purposes.

With all of these function blocks, we can attest that KEEL can be useful by different types of users who may expect to find specific features in a DM software.

In the following subsections we describe in detail the user profiles for whom KEEL is intended, its main features and the different integrated function blocks.

10.2.1 Main Features

KEEL is a software tool developed to ensemble and use different DM models. Although it was initially focused on the use of evolutionary algorithms for KDD, its continuous development has broadened the available ML paradigms for DM. We would like to note that this is the first software toolkit of this type containing a library of evolutionary algorithms with open source code in Java. The main features of KEEL are:

- Almost one hundred of data preprocessing algorithms proposed in specialized literature are included: data transformation, discretization, MVs treatment, noise filtering, instance selection and FS.
- More than two hundred of state-of-the-art techniques for classification, regression, subgroup discovery, clustering and association rules, ready to be used within the platform or to be extracted and integrated in any other particular project.
- Specialized modules for recent and difficult challenges in DM such as imbalanced learning and multiple instance learning.
- Being the initial key role of KEEL, EAs are presented in predicting models, pre-processing (evolutionary feature and instance selection) and post-processing (evolutionary tuning of fuzzy rules).
- It contains a statistical library to analyze algorithm results and comprises of a set of statistical tests for analyzing the normality and heteroscedasticity of the results, as well as performing parametric and non-parametric comparisons of the algorithms.
- Some algorithms have been developed using the Java Class Library for Evolutionary Computation (JCLEC) software [18].[14]
- A user-friendly interface is provided, oriented towards the analysis of algorithms.
- The software is designed for experiments containing multiple data sets and algorithms connected to each other to obtain the desired result. Experiments are independently script-generated from the user interface for an off-line run in the same or other machines.
- KEEL also allows for experiments in on-line mode, intended as an educational support for learning the operation of the algorithms included.

[14] http://jclec.sourceforge.net/.

- It contains a Knowledge Extraction Algorithms Library[15] with the incorporation of multiple evolutionary learning algorithms, together with classical learning approaches. The principal families of techniques included are:
  - *Evolutionary rule learning models*. Including different paradigms of evolutionary learning.
  - *Fuzzy systems*. Fuzzy rule learning models with a good trade-off between accuracy and interpretability.
  - *Evolutionary neural networks*. Evolution and pruning in ANNs, product unit ANNs, and RBFN models.
  - *Genetic programing*. Evolutionary algorithms that use tree representations for knowledge extraction.
  - *Subgroup discovery*. Algorithms for extracting descriptive rules based on patterns subgroup discovery.
  - *Data reduction* (*instance and feature selection and discretization*). EAs for data reduction.

KEEL integrates the library of algorithms in each of its function blocks. We have briefly presented its function blocks above but in the following subsections, we will describe the possibilities that KEEL offers in relation to data management, off-line experiment design and on-line educational design.

10.2.2 Data Management

The fundamental purpose of data preparation is to manipulate and transform raw data so that the information content enfolded in the data set can be exposed, or made more accessible [19]. Data preparation comprises of those techniques concerned with analyzing raw data so as to yield quality data, mainly including data collecting, data integration, data transformation, data cleaning, data reduction and data discretization [20]. Data preparation can be even more time consuming than DM, and can present similar challenges. Its importance lies in that the real-world data is impure (incomplete, noisy and inconsistent) and high-performance mining systems require quality data (the removal of anomalies or duplications). Quality data yields high-quality patterns (to recover missing data, purify data and resolve conflicts).

The *Data Management* module integrated in KEEL allows us to perform the data preparation stage independently of the remaining DM processes. This module is focused on the group of users denoted as domain experts. They are familiar with their data, they know the processes that produce the data and they are interested in reviewing to improve them or analyze them. On the other hand, domain users are those whose interests lies in applying processes to their own data and are usually not experts in DM.

[15] http://www.keel.es/software/prototypes/version1.0/AlgorithmsList.pdf.

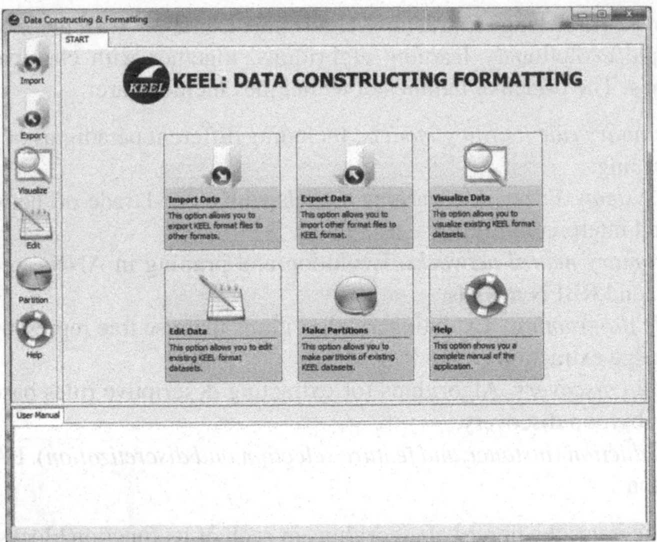

Fig. 10.1 Data management

Figure 10.1 shows an example window of the *Data Management* module in the section of *Data Visualization*. The module has seven sections, each of which is accessible through the buttons on the left side of the window. In the following, we will briefly describe them:

- *Creation of a new data set*: This option allows us to generate a new data set compatible with the other KEEL modules.
- *Import data to KEEL format*: Since KEEL works with a specific data format (similar to the ARFF format) in all its modules, this section allows us to convert various data formats to KEEL format, such as CSV, XML, ARFF, extracting data from data bases, etc.
- *Export data from KEEL format*: This option is the reverse of the previous one. It converts the data handled by KEEL procedures in other external formats to establish compatibility with other software tools.
- *Visualization of data*: This option is used to represent and visualize the data. With it, we can see a graphical distribution of each attribute and comparisons between two attributes.
- *Edition of data*: This area is dedicated to managing the data manually. The data set, once loaded, can be edited by modifying values, adding or removing rows and columns, etc.
- *Data Partition*: This zone allows us to make the partitions of data needed by the experiment modules to validate results. It supports *k*-FCV, 5 × 2-CV and hold-out validation with stratified partition.

- *Data Preparation*: This section allows us to perform automatic data preparation for DM, including cleaning, transformation and reduction of data. All techniques integrated in this section are also available in the experiments-related modules.

10.2.3 Design of Experiments: Off-Line Module

In the last few years, a large number of DM software tools have been developed for research purposes. Some of them are libraries that allow reductions in programming work when developing new algorithms: ECJ [21], JCLEC [18], learning classifier systems [22], etc. Others are DM suites that incorporate learning algorithms (some of them may use EAs for this task) and provide a mechanism to establish comparisons among them. Some examples are Weka [2], D2K [23], etc.

This module is a Graphical User Interface (GUI) that allows the design of experiments for solving various problems of regression, classification and unsupervised learning. Having designed the experiments, it generates the directory structure and files required for running them in any local machine with Java (see Fig. 10.2).

The experiments are graphically modeled, based on data flow and represented by graphs with node-edge connections. To design an experiment, we first have to indicate the type of validation (*k*-FCV [24] or 5×2-CV [25]) and the type of learning (regression, classification or unsupervised) to be used. Then, we have to select the data sources, drag the selected methods into the workspace and connect methods and data sets, combining the evolutionary learning algorithms with different pre-processing and post-processing techniques, if needed. Finally, we can add statistical tests to achieve a complete analysis of the methods being studied, and a report box to obtain a summary of the results. Notice that each component of the experiment is configured in separate dialogues that can be opened by double-clicking the respective node.

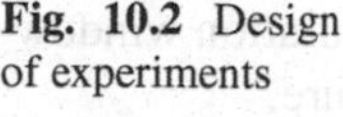

Fig. 10.2 Design of experiments

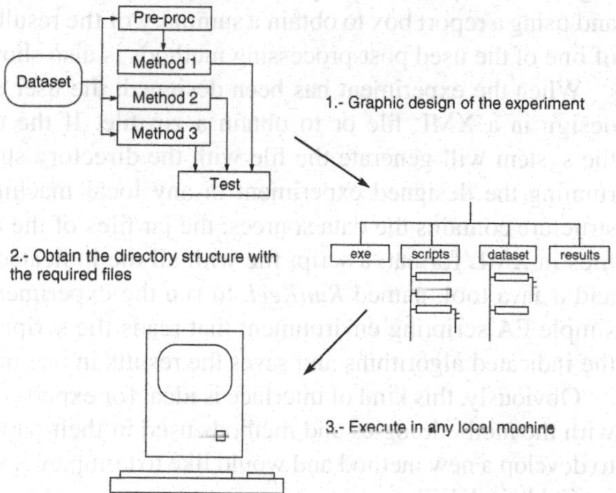

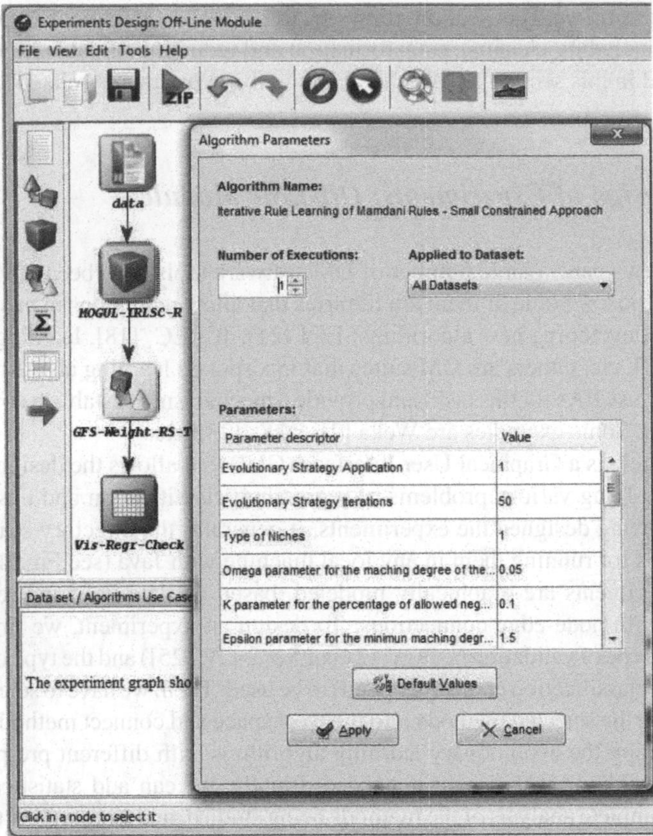

Fig. 10.3 Example of an experiment and the configuration window of a method

Figure 10.3 shows an example of an experiment following the MOGUL methodology and using a report box to obtain a summary of the results. The configuration window of one of the used post-processing methods is also shown in this figure.

When the experiment has been designed, the user can choose either to save the design in a XML file or to obtain a zip file. If the user chooses a zip file, then the system will generate the file with the directory structure and required files for running the designed experiment in any local machine with Java. This directory structure contains the data sources, the jar files of the algorithms, the configuration files in XML format, a script file with all the indicated algorithms in XML format, and a Java tool, named *RunKeel*, to run the experiment. *RunKeel* can be seen as a simple EA scripting environment that reads the script file in XML format, runs all the indicated algorithms and saves the results in one or several report files.

Obviously, this kind of interface is ideal for experts of specific areas who, familiar with the methodologies and methods used in their particular area of interest, intend to develop a new method and would like to compare it with the well-known methods available in KEEL.

10.2.4 Computer-Based Education: On-Line Module

There is a variety of terms used to describe the use of computers in education [26]. Computer-assisted instruction (CAI), computer-based education (CBE) and computer-based instruction (CBI) are the broadest terms and can refer to virtually any kind of computer use in educational environments. These terms may refer either to stand-alone computer learning activities or to computer activities which reinforce material introduced and taught by teachers.

Most of the software developed in DM and evolutionary computation domain is designed for research purposes (libraries, algorithms, specific applications, etc.). But there is some free software that is designed not only for research but also for educational purposes. These systems are easy to use due to the fact that they provide a GUI to assist user interaction with the system in all the tasks (selecting data, choosing parameters, running algorithms, visualize the results, etc.). Some examples of open source DM systems are Weka [2], Yale [27] and Tanagra [28].

This module is a GUI that allows the user to design an experiment (with one or more algorithms), run it and visualize the results on-line. The idea is to use this part of KEEL as a guideline to demonstrate the learning process of a certain model. This module has a similar structure to the previous one but only includes algorithms and options that are suitable for academic purposes.

When an experiment is designed the user can choose either to save the experiment in a XML file or to run it. If the user chooses to run it, then the system will show an auxiliary window to manage and visualize the execution of each algorithm. When the run finishes, this window will show the results obtained for each algorithm in separate tags, showing for example the confusion matrices for classification or the mean square errors for regression problems (see Fig. 10.4).

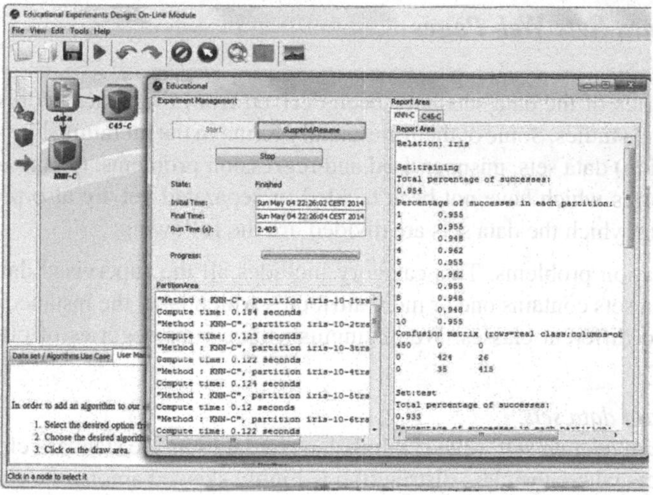

Fig. 10.4 Auxiliary window of an experiment with two algorithms

10.3 KEEL-Dataset

In this section we present the KEEL-dataset repository. It can be accessed through the main KEEL webpage.[16] The KEEL-dataset repository is devoted to the data sets in KEEL format which can be used with the software and provides:

- A detailed categorization of the considered data sets and a description of their characteristics. Tables for the data sets in each category have been also created.
- A descriptions of the papers which have used the partitions of data sets available in the KEEL-dataset repository. These descriptions include results tables, the algorithms used and additional material.

KEEL-dataset contains two main sections according to the previous two points. In the first part, the data sets of the repository are presented. They have been organized in several categories and sub-categories arranging them in tables. Each data set has a dedicated webpage in which its characteristics are presented. These webpages also provide the complete data set and the partitions ready to download.

On the other hand, the experimental studies section is a novel approach in these types of repositories. It provides a series of webpages for each experimental study with the data sets used and their results in different formats as well, ready to perform a direct comparison. Direct access to the paper's PDF for all the experimental studies included in this webpage is also provided.

In Fig. 10.5 the main webpage, in which these two main sections appear, is depicted.

In the rest of this section we will describe the two main sections of the KEEL-dataset repository webpage.

10.3.1 Data Sets Web Pages

The categories of the data sets have been derived from the topics addressed in the experimental studies. Some of them are usually found in the literature, like supervised (classification) data sets, unsupervised and regression problems. On the other hand, new categories which have not been tackled or separated yet are also present. The categories in which the data sets are divided are the following:

- Classification problems. This category includes all the supervised data sets. All these data sets contains one or more attributes which label the instances, mapping them into different classes. We distinguish three subcategories of classification data sets:
  - *Standard data sets.*
  - *Imbalanced data sets* [29–31]. Imbalanced data sets are standard classification data sets where the class distribution is highly skewed among the classes.

[16] http://keel.es/datasets.php.

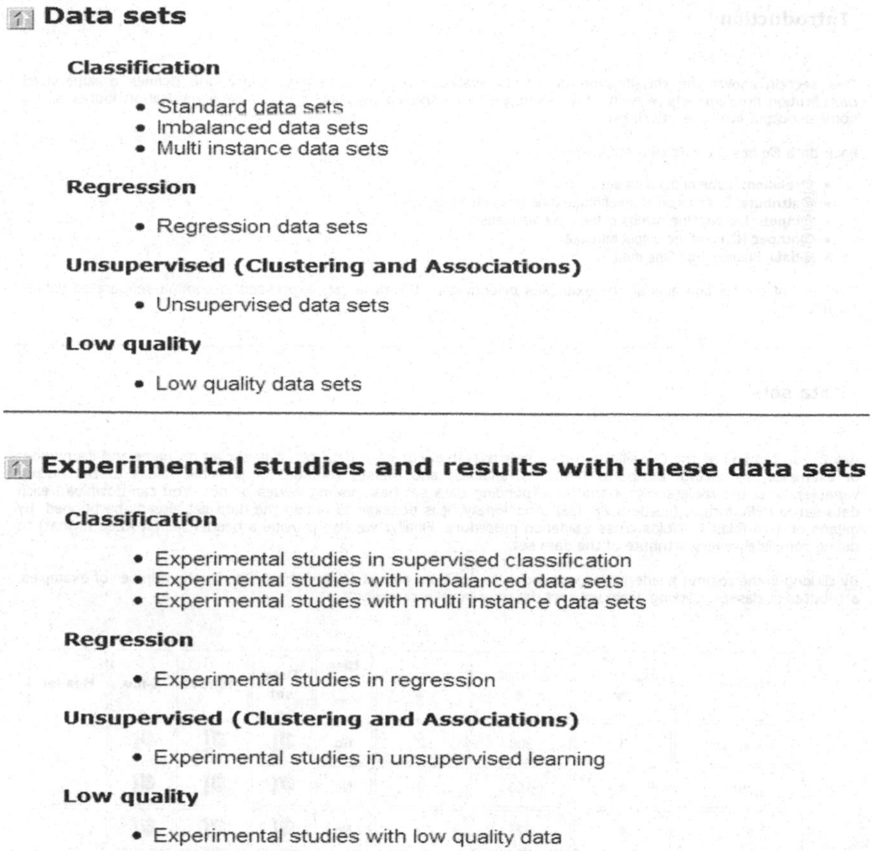

Data sets

Classification

- Standard data sets
- Imbalanced data sets
- Multi instance data sets

Regression

- Regression data sets

Unsupervised (Clustering and Associations)

- Unsupervised data sets

Low quality

- Low quality data sets

Experimental studies and results with these data sets

Classification

- Experimental studies in supervised classification
- Experimental studies with imbalanced data sets
- Experimental studies with multi instance data sets

Regression

- Experimental studies in regression

Unsupervised (Clustering and Associations)

- Experimental studies in unsupervised learning

Low quality

- Experimental studies with low quality data

Fig. 10.5 KEEL-dataset webpage (http://keel.es/datasets.php)

  – *Multi instance data sets* [32]. Multi-Instance data sets represent problems where there is a many-to-one relationship between feature vectors and their output attribute.

- Regression problems. These are data sets with a real valued output attribute, and the objective is to better approximate this output value using the input attributes.
- Unsupervised (Clustering and Associations) problems. Unsupervised data sets represent a set of data whose examples have been not labeled.
- Low quality data [33]. In this category the data sets which contain imprecise values in their input attributes are included, caused by noise or restrictions in the measurements. Therefore these low quality data sets can contain a mixture of crisp and fuzzy values. This is a unique category.

In Fig. 10.6 the webpage for the classification standard data sets is shown as an illustrative example of a particular category webpage. These webpages are structured in two main sections:

Introduction

This section shows the classification data sets avalaible in the repository. Every one defines a supervised classification problem, where each of its examples is composed by some nominal or numerical attributes and a nominal output attribute (its class).

Each data file has the following structure:

- **@relation:** Name of the data set
- **@attribute:** Description of an attribute (one for each attribute)
- **@inputs:** List with the names of the input attributes
- **@output:** Name of the output attribute
- **@data:** Starting tag of the data

The rest of the file contains all the examples belonging to the data set, expressed in comma sepparated values format.

Data sets

Below you can find all the Classification data sets available. For each data set, it is shown its name and its number of examples (instances), attributes (features,variables) and classes (number of possible values of the output variable). Also, the table shows if the corresponding data set has missing values or not. You can download each data set in KEEL format (inside a ZIP file). Additionally, it is possible to obtain the data set already partitioned, by means of a 10-folds / 5-folds cross validation procedure. Finally, we also provide a header file (in KEEL format) to define completely every attribute of the data set.

By clicking in the column headers, you can order the table by names (alphabetically) or by the number of examples, attributes or classes. Clicking again will sort the rows in reverse order.

Name	#Attributes	#Examples	#Classes	Miss Val.	Data set	10-fcv	5-fcv	Header
haberman	3	306	2	No				
iris	4	150	3	No				
balance	4	625	3	No				

Fig. 10.6 Fraction of Keel-dataset standard data sets' webpage

- First, the structure of the header of this type of Keel data set file is pointed out. This description contains the tags used to identify the different attributes, the name of the data set and indicates the starting point of the data.
- The second part is a enumeration of the different data sets contained in the webpage. This enumeration is presented in a table. The table shows the characteristics of all the data sets: the name of the data set, number of attributes, number of examples and number of classes (if applicable). Moreover the possibility of downloading the entire data set or different kind of partitions in Keel format in a ZIP file is presented. A header file is also available with particular information of the data set.

The tables' columns can be also sorted attending to the different data set's characteristics, like the number of attributes or examples.

Clicking on the name of the data set in the table will open the specific webpage for this data set. This webpage is composed of tables which gather all information available on the data set.

- The first table will always contain the general information of the data set: name, number of attributes, number of instances, number of classes, presence of MVs, etc.
- The second table contains the relation of attributes of the data set. For each attribute, the domain of the values is given. If it is a numerical attribute, the minimum and maximum values of the domain are presented. In the case of nominal attributes, the complete set of values is shown. The class attribute (if applicable) is stressed with a different color.

Additional information of the data set is also included, indicating its origin, applications and nature. In a second part of the webpage, the complete data set and a number of partitions can be downloaded in Keel format.

10.3.2 Experimental Study Web Pages

This section contains the links to the different experimental studies for the respective data set categories. For each category, a new webpage has been built. See Fig. 10.7 for the webpage devoted to the experimental studies with standard classification data sets. These webpages contain published journal publications which use the correspondent kind of data sets in the repository. The papers are grouped by the publication year. Each paper can contain up to four links:

Introduction

This section shows some relevant research papers in which some of the classification data sets avalaible in KEEL-dataset have been employed.

For each study, we provide its reference (plain text and BibTeX formats), abstract and summary. A pdf version the article can also be downloaded. Additionally, we offer complementary material about the experimental studies carried up: Algorithms tested, data sets employed and results obtained (XLS and CVS formats).

Experimental studies and results with these data sets

Jump to year: 2010 **(1)**

Year 2010 (1):

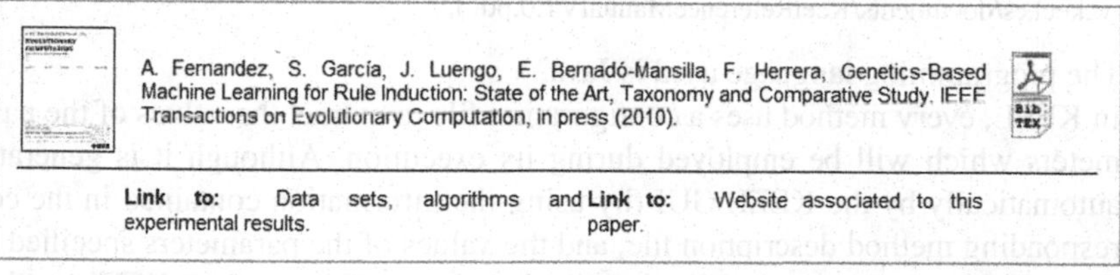

A. Fernandez, S. García, J. Luengo, E. Bernadó-Mansilla, F. Herrera, Genetics-Based Machine Learning for Rule Induction: State of the Art, Taxonomy and Comparative Study. IEEE Transactions on Evolutionary Computation, in press (2010).
BIB TEX
Link to: Data sets, algorithms and experimental results.
Link to: Website associated to this paper.

Fig. 10.7 Keel-dataset experimental studies with standard classification data sets webpage

- The first link is the PDF file of the paper.
- The second link is the Bibtex reference of the paper.
- The bottom-left link *Data sets, algorithms and experimental results* is always present. It references to the particular Keel-dataset webpage for the paper.
- The bottom-right link *Website associated to this paper* is only present for some papers which have a particular and external webpage related to them.

The particular Keel-dataset for the paper presents the relevant information of the publication. The abstract of the paper, an outline and the details of the experimental study are included. These details consist of the names of the algorithms analyzed, the list of data sets used and the results obtained. Both data sets used and the complete results of the paper are available for download in separate ZIP files. Moreover, the results are detailed and listed in CSV and XLS (Excel) formatted files. In Fig. 10.8 an example of the webpage for a specific publication with all these fields is shown.

10.4 Integration of New Algorithms into the KEEL Tool

In this section the main features that any researcher must take into account to integrate a new algorithm into the KEEL software tool are described. Next, a simple codification example is provided in order to clarify the integration process.

10.4.1 Introduction to the KEEL Codification Features

This section is devoted to describing in detail how to implement or to import an algorithm into the KEEL software tool. The KEEL philosophy tries to include the least possible constraints for the developer, in order to ease the inclusion of new algorithms. Thus, it is not necessary to follow the guidelines of any design pattern or framework in the development of a new method. In fact, each algorithm has its source code in a single folder and does not depend on a specific structure of classes, making the integration of new methods straightforward.

We enumerate the list of details to take into account before codifying a method for the KEEL software, which is also detailed at the KEEL Reference Manual (http://www.keel.es/documents/KeelReferenceManualV1.0.pdf).

- The programming language used is Java.
- In KEEL, every method uses a configuration file to extract the values of the parameters which will be employed during its execution. Although it is generated automatically by the KEEL GUI (by using the information contained in the corresponding method description file, and the values of the parameters specified by the user), it is important to fully describe its structure because any KEEL method must be able to read it completely, in order to get the values of its parameters specified in each execution.

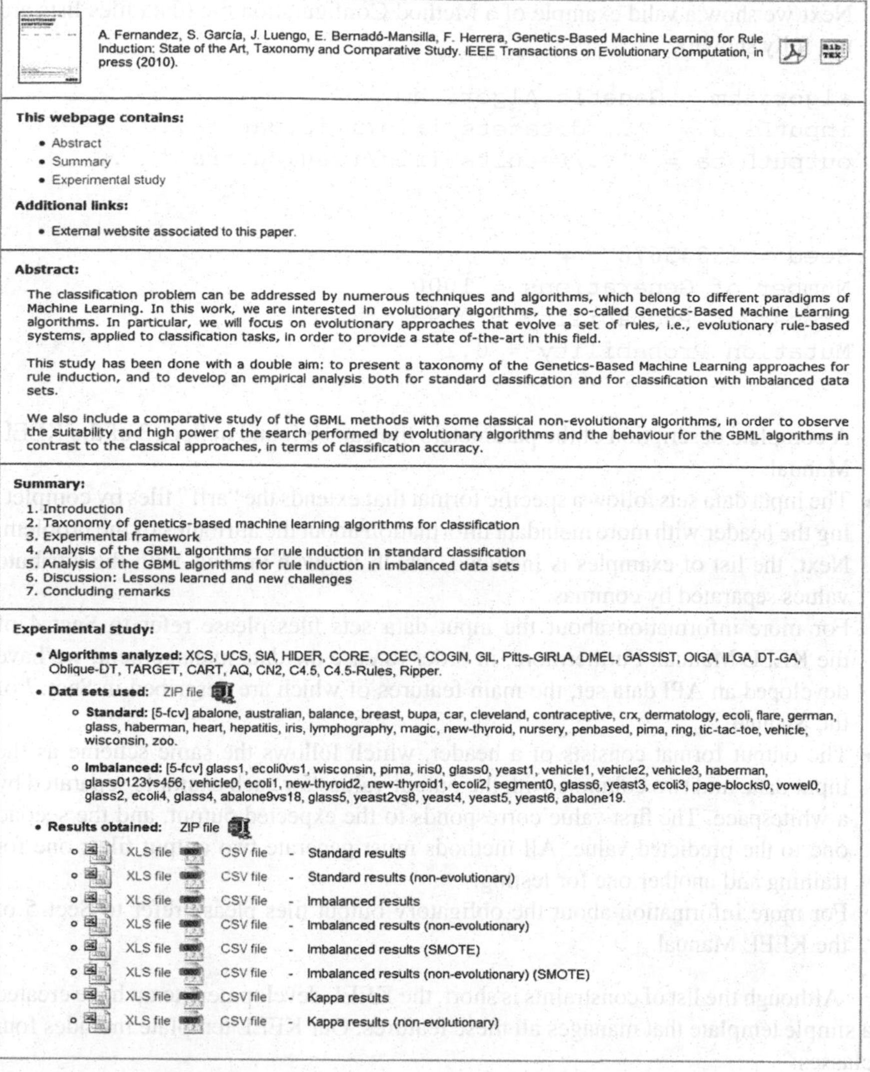

A. Fernandez, S. García, J. Luengo, E. Bernadó-Mansilla, F. Herrera, Genetics-Based Machine Learning for Rule Induction: State of the Art, Taxonomy and Comparative Study. IEEE Transactions on Evolutionary Computation, in press (2010).

This webpage contains:

- Abstract
- Summary
- Experimental study

Additional links:

- External website associated to this paper.

Abstract:

The classification problem can be addressed by numerous techniques and algorithms, which belong to different paradigms of Machine Learning. In this work, we are interested in evolutionary algorithms, the so-called Genetics-Based Machine Learning algorithms. In particular, we will focus on evolutionary approaches that evolve a set of rules, i.e., evolutionary rule-based systems, applied to classification tasks, in order to provide a state of-the-art in this field.

This study has been done with a double aim: to present a taxonomy of the Genetics-Based Machine Learning approaches for rule induction, and to develop an empirical analysis both for standard classification and for classification with imbalanced data sets.

We also include a comparative study of the GBML methods with some classical non-evolutionary algorithms, in order to observe the suitability and high power of the search performed by evolutionary algorithms and the behaviour for the GBML algorithms in contrast to the classical approaches, in terms of classification accuracy.

Summary:

1. Introduction
2. Taxonomy of genetics-based machine learning algorithms for classification
3. Experimental framework
4. Analysis of the GBML algorithms for rule induction in standard classification
5. Analysis of the GBML algorithms for rule induction in imbalanced data sets
6. Discussion: Lessons learned and new challenges
7. Concluding remarks

Experimental study:

- **Algorithms analyzed:** XCS, UCS, SIA, HIDER, CORE, OCEC, COGIN, GIL, Pitts-GIRLA, DMEL, GASSIST, OIGA, ILGA, DT-GA, Oblique-DT, TARGET, CART, AQ, CN2, C4.5, C4.5-Rules, Ripper.
- **Data sets used:** ZIP file
  - **Standard:** [5-fcv] abalone, australian, balance, breast, bupa, car, cleveland, contraceptive, crx, dermatology, ecoli, flare, german, glass, haberman, heart, hepatitis, iris, lymphography, magic, new-thyroid, nursery, penbased, pima, ring, tic-tac-toe, vehicle, wisconsin, zoo.
  - **Imbalanced:** [5-fcv] glass1, ecoli0vs1, wisconsin, pima, iris0, glass0, yeast1, vehicle1, vehicle2, vehicle3, haberman, glass0123vs456, vehicle0, ecoli1, new-thyroid2, new-thyroid1, ecoli2, segment0, glass6, yeast3, ecoli3, page-blocks0, vowel0, glass2, ecoli4, glass4, abalone9vs18, glass5, yeast2vs8, yeast4, yeast5, yeast6, abalone19.
- **Results obtained:** ZIP file
  - XLS file CSV file - Standard results
  - XLS file CSV file - Standard results (non-evolutionary)
  - XLS file CSV file - Imbalanced results
  - XLS file CSV file - Imbalanced results (non-evolutionary)
  - XLS file CSV file - Imbalanced results (SMOTE)
  - XLS file CSV file - Imbalanced results (non-evolutionary) (SMOTE)
  - XLS file CSV file - Kappa results
  - XLS file CSV file - Kappa results (non-evolutionary)

Fig. 10.8 Keel-dataset example of an experimental study dedicated webpage

Each configuration file has the following structure:

- *algorithm*: Name of the method.
- *inputData*: A list of the input data files of the method.
- *outputData*: A list of the output data files of the method.
- *parameters*: A list of parameters of the method, containing the name of each parameter and its value (one line is used for each one).

Next we show a valid example of a Method Configuration file (data files lists are not fully shown):

```
algorithm = Genetic Algorithm
inputData = ``../datasets/iris/iris.dat'' ...
outputData = ``../results/iris/result0.tra'' ...


Seed = 12345678
Number of Generations = 1000
Crossover Probability = 0.9
Mutation Probability = 0.1
...
```

A complete description of the parameters file can be found in Sect. 3 of the KEEL Manual.

- The input data sets follow a specific format that extends the "arff" files by completing the header with more metadata information about the attributes of the problem. Next, the list of examples is included, which is given in rows with the attribute values separated by commas.
  For more information about the input data sets files please refer to Sect. 4 of the KEEL Manual. Furthermore, in order to ease the data management, we have developed an API data set, the main features of which are described in Sect. 7 of the Manual.
- The output format consists of a header, which follows the same scheme as the input data, and two columns with the output values for each example separated by a whitespace. The first value corresponds to the expected output, and the second one to the predicted value. All methods must generate two output files: one for training and another one for testing.
  For more information about the obligatory output files please refer to Sect. 5 of the KEEL Manual.

Although the list of constraints is short, the KEEL development team have created a simple template that manages all these features. Our KEEL template includes four classes:

1. **Main**: This class contains the main instructions for launching the algorithm. It reads the parameters from the file and builds the "algorithm object".

```
public class Main {

  private parseParameters parameters;

  private void execute(String confFile) {
       parameters = new parseParameters();
       parameters.parseConfigurationFile(confFile);
       Algorithm method = new Algorithm(parameters);
       method.execute();
  }
```

```
    public static void main(String args[]) {
        Main program = new Main();
        System.out.println("Executing Algorithm.");
        program.execute(args[0]);
    }
}
```

2. **ParseParameters**: This class manages all the parameters, from the input and output files, to every single parameter stored in the parameters file.

```
public class parseParameters {

    private String algorithmName;
    private String trainingFile, validationFile, testFile;
    private ArrayList <String> inputFiles;
    private String outputTrFile, outputTstFile;
    private ArrayList <String> outputFiles;
    private ArrayList <String> parameters;

    public parseParameters() {
        inputFiles = new ArrayList<String>();
        outputFiles = new ArrayList<String>();
        parameters = new ArrayList<String>();

    }

    public void parseConfigurationFile(String fileName) {
        StringTokenizer line;
        String file = Files.readFile(fileName);

        line = new StringTokenizer(file, "\n\r");
        readName(line);
        readInputFiles(line);
        readOutputFiles(line);
        readAllParameters(line);

    };
    ...
}
```

3. **myDataset**: This class is an interface between the classes of the API data set and the algorithm. It contains the basic options related to data access.

```
public class myDataset {

    private double[][] X;
    private double[] outputReal;
    private String[] output;

    private int nData;
    private int nVars;
    private int nInputs;

    private InstanceSet IS;

    public myDataset() {
        IS = new InstanceSet();
    }

    public double[] getExample(int pos) {
        return X[pos];
```

```
    }

    public void readClassificationSet(String datasetFile,
            boolean train) throws IOException {
        try {
            IS.readSet(datasetFile, train);
            nData = IS.getNumInstances();
            nInputs = Attributes.getInputNumAttributes();
            nVars = nInputs + Attributes.getOutputNumAttributes();
        ...
    }
}
```

4. **Algorithm**: This class is devoted to storing the main variables of the algorithm and naming the different procedures for the learning stage. It also contains the functions for writing the obligatory output files.

```
public class Algorithm {

    myDataset train, val, test;
    String outputTr, outputTst;
    private boolean somethingWrong = false;

    public Algorithm(parseParameters parameters) {

        train = new myDataset();
        val = new myDataset();
        test = new myDataset();
        try {
            System.out.println("\nReading the training set:" +
                        parameters.getTrainingInputFile());
            train.readClassificationSet(parameters.getTrainingInputFile(),
                        true);
            System.out.println("\nReading the validation set:" +
                        parameters.getValidationInputFile());
            val.readClassificationSet(parameters.getValidationInputFile(),
                        false);
            System.out.println("\nReading the test set:" +
                        parameters.getTestInputFile());
            test.readClassificationSet(parameters.getTestInputFile(),
                        false);
        } catch (IOException e) {
            System.err.println("There was a problem while reading
                        the input data sets:" + e);
            somethingWrong = true;
        }

        outputTr = parameters.getTrainingOutputFile();

    ...
    }
}
```

The template can be downloaded by clicking on the link http://www.keel.es/software/KEEL_template.zip, which additionally supplies the user with the whole API data set together with the classes for managing files and the random number generator.

Most of the functions of the classes presented above are self-explanatory and fully documented to help the developer understand their use. Nevertheless, in the next section we will explain in detail how to encode a simple algorithm within the KEEL software tool.

10.5 KEEL Statistical Tests

Nowadays, the use of statistical tests to improve the evaluation process of the performance of a new method has become a widespread technique in the field of DM [34–36]. Usually, they are employed inside the framework of any experimental analysis to decide when an algorithm is better than other one. This task, which may not be trivial, has become necessary to confirm when a new proposed method offers a significant improvement over the existing methods for a given problem.

Two kinds of tests exist: parametric and non-parametric, depending on the concrete type of data employed. As a general rule, a non-parametric test is less restrictive than a parametric one, although it is less robust than a parametric when data is well conditioned.

Parametric tests have been commonly used in the analysis of experiments in DM. For example, a common way to test whether the difference between the results of two algorithms is non-random is to compute a paired t-test, which checks whether the average difference in their performance over the data sets is significantly different from zero. When comparing a set of multiple algorithms, the common statistical method for testing the differences between more than two related sample means is the repeated-measures ANOVA (or within-subjects ANOVA) [37]. Unfortunately, parametric tests are based on assumptions which are most probably violated when analyzing the performance of computational intelligence and DM algorithms [38–40]. These assumpitions are known as independence, normality and homoscedasticity.

Nonparametric tests can be employed in the analysis of experiments, providing the researcher with a practical tool to use when the previous assumptions can not be satisfied. Although they are originally designed for dealing with nominal or ordinal data, it is possible to conduct ranking based transformations to adjust the input data to the test requirements. Several nonparemetric methods for pairwise and multiple comparison are available to contrast adequately the results obtained in any Computational Intelligence experiment. A wide description about the topic with examples, cases of studies, bibliographic recommendations can be found in the SCI2S thematic public website on *Statistical Inference in Computational Intelligence and Data Mining*.[17]

KEEL is one of the fewest DM software tools that provides the researcher with a complete set of statistical procedures for pairwise and multiple comparisons. Inside the KEEL environment, several parametric and non-parametric procedures have been coded, which should help to contrast the results obtained in any experiment performed with the software tool. These tests follow the same methodology that the rest of elements of KEEL, facilitating both its employment and its integration inside a complete experimental study.

[17] http://sci2s.ugr.es/sicidm/.

Table 10.1 Statistical procedures available in KEEL

Procedure	References	Description
5x2cv-f test	[25]	Approximate f statistical test for 5x2-CV
T test	[41]	Statistical test based on the Student's t distribution
F test	[42]	Statistical test based on the Snedecor's F distribution
Shapiro-Wilk test	[43]	Variance test for normality
Mann-Whitney U test	[44]	U statistical test of difference of means
Wilcoxon test	[45]	Nonparametric pairwise statistical test
Friedman test	[46]	Nonparametric multiple comparisons statistical test
Iman-Davenport test	[47]	Derivation from the Friedman's statistic (less conservative)
Bonferroni-Dunn test	[48]	Post-Hoc procedure similar to Dunnet's test for ANOVA
Holm test	[49]	Post-Hoc sequential procedure (most significant first)
Hochberg test	[50]	Post-Hoc sequential procedure (less significant first)
Nemenyi test	[51]	Comparison with all possible pairs
Hommel test	[52]	Comparison with all possible pairs (less conservative)

Table 10.1 shows the procedures existing in the KEEL statistical package. For each test, a reference and a brief description is given (an extended description can be found in the *Statistical Inference in Computational Intelligence and Data Mining* website and in the KEEL website[18]).

10.5.1 Case Study

In this section, we present a case study as an example of the functionality and process of creating an experiment with the KEEL software tool. This experimental study is focused on the comparison between the new algorithm imported (SGERD) and several evolutionary rule-based algorithms, and employs a set of supervised classification domains available in KEEL-dataset. Several statistical procedures available in the KEEL software tool will be employed to contrast the results obtained.

10.5.1.1 Algorithms and Classification Problems

Five representative evolutionary rule learning methods have been selected to carry out the experimental study: Ant-Miner, CO-Evolutionary Rule Extractor (CORE), HIerarchical DEcision Rules (HIDER), Steady-State Genetic Algorithm for Extracting Fuzzy Classification Rules From Data (SGERD) and Tree Analysis with Randomly Generated and Evolved Trees (TARGET) methodology. Table 10.2 shows their references and gives a brief description of each one.

[18] http://www.keel.es.

Table 10.2 Algorithms tested in the experimental study

Method	Reference	Description
Ant-Miner	[53]	An Ant Colony System based using a heuristic function based In the entropy measure for each attribute-value
CORE	[54]	A coevolutionary method which employs as fitness measure a Combination of the true positive rate and the false positive rate
HIDER	[55]	A method which iteratively creates rules that cover Randomly selected examples of the training set
SGERD	[56]	A steady-state GA which generates a prespecified number Of rules per class following a GCCL approach
TARGET	[57]	A GA where each chromosome represents a complete decision tree

On the other hand, we have used 24 well-known classification data sets (they are publicly available on the KEEL-dataset repository web page,[19] including general information about them, partitions and so on) in order to check the performance of these methods. Table 10.3 shows their main characteristics where $\#Ats$ is the number of attributes, $\#Ins$ is the number of instances and $\#Cla$ is the number of Classes. For each data set the number of examples, attributes and classes of the problem described are shown. We have employed a 10-FCV procedure as a validation scheme to perform the experiments.

Table 10.3 Data sets employed in the experimental study

Name	#Ats	#Ins	#Cla	Name	#Ats	#Ins	#Cla
HAB	3	306	2	Wisconsin	9	699	2
IRI	4	150	3	Tic-tac-toe	9	958	2
BAL	4	625	3	Wine	13	178	3
NTH	5	215	3	Cleveland	13	303	5
MAM	5	961	2	Housevotes	16	435	2
BUP	6	345	2	Lymphography	18	148	4
MON	6	432	2	Vehicle	18	846	4
CAR	6	1,728	4	Bands	19	539	2
ECO	7	336	8	German	20	1,000	2
LED	7	500	10	Automobile	25	205	6
PIM	8	768	2	Dermatology	34	366	6
GLA	9	214	7	Sonar	60	208	2

[19] http://www.keel.es/datasets.php.

10.5.1.2 Setting up the Experiment Under KEEL Software

To do this experiment in KEEL, first of all we click on the Experiment option in the main menu of the KEEL software tool, define the experiment as a Classification problem and use a 10-FCV procedure to analyze the results. Next, the first step of the experiment graph setup is to choose the data sets to be used in Table 10.3. The partitions in KEEL are static, meaning that further experiments carried out will stop being dependent on particular data partitions.

The graph in Fig. 10.9 represents the flow of data and results from the algorithms and statistical techniques. A node can represent an initial data flow (group of data sets), a pre-process/post-process algorithm, a learning method, test or a visualization of results module. They can be distinguished easily by the color of the node. All their parameters can be adjusted by clicking twice on the node. Notice that KEEL incorporates the option of configuring the number of runs for each probabilistic algorithm, including this option in the configuration dialog of each node (3 in this case study). Table 10.4 shows the parameter's values selected for the algorithms employed in this experiment (they have been taken from their respective papers following the indications given by the authors).

The methods present in the graph are connected by directed edges, which represent a relationship between them (data or results interchange). When the data is interchanged, the flow includes pairs of train-test data sets. Thus, the graph in this specific example describes a flow of data from the 24 data sets to the nodes of the

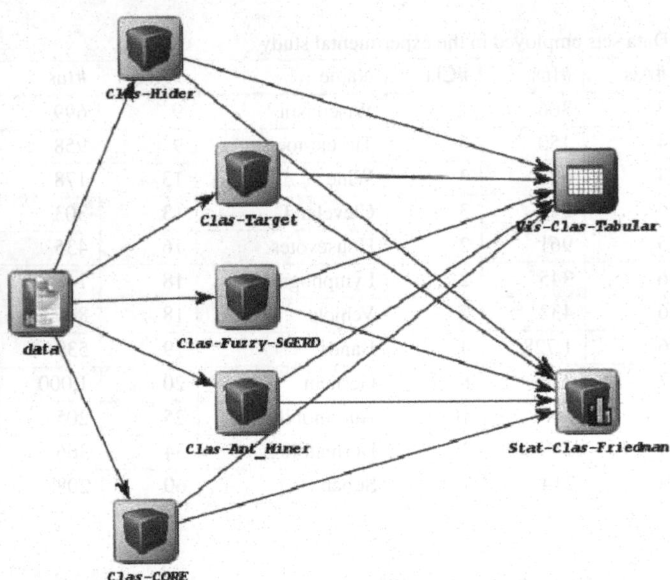

Fig. 10.9 Graphical representation of the experiment in KEEL

Table 10.4 Parameter' values employed in the experimental study

Algorithm	Parameters
Ant-Miner	Number of ants: 3000, Maximum uncovered samples: 10, Maximum samples by rule: 10
	Maximum iterations without converge: 10
CORE	Population size: 100, Co-population size: 50, Generation limit: 100
	Number of co-populations: 15, Crossover rate: 1.0
	Mutation probability: 0.1, Regeneration probability: 0.5
HIDER	Population size: 100, Number of generations: 100, Mutation probability: 0.5
	Cross percent: 80, Extreme mutation probability: 0.05, Prune examples factor: 0.05
	Penalty factor: 1, Error coefficient: 1
SGERD	Number of Q rules per class: Computed heuristically, Rule evaluation criteria = 2
TARGET	Probability of splitting a node: 0.5, Number of total generations for the GA: 100
	Number of trees generated by crossover: 30, Number of trees generated by mutation: 10
	Number of trees generated by clonation: 5, Number of trees Generated by immigration: 5

five learning methods used (Clas-AntMiner, Clas-SGERD, Clas-Target, Clas-Hider and Clas-CORE).

After the models are trained, the instances of the data set are classified. These results are the inputs for the visualization and test modules. The module Vis-Clas-Tabular receives these results as input and generates output files with several performance metrics computed from them, such as confusion matrices for each method, accuracy and error percentages for each method, fold and class, and a final summary of results. Figure 10.9 also shows another type of results flow, the node Stat-Clas-Friedman which represents the statistical comparison, results are collected and a statistical analysis over multiple data sets is performed by following the indications given in [38].

Once the graph is defined, we can set up the associated experiment and save it as a zip file for an off-line run. Thus, the experiment is set up as a set of XML scripts and a JAR program for running it. Within the results directory, there will be directories used for housing the results of each method during the run. For example, the files allocated in the directory associated to an interval learning algorithm will contain the knowledge or rule base. In the case of a visualization procedure, its directory will house the results files. The results obtained by the analyzed methods are shown in the next section, together with the statistical analysis.

10.5.1.3 Results and Analysis

This subsection describes and discusses the results obtained from the previous experiment configuration. Tables 10.5 and 10.6 show the results obtained in training and test stages, respectively. For each data set, the average and standard deviations in accuracy obtained by the module Vis-Clas-Tabular are shown, with the best results stressed in **boldface**.

Focusing on the test results, the average accuracy obtained by Hider is the highest one. However, this estimator does not reflect whether or not the differences among the methods are significant. For this reason, we have carried out an statistical analysis based on multiple comparison procedures (see http://sci2s.ugr.es/sicidm/ for a full

Table 10.5 Average results and standard deviations of training accuracy obtained

Data set	Ant Miner		CORE		HIDER		SGERD		TARGET	
	Mean	SD	Mean	SD	Mean	SD	Mean	SD	Mean	SD
HAB	**79.55**	1.80	76.32	1.01	76.58	1.21	74.29	0.81	74.57	1.01
IRI	97.26	0.74	95.48	1.42	**97.48**	0.36	97.33	0.36	93.50	2.42
BAL	73.65	3.38	68.64	2.57	75.86	0.40	76.96	2.27	**77.29**	1.57
NTH	**99.17**	0.58	92.66	1.19	95.97	0.83	90.23	0.87	88.05	2.19
MAM	81.03	1.13	79.04	0.65	**83.60**	0.75	74.40	1.43	79.91	0.65
BUP	**80.38**	3.25	61.93	0.89	73.37	2.70	59.13	0.68	68.86	0.89
MON	97.22	0.30	87.72	7.90	97.22	0.30	80.56	0.45	**97.98**	7.90
CAR	77.95	1.82	**79.22**	1.29	70.02	0.02	67.19	0.08	77.82	0.29
ECO	87.90	1.27	67.03	3.69	**88.59**	1.77	73.02	0.86	66.22	4.69
LED	59.42	1.37	28.76	2.55	**77.64**	0.42	40.22	5.88	34.24	3.55
PIM	71.86	2.84	72.66	2.62	**77.82**	1.16	73.71	0.40	73.42	2.62
GLA	81.48	6.59	54.26	1.90	**90.09**	1.64	53.84	2.96	45.07	0.90
WIS	92.58	1.65	94.71	0.64	**97.30**	0.31	93.00	0.85	96.13	0.64
TAE	69.62	2.21	69.46	1.20	69.94	0.53	69.94	0.53	**69.96**	2.20
WIN	**99.69**	0.58	99.06	0.42	97.19	0.98	91.76	1.31	85.19	1.58
CLE	60.25	1.35	56.30	1.97	**82.04**	1.75	46.62	2.23	55.79	2.97
HOU	94.28	1.84	**96.98**	0.43	**96.98**	0.43	**96.98**	0.43	**96.98**	0.43
LYM	77.11	5.07	65.99	5.43	**83.70**	2.52	77.48	3.55	75.84	4.43
VEH	59.52	3.37	36.49	3.52	**84.21**	1.71	51.47	1.19	51.64	2.52
BAN	67.61	3.21	66.71	2.01	**87.13**	2.15	63.84	0.74	71.14	2.01
GER	71.14	1.19	70.60	0.63	**73.54**	0.58	67.07	0.81	70.00	1.37
AUT	69.03	8.21	31.42	7.12	**96.58**	0.64	52.56	1.67	45.66	6.12
DER	86.18	5.69	31.01	0.19	**94.91**	1.40	72.69	1.04	66.24	1.81
SON	74.68	0.79	53.37	0.18	**98.29**	0.40	75.69	1.47	76.87	1.18
Average	79.52	2.51	68.16	2.14	**86.09**	1.04	71.76	1.37	72.43	2.33

Table 10.6 Average results and standard deviations of test accuracy obtained

Data set	Ant Miner		CORE		HIDER		SGERD		TARGET	
	Mean	SD	Mean	SD	Mean	SD	Mean	SD	Mean	SD
HAB	72.55	5.27	72.87	4.16	**75.15**	4.45	74.16	2.48	71.50	2.52
IRI	96.00	3.27	92.67	4.67	**96.67**	3.33	**96.67**	3.33	92.93	4.33
BAL	70.24	6.21	70.08	7.11	69.60	3.77	75.19	6.27	**75.62**	7.27
NTH	**90.76**	6.85	**90.76**	5.00	90.28	7.30	88.44	6.83	86.79	5.83
MAM	81.48	7.38	77.33	3.55	**82.30**	6.50	74.11	5.11	79.65	2.11
BUP	57.25	7.71	61.97	4.77	65.83	10.04	57.89	3.41	**65.97**	1.41
MON	**97.27**	2.65	88.32	8.60	**97.27**	2.65	80.65	4.15	96.79	5.15
CAR	77.26	2.59	**79.40**	3.04	70.02	0.16	67.19	0.70	77.71	2.70
ECO	58.58	9.13	64.58	4.28	**75.88**	6.33	72.08	7.29	65.49	4.29
LED	55.32	4.13	27.40	4.00	**68.20**	3.28	40.00	6.75	32.64	6.75
PIM	66.28	4.26	73.06	6.03	73.18	6.19	**73.71**	3.61	73.02	6.61
GLA	53.74	12.92	45.74	9.36	**64.35**	12.20	48.33	5.37	44.11	5.37
WIS	90.41	2.56	92.38	2.31	**96.05**	2.76	92.71	3.82	95.75	0.82
TAE	64.61	5.63	**70.35**	3.77	69.93	4.73	69.93	4.73	69.50	2.73
WIN	92.06	6.37	**94.87**	4.79	82.61	6.25	87.09	6.57	82.24	7.57
CLE	**57.45**	5.19	53.59	7.06	55.86	5.52	44.15	4.84	52.99	1.84
HOU	93.56	3.69	**97.02**	3.59	**97.02**	3.59	**97.02**	3.59	96.99	0.59
LYM	73.06	10.98	65.07	15.38	72.45	10.70	72.96	13.59	**75.17**	10.59
VEH	53.07	4.60	36.41	3.37	**63.12**	4.48	51.19	4.85	49.81	5.85
BAN	59.18	6.58	64.23	4.23	62.15	8.51	62.71	4.17	**67.32**	6.17
GER	66.90	3.96	69.30	1.55	**70.40**	4.29	66.70	1.49	70.00	0.49
AUT	53.74	7.79	32.91	6.10	**62.59**	13.84	50.67	10.27	42.82	13.27
DER	81.16	7.78	31.03	1.78	**87.45**	3.26	69.52	4.25	66.15	4.25
SON	71.28	5.67	53.38	1.62	52.90	2.37	73.45	7.34	**74.56**	8.34
Average	72.22	5.97	66.86	5.01	**75.05**	5.69	70.27	5.20	71.06	4.87

description), by including a node called Stat-Clas-Friedman in the KEEL experiment. Here, we include the information provided by this statistical module:

- Table 10.7 shows the obtained average rankings across all data sets following the Friedman procedure for each method. They will be useful to calculate the p-value and to detect significant differences between the two methods.
- Table 10.8 depicts the results obtained from the use of the Friedman and Iman-Davenport test. Both, the statistics and p-values are shown. As we can see, a level of significance $\alpha = 0.10$ is needed in order to consider that differences among the methods exist. Note also that the p-value obtained by the Iman-Davenport test is lower than that obtained by Friedman, this is always true.

Table 10.7 Average rankings of the algorithms by Friedman procedure

Algorithm	Ranking
AntMiner	3.125
CORE	3.396
Hider	**2.188**
SGERD	3.125
Target	3.167

Table 10.8 Results of the Friedman and Iman-Davenport tests

Friedman value	*p*-value	Iman-Davenport value	*p*-value
8.408	0.0777	2.208	0.0742

Table 10.9 Adjusted *p*-values. Hider is the control algorithm

I	Algorithm	Unadjusted p	p_{Holm}	p_{Hoch}
1	CORE	0.00811	0.032452	0.03245
2	Target	0.03193	0.09580	0.03998
3	AntMiner	0.03998	0.09580	0.03998
4	SGERD	0.03998	0.09580	0.03998

- Finally, in Table 10.9 the adjusted p-values are shown considering the best method (Hider) as the control algorithm and using the three post-hoc procedures explained above. The following analysis can be made:
  - The procedure of Holm verifies that Hider is the best method with $\alpha = 0.10$, but it only outperforms CORE considering $\alpha = 0.05$.
  - The procedure of Hochberg checks the supremacy of Hider with $\alpha = 0.05$. In this case study, we can see that the Hochberg method is the one with the highest power.

10.6 Summarizing Comments

In this chapter we have introduced a series of non-commercial Java software tools, and focused on a particular one named KEEL, that provides a platform for the analysis of ML methods applied to DM problems. This tool relieves researchers of much technical work and allows them to focus on the analysis of their new learning models in comparison with the existing ones. Moreover, the tool enables researchers with little knowledge of evolutionary computation methods to apply evolutionary learning algorithms to their work.

We have shown the main features of this software tool and we have distinguished three main parts: a module for data management, a module for designing experiments with evolutionary learning algorithms, and a module educational goals. We have also shown some case studies to illustrate functionalities and the experiment set up processes.

Apart from the presentation of the main software tool, three other complementary aspects of KEEL have been also described:

- KEEL-dataset, a data set repository that includes the data set partitions in the KEEL format and shows some results obtained in these data sets. This repository can free researchers from merely "technical work" and facilitate the comparison of their models with the existing ones.
- Some basic guidelines that the developer may take into account to facilitate the implementation and integration of new approaches within the KEEL software tool. We have shown the simplicity of adding a simple algorithm (SGERD in this case) into the KEEL software with the aid of a Java template specifically designed for this purpose. In this manner, the developer only has to focus on the inner functions of their algorithm itself and not on the specific requirements of the KEEL tool.
- A module of statistical procedures which let researchers contrast the results obtained in any experimental study using statistical tests. This task, which may not be trivial, has become necessary to confirm when a new proposed method offers a significant improvement over the existing methods for a given problem.

References

1. Han, J., Kamber, M., Pei, J.: Data mining: Concepts and techniques, second edition (The Morgan Kaufmann series in data management systems). Morgan Kaufmann, San Francisco (2006)
2. Witten, I.H., Frank, E.: Data mining: practical machine learning tools and techniques, second edition (Morgan Kaufmann series in data management systems). Morgan Kaufmann Publishers Inc., San Francisco (2005)
3. Demšar, J., Curk, T., Erjavec, A., Gorup, Črt, Hočevar, T., Milutinovič, M., Možina, M., Polajnar, M., Toplak, M., Starič, A., Štajdohar, M., Umek, L., Žagar, L., Žbontar, J., Žitnik, M., Zupan, B.: Orange: Data mining toolbox in python. J. Mach. Learn. Res. **14**, 2349–2353 (2013)
4. Abeel, T., de Peer, Y.V., Saeys, Y.: Java-ML: A machine learning library. J. Mach. Learn. Res. **10**, 931–934 (2009)
5. Hofmann, M., Klinkenberg, R.: RapidMiner: Data mining use cases and business analytics applications. Chapman and Hall/CRC, Florida (2013)
6. Williams, G.J.: Data mining with rattle and R: The art of excavating data for knowledge discovery. Use R!. Springer, New York (2011)
7. Sonnenburg, S., Braun, M., Ong, C., Bengio, S., Bottou, L., Holmes, G., LeCun, Y., Müller, K.R., Pereira, F., Rasmussen, C., Rätsch, G., Schölkopf, B., Smola, A., Vincent, P., Weston, J., Williamson, R.: The need for open source software in machine learning. J. Mach. Learn. Res. **8**, 2443–2466 (2007)
8. Alcalá-Fdez, J., Sánchez, L., García, S., del Jesus, M., Ventura, S., Garrell, J., Otero, J., Romero, C., Bacardit, J., Rivas, V., Fernández, J., Herrera, F.: KEEL: A software tool to assess evolutionary algorithms to data mining problems. Soft Comput. **13**(3), 307–318 (2009)
9. Derrac, J., García, S., Herrera, F.: A survey on evolutionary instance selection and generation. Int. J. Appl. Metaheuristic Comput. **1**(1), 60–92 (2010)
10. Kudo, M., Sklansky, J.: Comparison of algorithms that select features for pattern classifiers. Pattern Recognit. **33**(1), 25–41 (2000)
11. Quinlan, J.R.: C4.5: programs for machine learning. Morgan Kaufmann Publishers, San Francisco (1993)

12. Schölkopf, B., Smola, A.J.: Learning with kernels : support vector machines, regularization, optimization, and beyond. Adaptive computation and machine learning. MIT Press, Cambridge (2002)
13. Frenay, B., Verleysen, M.: Classification in the presence of label noise: A survey. Neural Netw. Learn. Syst., IEEE Trans. **25**(5), 845–869 (2014)
14. Garcia, E.K., Feldman, S., Gupta, M.R., Srivastava, S.: Completely lazy learning. IEEE Trans. Knowl. Data Eng. **22**(9), 1274–1285 (2010)
15. Alcalá, R., Alcalá-Fdez, J., Casillas, J., Cordón, O., Herrera, F.: Hybrid learning models to get the interpretability-accuracy trade-off in fuzzy modeling. Soft Comput. **10**(9), 717–734 (2006)
16. Rivas, A.J.R., Rojas, I., Ortega, J., del Jesús, M.J.: A new hybrid methodology for cooperative-coevolutionary optimization of radial basis function networks. Soft Comput. **11**(7), 655–668 (2007)
17. Bernadó-Mansilla, E., Ho, T.K.: Domain of competence of xcs classifier system in complexity measurement space. IEEE Trans. Evol. Comput. **9**(1), 82–104 (2005)
18. Ventura, S., Romero, C., Zafra, A., Delgado, J.A., Hervas, C.: Jclec: A java framework for evolutionary computation. Soft Comput. **12**(4), 381–392 (2007)
19. Pyle, D.: Data preparation for data mining. Morgan Kaufmann Publishers Inc., San Francisco (1999)
20. Zhang, S., Zhang, C., Yang, Q.: Data preparation for data mining. Appl. Artif. Intel. **17**(5–6), 375–381 (2003)
21. Luke, S., Panait, L., Balan, G., Paus, S., Skolicki, Z., Bassett, J., Hubley, R., Chircop, A.: ECJ: A Java based evolutionary computation research system. http://cs.gmu.edu/eclab/projects/ecj
22. Meyer, M., Hufschlag, K.: A generic approach to an object-oriented learning classifier system library. J. Artif. Soc. Soc. Simul. 9(3) (2006) http://jasss.soc.surrey.ac.uk/9/3/9.html
23. Llorá, X.: E2k: Evolution to knowledge. SIGEVOlution **1**(3), 10–17 (2006)
24. Kohavi, R.: A study of cross-validation and bootstrap for accuracy estimation and model selection. In: Proceedings of the 14th International Joint Conference on Artificial Intelligence. IJCAI'95, vol. 2, pp. 1137–1143. Morgan Kaufmann Publishers Inc., San Francisco, CA, USA (1995)
25. Dietterich, T.G.: Approximate statistical tests for comparing supervised classification learning algorithms. Neural Comput. **10**(7), 1895–1923 (1998)
26. Ortega, M., Bravo, J. (eds.): Computers and education in the 21st century. Kluwer, Dordrecht (2000)
27. Mierswa, I., Wurst, M., Klinkenberg, R., Scholz, M., Euler, T.: Yale: Rapid prototyping for complex data mining tasks. In: Ungar, L., Craven, M., Gunopulos, D., Eliassi-Rad, T. (eds.) KDD '06: Proceedings of the 12th ACM SIGKDD International Conference on Knowledge Discovery and Data Mining, pp. 935–940. NY, USA, New York (2006)
28. Rakotomalala, R.: Tanagra : un logiciel gratuit pour l'enseignement et la recherche. In: S. Pinson, N. Vincent (eds.) EGC, Revue des Nouvelles Technologies de l'Information, pp. 697–702. Cpadus-ditions (2005)
29. Batista, G.E.A.P.A., Prati, R.C., Monard, M.C.: A study of the behaviour of several methods for balancing machine learning training data. SIGKDD Explor. **6(1)**, 20–29 (2004)
30. He, H., Garcia, E.A.: Learning from imbalanced data. IEEE Trans. Knowl. Data Eng. **21**(9), 1263–1284 (2009)
31. Sun, Y., Wong, A.K.C., Kamel, M.S.: Classification of imbalanced data: A review. Int. J. Pattern Recognit. Artif. Intel. **23**(4), 687–719 (2009)
32. Dietterich, T., Lathrop, R., Lozano-Perez, T.: Solving the multiple instance problem with axis-parallel rectangles. Artif. Intell. **89**(1–2), 31–71 (1997)
33. Sánchez, L., Couso, I.: Advocating the use of imprecisely observed data in genetic fuzzy systems. IEEE Trans. Fuzzy Syst. **15**(4), 551–562 (2007)
34. Děmsar, J.: Statistical comparisons of classifiers over multiple data sets. J. Mach. Learn. Res. **7**, 1–30 (2006)
35. García, S., Fernández, A., Luengo, J., Herrera, F.: Advanced nonparametric tests for multiple comparisons in the design of experiments in computational intelligence and data mining: Experimental analysis of power. Inf. Sci. **180**(10), 2044–2064 (2010)

36. García, S., Herrera, F.: An extension on statistical comparisons of classifiers over multiple data sets for all pairwise comparisons. J. Mach. Learn. Res. **9**, 2579–2596 (2008)
37. Fisher, R.A.: Statistical methods and scientific inference (2nd edition). Hafner Publishing, New York (1959)
38. García, S., Fernández, A., Luengo, J., Herrera, F.: A study of statistical techniques and performance measures for genetics-based machine learning: Accuracy and interpretability. Soft Comput. **13**(10), 959–977 (2009)
39. García, S., Molina, D., Lozano, M., Herrera, F.: A study on the use of non-parametric tests for analyzing the evolutionary algorithms' behaviour: A case study on the CEC 2005 special session on real parameter optimization. J. Heuristics **15**, 617–644 (2009)
40. Luengo, J., García, S., Herrera, F.: A study on the use of statistical tests for experimentation with neural networks: Analysis of parametric test conditions and non-parametric tests. Expert Syst. with Appl. **36**, 7798–7808 (2009)
41. Cox, D., Hinkley, D.: Theoretical statistics. Chapman and Hall, London (1974)
42. Snedecor, G.W., Cochran, W.C.: Statistical methods. Iowa State University Press, Ames (1989)
43. Shapiro, S.S.: M.W.: An analysis of variance test for normality (complete samples). Biometrika **52**(3–4), 591–611 (1965)
44. Mann, H.B., Whitney, D.R.: On a test of whether one of two random variables is stochastically larger than the other. Ann. Math. Stat **18**, 50–60 (1947)
45. Wilcoxon, F.: Individual comparisons by ranking methods. Biometrics **1**, 80–83 (1945)
46. Friedman, M.: The use of ranks to avoid the assumption of normality implicit in the analysis of variance. J. the Am. Stat. Assoc. **32**(200), 675–701 (1937)
47. Iman, R., Davenport, J.: Approximations of the critical region of the friedman statistic. Commun. Stat. **9**, 571–595 (1980)
48. Sheskin, D.: Handbook of parametric and nonparametric statistical procedures. Chapman and Hall/CRC, Boca Raton (2006)
49. Holm, S.: A simple sequentially rejective multiple test procedure. Scand. J. Stat. **6**, 65–70 (1979)
50. Hochberg, Y.: A sharper bonferroni procedure for multiple tests of significance. Biometrika **75**, 800–803 (1988)
51. Nemenyi, P.B.: Distribution-free multiple comparisons, ph.d. thesis (1963)
52. Bergmann, G., Hommel, G.: Improvements of general multiple test procedures for redundant systems of hypotheses. In: Bauer, G.H.P., Sonnemann, E. (eds.) Multiplc hypotheses testing, pp. 100–115. Springer, Berlin (1988)
53. Parpinelli, R., Lopes, H., Freitas, A.: Data mining with an ant colony optimization algorithm. IEEE Trans. Evol. Comput. **6**(4), 321–332 (2002)
54. Tan, K.C., Yu, Q., Ang, J.H.: A coevolutionary algorithm for rules discovery in data mining. Int. J. Syst. Sci. **37**(12), 835–864 (2006)
55. Aguilar-Ruiz, J.S., Giráldez, R., Riquelme, J.C.: Natural encoding for evolutionary supervised learning. IEEE Trans. Evol. Comput. **11**(4), 466–479 (2007)
56. Mansoori, E., Zolghadri, M., Katebi, S.: SGERD: A steady-state genetic algorithm for extracting fuzzy classification rules from data. IEEE Trans. Fuzzy Syst. **16**(4), 1061–1071 (2008)
57. Gray, J.B., Fan, G.: Classification tree analysis using TARGET. Comput. Stat. Data Anal. **52**(3), 1362–1372 (2008)

Index

S. García et al., *Data Preprocessing in Data Mining*,
Intelligent Systems Reference Library 72, DOI 10.1007/978-3-319-10247-4

Printed in the United States
By Bookmasters